WORKS ISSUED BY
THE HAKLUYT SOCIETY

THE JOURNAL OF
JEAN-FRANÇOIS DE GALAUP DE LA PÉROUSE
VOLUME II

SECOND SERIES
NO. 180

HAKLUYT SOCIETY

COUNCIL AND OFFICERS, 1994-95

PRESIDENT
Professor P. E. H. Hair

VICE-PRESIDENTS

Dr T. E. Armstrong
Professor C. F. Beckingham, FBA
Professor C. R. Boxer, FBA

Professor Eila M. J. Campbell
Professor D. B. Quinn, HON.FBA
Professor Glyndwr Williams

Sir Harold Smedley, KCMG, MBE

COUNCIL *(with date of election)*

Professor R. C. Bridges (1990)
Professor W. E. Butler (1992)
Tony Campbell (1990)
Lt. Cdr. A. C. F. David (1990)
S. M. Easton (co-opted)
A. J. Farrington (1992)
Dr F. F. R. Fernandez-Armesto (1992)
Dr John B. Hattendorf (1994)

Dr Peter Jackson (1993)
J. C. H. King (1991)
Dr Robin Law (1993)
Anthony Payne (1991)
A. N. Ryan (1991)
Royal Geographical Society
 (Mrs Dorothy Middleton)
Mrs Ann Shirley (1994)
M. F. Strachan, CBE, FRSE (1994)

Dr Helen Wallis, OBE (1993)

TRUSTEES
Sir Geoffrey Ellerton, CMG, MBE
H. H. L. Smith
G. H. Webb, CMG, OBE
Professor Glyndwr Williams

HONORARY TREASURER
David Darbyshire

HONORARY SECRETARIES
Dr W. F. Ryan
Warburg Institute, University of London, Woburn Square,
London WC1H 0AB
Mrs Sarah Tyacke
(Keeper of Public Records)
Public Record Office, Ruskin Avenue, Kew, Richmond, Surrey TW9 4DU

HONORARY SECRETARIES FOR OVERSEAS
Australia: Ms Maura O'Connor, Curator of Maps, National Library of Australia, Canberra, A.C.T. 2601
Canada: Professor J. B. Bird, McGill University, Montreal, Quebec
New Zealand: J. E. Traue, Department of Librarianship, Victoria University of Wellington, P.O. Box 600, Wellington
South Africa: Dr F. R. Bradlow, 401/402 French Bank Building, 4 Church Square, Cape Town 8001
U.S.A.: Dr Norman Fiering, The John Carter Brown Library, Brown University, Box 1894, Providence, Rhode Island 02912
Professor Norman J. W. Thrower, Department of Geography, University of California, Los Angeles, 405 Hilgard Avenue, Los Angeles, California 90024–1698
Western Europe: Paul Putz, 54 rue Albert I, 1117 Luxembourg, Luxembourg.

ADMINISTRATIVE ASSISTANT
Mrs Fiona Easton
Postal address only: Hakluyt Society, c/o The Map Library,
The British Library, Great Russell Street, London WC1B 3DG
(to whom queries and application for membership may be made)
Telephone 01986–788359   Fax 01986–788181

View of Petropavlovsk, Kamchatka, by Blondela. SHM 352:23.

The Journal
of
Jean-François de Galaup de la Pérouse
1785–1788

VOLUME II

Translated and edited by
JOHN DUNMORE

THE HAKLUYT SOCIETY
LONDON
1995

© The Hakluyt Society 1995

ISBN 0 904180 39 5
ISSN 0072 9396

Typeset by Waveney Typesetters, Norwich
Printed in Great Britain at
the University Press, Cambridge

SERIES EDITORS
W. F. RYAN and SARAH TYACKE

British Library Cataloguing-in-Publication Data
A catalogue record for this book is
available from the British Library

Published by the Hakluyt Society
c/o The Map Library
British Library, Great Russell Street
London WC1B 3DG

CONTENTS

List of Illustrations and Maps	vi
THE JOURNAL (Chapters XIV–XX)	233
Appendices	
I. Selected Correspondence	451
II. The Muster Roll	543
III. The Death of Father Receveur	564
IV. Principal Monuments erected to La Pérouse	570
Sources for the Journal	571
Select Bibliography	577
Index	589

LIST OF ILLUSTRATIONS AND MAPS

View of Petropavlosk — *Frontispiece*

11. View of Cavite — *between pages* 250 and 251
12. Chart of Hoapinsu

13. Chart of part of Quelpaert Island — *between pages* 264 and 265
14. Charts of the islands of Botol and Kumi

15. Chart of part of the Korean archipelago — *between pages* 268 and 269
16. Chart of part of Dagelet Island

17. Chart of Ternay Bay — *between pages* 280 and 281
18. La Pérouse and officers with inhabitants of Langle Bay

19. Chart of Castries Bay — *between pages* 304 and 305
20. French officers among local tombs

21. Chart of D'Estaing Bay — *between pages* 322 and 323
22. Chart of Langle Bay

23. Navigators archipelago (Samoas) — *between pages* 387 and 388

24. Chart of part of Maouna — *between pages* 416 and 417
25. Massacre Cove, Maouna

26. Map of the Expedition through the Pacific — *between pages* 448 and 449
27. Map of the Expedition in Alaska
28. Map of the Expedition in North-East Asia
29. Proposed route from Botany Bay to the Ile de France (Mauritius)
30. Presumed route from Botany Bay

CHAPTER XIV[1]

Arrival at Cavite – our reception by the Castilan or King's Lieutenant who is in charge of this place. Our vessels had been seen from Manila and the Governor had sent the Captn of the bay to meet us. Mr Boutin, lieutenant de vaisseau is sent to the Governor-General; welcome extended to this officer. Orders issued to Cavite to provide all our needs. Details on this town and its arsenal. Visit by the commanders of the two frigates and several officers to the Governor-General. Description of Manila and district. Its population, disadvantage resulting from the form of government established there; penances witnessed during Holy Week; tax on tobacco; foundation of the new Philippines Company. Reflections of this new development. Details on the southern Philippines which do not recognise Spanish sovereignty. Spain has only one military garrison on Mindanao with 150 men. Constant warfare with the Moors or Moslems of these various islands. Two-day stay in Manila at the home of Mr Sébir, a French merchant; invaluable services rendered by him; his intellect; his plans when he settled in Manila. Excellent services rendered by this colony's administrator. Military situation of the island of Luzon.

We had hardly dropped anchor at the entrance to the port of Cavite when an officer came on behalf of the Commandant of that place to request us not to communicate with the land until instructions were received from the Governor-General to whom the king's lieutenant proposed to send a message as soon as he knew the reasons for our call. We promptly replied that we sought food and permission to repair our frigates in order to continue our voyage as soon as possible, but before this Spanish officer left, the commander of the

[1] This chapter is preceded by the title 'La Pérouse manuscript to form the third volume of his Voyage'. It is written by a different hand from earlier chapters. The spelling is more consistent. Annotations in a different handwriting would seem to indicate that the chapter was revised before being sent on to Paris.

bay[1] arrived from Manila from where our ships had been sighted. He told us that our arrival in the China Seas was known there and that letters from the Spanish ambassador had notified the Governor-General several months ago that we were likely to call. This officer added that the season made it possible to anchor in front of Manila, where we would find all the resources and the entertainment the Philippines could provide, but we were at anchor in front of an arsenal within a musket shot from land, and we possibly were discourteous enough to let the officer know that nothing could make up for such advantages. He was good enough to allow Mr Boutin, *lieutenant de vaisseau*, to go in his boat to notify the Governor-General of our arrival and ask him to issue instructions that our various requests should be met not later than 5 April, the later plan of our voyage requiring the two frigates to be under sail before the tenth of that month. Mr de Basco,[2] brigadier of the naval forces, Governor-General of Manila, extended the warmest welcome to the officers I sent him and gave firm orders that nothing should delay our departure.

He also wrote to the Castilon to allow us to communicate with the town and provide us with all the assistance and facilities at his disposal. Mr Boutin's return with Mr de Basco's despatches made us all citizens of Cavite, as our ships were so close to land that we could go ashore and return whenever we wished. We rented several houses to work on our sails, salt our provisions, build two boats, house our naturalists and our geographer-surveyors, and the kindly commandant lent us his to set up our observatory. We enjoyed as

---

[1] Marginal note: 'In Spain the commander of the bay is head of customs; he has a military rank: in Manila he has the rank of captain.'

[2] José de Basco y Vargas was not a nobleman but a *señor*. His commoner background created difficulties for him from the moment he landed in Manila in July 1778. The local Spanish families did not relish the idea of bowing before a former naval captain promoted to governor of the Philippines. The King refused to remove him and they tried to replace him with Don Pedro de Sarrio, *Teniente del Rey*, who had previously acted as interim governor. The coup failed and Vargas then carried out a number of major reforms of the colony's agriculture and administration. He founded in 1781 the *Sociedade económica de amigos del país*. Although La Pérouse criticised the tobacco monopoly he introduced, it helped to reduce imports and stimulated the local industry while increasing local revenues: the downside was administrative corruption. He created a wine monopoly in 1786, reformed the justice system and education and allowed Chinese to settle in the district of Parian. He conquered the Batan Islands and was named 'Conde de la Conquista de las Batanas'. He sailed for Spain in November 1787, was raised to the rank of rear-admiral and appointed governor of Cartagena.

much freedom as if we had been in the countryside, and we could find in the arsenal and the market place the same resources we would have found in one of the best ports in Europe. Cavite,[1] three leagues south-west of Manila, was formerly a fairly important settlement, but in the Philippines as in Europe the great cities suck in so to speak the smaller ones, and all that remains today is the commander of the arsenal and a *contador*, two port lieutenants, the castilon in charge of the establishment with 150 men belonging to the garrison and the officers attached to this soldiery.

All the other residents are of mixed blood or Indians attached to the arsenal, making up, together with their families which ordinarily are very large, a population of some four thousand souls living in the town or in the suburb of St Roch. There are two parishes with three monasteries, each with two religious although thirty could very comfortably live in them. The Jesuits[2] formerly owned a very fine house here which the New Company took over; mostly one now sees only ruins, the old stone buildings having been abandoned or taken over by the Indians who do not repair them, and Cavite, the second city of the Philippines, capital of a province which bears its name, is today nothing more than a wretched village with no other Spanish than the military officers or administrators, but although the town is only a mass of ruins the same cannot be said of the port where Mr de Bermudès, brigadier in the naval forces, who is in charge, has established an order which makes one regret that his talents are exercised on such a small scene; all his workers are Indians and he has exactly the same workshops as in our European arsenals. This officer who is of the same rank as the governor-general considers no detail to be beneath him and his conversation proved to us that none is beyond his ability. Everything we asked of him was carried out with the utmost graciousness. The forges, the pulley shop, the rope works laboured for several days for our frigates. Mr de Bermudès anticipated our

---

[1] Cavite's importance was related to its arsenal and shipyards. Manila, on the other hand, had continued to grow as trade and the administrative structure of the Philippines developed. The fortified town of Cavite, dominated by Fort San Felipe, was situated on an island in Cavite Bay, with San Roque on a peninsula further west; Cavite Viejo was a village on the south side of the bay on the Luzon mainland.

[2] The Jesuits had established themselves in Cavite in 1614, largely at the request of the Archbishop of Manila, in order to minister to the port workers. The church of Nuestra Señora de Loreto was opened in 1632, followed a little later by the Jesuit college which is no doubt the residence mentioned by La Pérouse.

wishes, and his friendship was all the more flattering in that one could tell from his character that he did not offer it easily and that his principles were marked by a certain austerity which possibly had not helped his career in the forces. As we could not hope to find anywhere in the world such a convenient port, Mr de Langle and I decided to have our entire rigging checked and our yards stripped. These precautions did not involve any loss of time because we had to wait at least a month before the various supplies we had requested from the administrator of Manila could arrive on board.

Two days after our arrival at Cavite we went with Mr de Langle and several officers to Manila. The crossing took only two and a half hours in our boats in which we had taken a number of soldiers on account of the Moors who often infest Manila Bay; our first call was on the Governor who kept us to dinner and gave orders to his captain of the guards to lead us to the archbishop, the administrator and the various *oidors*.[1] It was not the least tiring day of our campaign: it was extremely hot and we were walking in a country where the least of the citizens always travels in a carriage, but there were none to be hired, unlike in Batavia, and without Mr Sebir, a French merchant who was by chance told of our arrival and who sent us his coach, we would have been forced to give up the various calls we had planned.

Including its suburbs, the city of Manila is very extensive; its population is estimated at 38 thousand souls, among them fewer than a thousand or twelve hundred Spanish, the others being mixed bloods, Indians or Chinese who apply themselves to every accomplishment and every kind of work. The less wealthy Spanish households have one or more carriages; two of the finest horses available cost thirty piastres, their fodder and the driver's wages six piastres a month, so there is no place where a coach is cheaper to run and at the same time none where one is more necessary. The surroundings of Manila are charming; the finest river runs level with the ground and splits into several channels, the two main ones leading to the famous Laguna or Bay Lake[2] which is situated seven

[1] An *oidor* was a member of the *audiencia* or Supreme Court whose functions were both administrative and judicial.
[2] The Pasig River which divided the old town in two has its source in the Laguna de Bay, a large lake south-east of the town. Rebuilt in 1645 following a series of earthquakes, Manila was divided into quarters, Manila proper being on the left bank, Binondo and Tondo on the right bank, various suburbs on the islands of San Miguel and Isleta, and the Chinese quarter set apart at Parian.

CHAPTER XIV

leagues inland and lined with more than a hundred Indian villages and the most fertile of soils.

Manila, built along the bay that bears its name, the shore of which stretches for over twenty-five leagues, and at the mouth of a river which is navigable as far as the lake from which it issues, is possibly the most happily situated town in the world. Every type of food grows in the greatest abundance and is extremely cheap, but clothes, European smallwares and furniture fetch excessive prices; the lack of manufactures, the prohibitions and annoyances of every kind that affect trade mean that goods from India and China are at least as expensive as in Europe, and this colony where the various taxes bring nearly eight hundred thousand piastres into the coffers of the Treasury still costs Spain fifteen hundred thousand francs annually, which sum is sent to it from Mexico. The immense possessions of the Spanish in America have not allowed that government to deal with the Philippines; they are still like those lands belonging to great noblemen which remain fallow but could make a fortune for several families.

But I do not hesitate to assert that a very great nation that had no other colony than the Philippines and established there the form of government that was the most appropriate for it could contemplate without any feeling of envy all the European settlements of Africa and America.

Three million people reside in these islands, approximately a third of them in Luzon; these people impressed me as being in no way inferior to those of Europe; they cultivate their lands with intelligence, work as carpenters, joiners, blacksmiths, goldsmiths, spinners, masons. I have travelled to their villages and found them kind, hospitable and courteous, and although the Spanish speak of them and treat them with contempt I noticed that they blamed them for all the shortcomings of the administration they have set up among them; it is known that the greed for gold and the desire for conquests which animated the Spanish and the Portuguese two centuries ago led adventurers from these nations to travel through the various seas and islands of both hemispheres for no other reason than to find that precious metal.

No doubt a few goldbearing rivers and the proximity of spices were at the heart of the first Spanish settlements, but the results did not come up to expectations, and the original avarice was replaced by religious enthusiasm; legions of religious from every order were sent to preach Christianity, and the harvest was so abundant that soon there were eight or nine hundred thousand Christians in the different islands. Had this zeal been leavened with a little philosophy the system would have guaranteed Spanish supremacy and made these settlements useful to the home country, but the aim was always to make Christians and saints and never citizens. The people were split into parishes subjected to most detailed and outlandish practices; every transgression, every sin, is still punished by the whip – there is a tariff for absences from prayers and mass which is administered to men and to women at the church door by order of the priest; feasts, meetings of confraternities and private devotions take up an inordinate time, and, because in hot countries people get more excited than in more temperate climates, I saw during Holy Week masked penitents dragging chains through the streets, their legs and loins wrapped in a bundle of thorns, being lashed every time they stop in front of all the church porches or chapels, and in a word undergoing penances as harsh as those of Indian fakirs. These practices, more likely to create enthusiasts than truly devout people, have now been banned by the Archbishop of Manila, but they are probably encouraged, if not actually required, by certain confessors.

   This monachist regime which excites the soul and over-persuades these people (already lazy as the result of the climate and the lack of necessity) that life is only a transition and that the goods of this world are superfluities, combines with the impossibility of getting a satisfactory price for the fruits of the earth, which could make up for the labour involved; so once the inhabitants have the rice, sugar and vegetables that are necessary for their sustenance the rest loses all value; in such circumstances one has seen sugar selling for less than a *sol*[1] a pound and rice remains unharvested on the soil. I believe that the most knowledgeable group of men would be hard put to conjure up a more absurd form of government than the one which has ruled these colonies for two centuries; the port of Manila which should be free and open to all nations has until recently been closed to Europeans and merely half-opened to a few Moors, Armenians or Portuguese from Goa. The most despotic authority is entrusted to the governor. The *audiencia* that should act as a

---

[1] A *sol*, or *sou*, was a twentieth of the old French pound, the *livre*, the value of which at the time was considerably lower than the English pound.

moderating influence is powerless before the will of the King's representative; he can not only *de jure* but *de facto* expel or receive or confiscate the goods of foreigners who come to Manila in the hope of some gain and who venture there solely in the expectation of very high profits that are ruinous for the consumers.

There is no freedom; inquisitors and monks watch over consciences, *oidors* over private matters, and the government over the most innocent proceedings; a stroll inland on the island, a conversation, are its business – to sum up, the finest and most delightful country in the world is certainly the last one a free man would want to live in. I saw in Manila that honest and virtuous governor of the Marianas, that Mr Tobias who was too highly praised by the Abbé Rainal for his own good,[1] being persecuted by the monks whose wife they turned against him by depicting him as ungodly; she has asked for a separation in order not to have to live with a reprobate. All the fanatics cheered this decision. Mr Tobias is lieutenant-colonel of the regiment controlling the Manila garrison; he is recognised as one of the best officers in the country, but the governor has ordered that all his pay be handed over to his wife and has left him a mere twenty-six piastres for his subsistence and his son's. This worthy soldier was reduced to despair and waiting for an opportunity to escape from the colony to go and throw himself at the King's feet and beg him for justice. A very wise but unhappily ineffective legislation was supposed to moderate such excessive power – it allows each citizen to lay a complaint against the retiring governor before his successor, but it is in the latter's own interest to overlook everything that his predecessor is reproached for, and a citizen bold enough to complain runs the risk of new and worse vexations.

---

[1] Guillaume Raynal's *Histoire philosophique et politique du commerce et des établissements européens dans les deux Indes* was published in 1772. In book VI, ch. 22, he outlined the history of Guam and the Marianas which had stagnated until Tobias's arrival; he described him as 'an active man, humane, enlightened, [who] understood at last that the population would not recover, that it would even continue to fall, unless he succeeded in turning his island into an agricultural unit. This worthy idea led him to become himself a farmer. Following his example, the natives cleared the land the ownership of which he had guaranteed them....May this worthy and respectable Spaniard obtain one day that which would complete his happiness: the consolation of seeing a lessening in the passion these beloved children of his have for coconut wine and an increase in their application to work.' (Abridged edn by Yves Benot, Paris, 1981, pp. 104–5.) The misfortunes that later befell this administrator whose ideals were ahead of his time are due to jealousy, corruption, bigotry and marital troubles.

The most outrageous distinctions are made and maintained with the greatest strictness. The number of horses that may draw each carriage is determined for each social class; coachmen must give way in front of a larger number of horses, and by the mere whim of an *oidor* all the carriages which are unlucky enough to be going in the same direction must line up behind his coach; so many evils in the administration, so many vexations that result from them nevertheless have not altogether destroyed the country's advantages; peasants still have a look of happiness we do not find in our European villages, their homes are admirably neat and shaded by areca palm-trees[1] and other local fruit trees. The tax payable by the head of each family is quite moderate – it is five and a half *reales*, including the tax payable to the Church which is collected by the King, and all the bishops, canons and priests are paid by him but they have instituted a set of fees which make up for the slenderness of their wages.

A terrible and almost unbearable plague has appeared in recent years which militates against this remnant of happiness. It is a tax on tobacco; these people have such an immoderate passion for the smoke of this narcotic that there is not a single moment when an Indian man or woman does not have a *cigaro*[2] between the lips; children have hardly got out of their crib when they take up the habit. Luzon tobacco was the best in Asia; everyone grew some around his house for his own needs, and the small number of foreign vessels allowed to call at Manila took some to every part of India.

A prohibitory law has just been promulgated. All the tobacco plants belonging to individuals have been pulled up and cultivation restricted to fields where tobacco is grown for the King's profits. The price has been fixed at half a piastre a pound and although consumption has fallen drastically a workman's daily pay is not enough to buy the tobacco his family smokes in one day; everyone agrees that a tax of two piastres added to the capitation tax would have earned the Treasury an amount equal to the proceeds of the

---

[1] There are several varieties of areca palms in the Philippines, the best known of which is probably the betel-producing *Areca catechu*.

[2] The term 'cigar' was not widely known in France at the time and Milet-Mureau felt that he needed to enlighten his 1797 readers with a marginal note explaining that it was made with rolled-up tobacco leaves that one then smoked. One finds in France at the time the term *cigale* by association with the word for grasshopper.

tobacco sales and would not have given rise to the same disorders. Serious uprisings have occurred everywhere in the island, soldiers have been called out to quell them, an army of tax collectors is employed to prevent smuggling and compel consumers to apply to the royal offices: several have been murdered but quickly avenged by the tribunals which bother with far fewer formalities when dealing with Indians than with other citizens.

There is still a remnant of discontent which the slightest spark could turn into a very real danger, and there can be no doubt that an enemy country intent on conquest would find an army of Indians placing themselves under its orders the moment it set foot on the island and brought them weapons. This picture of the present state of Manila is quite different from what could be depicted within a few years if the Spanish government at last adopted a better constitution. The soil welcomes the most precious crops and in the island of Luzon nine hundred thousand people of both sexes can be encouraged to till it. This climate enables silk to be harvested ten times a year whereas China's hardly allows one to hope for two crops.

Cotton, indigo, sugar canes and coffee grow wild in the midst of the natives who ignore them; there is every indication that spices would not be inferior to those of the Moluccas. Total free trade for all nations would ensure a certain sale which should then encourage the cultivation of all these plants; a modest duty on all exports would, within a few years, be enough to meet all the government's expenses; religious freedom and a few privileges if granted to the Chinese would attract into the island a hundred thousand inhabitants from the eastern provinces of that empire, whom the tyranny of the mandarins is driving towards the sea, and if the Spanish added to these advantages the conquest of Macao, their settlements in Asia and the benefits their commerce would draw from them would certainly exceed those of the Dutch in the Moluccas and Java. The foundation of the New Company of the Philippines[1]

---

[1] The Real Compañia de Filipinas was founded by King Carlos III in 1785. It was granted the right to trade directly with Spain and to deal with China and the Indian states. Formerly, goods were carried by the galleon which made an annual crossing from Manila to Mexico. As this had ensured a highly lucrative monopoly for certain traders, the arrival of the Royal Company and the prospect of Manila eventually becoming a free port aroused bitter hostility. An attempt back in 1765–77 to establish direct trade links between Spain and the Philippines – the voyage of the *Buen Consejo* – had given rise to the same opposition, but royal support for freer

proves that the government's attention has finally turned towards this part of the world; it has adopted, albeit in part, the plan of Cardinal Alberoni;[1] this minister had felt that Spain, having no manufactures, would do better if her precious metals helped the nations of Asia to become wealthier, instead of those of Europe who were her rivals, whose trade she enriched and whose strength she increased by consuming their industrial products. Accordingly he believed it was his duty to make Manila into a trading centre open to every country, and he wanted to invite shipowners from the various provinces of Spain to go there to obtain the cloth and other material from China and India which the colonies and the metropolis required.

We know that this minister had more intelligence than knowledge. He knew Europe fairly well, but he had not the slightest idea about Asia. The items most in demand in Spain and her colonies are those of the Coromandel coast and Bengal: it is certainly as easy to ship them to Cadiz as to Manila which is a considerable distance away and subject to monsoons that expose such navigation to substantial risks and delays, so that the difference between prices in Manila and in India must be at least fifty per cent, and if to this price one adds the extremely high cost of organising shipping in Spain for such a faraway country, one realises that goods transiting through Manila need to be sold at a very high price in continental Spain, and even more so in Spanish America, and that countries which, like England, Holland and France, trade direct will always be able to engage in smuggling and make great profits. It was nevertheless this ill-conceived plan which lay at the basis of the New Company's programme, but with restrictions and prejudices that make it a hundred times worse than the Italian minister's and such that it seems impossible to me that this company can survive for four years, even though its monopoly has so to speak swallowed up the

---

trade was now proving a real threat to the old guard; they were saved by the European conflict of 1793–6 which partly isolated the Philippines and seriously affected eastern trade. See on these struggles Manuel Azcarraga y Palmero, *La Libertad del comercio en las islas Filipinas*, Madrid, 1871, M.L. Diaz-Trechuelo Spinola, *La Real Compañia de Filipinas*, Seville, 1965, and W.L. Schurz, *The Manila Galleon*, New York, 1939.

[1] Giulio Alberoni (1664–1752) became Spanish chief minister in 1715. An energetic reformer, he transformed the old administrative structures and took prompt steps to develop Spain's external trade. He was dismissed in 1719.

whole of the nation's trade in its American colonies; the so-called Manila Fair where the New Company is to obtain its supplies is only open to Indian countries, as if one was afraid that sellers might compete with each other and cloth from the coast and Bengal might be available too cheaply.

It has not anyhow been noticed that these supposed Armenian or Goanese flags merely cover English goods, and as these different disguises involve additional costs, the consumers have to pay, and the difference in price between India and Manila is not fifty per cent, but sixty and even eighty.

This flaw is compounded by the company's exclusive right to purchase the products of Luzon where trade is not stimulated by any competition on the part of buyers, so that it will always remain in the state of inertia that has stifled it for two centuries. Enough authors have commented on Manila's civil and military administration, and I felt that I should make the town known in this new context which, following the foundation of the New Company, may be of some interest in a century when everyone who is called upon to hold some position in the state is aware of the theory of commerce.[1]

The Spanish have a few settlements on various islands to the south of Luzon, but they are only really on sufferance there and the situation in which their subjects in Luzon are placed in no way commits the inhabitants of other islands to recognise their sovereignty. On the contrary they are constantly at war, and these so-called Moors I have already mentioned, who infest their coastline, make such frequent landings and carry off into slavery Indians of both sexes who have accepted Spanish domination, are inhabitants of Mindanao, Mindoro and Panay who recognise no other authority than that of their local sovereigns who are as incorrectly termed sultans as they themselves are wrongly called Moors. They are in reality Malays and took up Mohammedanism at about the same time as Christianity was first brought to Manila. The Spanish called

---

[1] La Pérouse's strong support for greater freedom in international trade reflects the ideas of the Physiocrats, a group of French economists led by François Quesnay (1694–1774) who were highly influential in the years leading up to the French Revolution. They favoured free trade and laisser-faire, as well as a single tax – a view reflected in La Pérouse's comments on an economic policy for the Philippines. Their belief that all wealth originates in agriculture meant that their influence rapidly waned as the Industrial Revolution began to transform western Europe and give rise to new economic theories.

them Moors and their rulers sultans because they shared the religion of people from Africa with whom they were also at war for so many centuries. The only Spanish military outpost in the southern Philippines is Somboango[1] on Mindanao Island, where they maintain a garrison of 150 men commanded by a military governor appointed by the governor of Manila. On the other islands there are only a few villages defended by some wretched guns that have to be serviced by militias and commanded by *alcaldes* chosen by the Governor General, who can be selected from any class of citizens and who are not soldiers. The real rulers of the various islands where the Spanish villages are situated would have soon destroyed them, had it not been in their own interest to preserve them. These so-called Moors maintain peace in their own territories, but they send out their subjects to carry out piratical raids on the coasts of Luzon, and the *alcaldes* buy a very large number of slaves taken by these pirates, which saves them the trouble of taking them to Batavia where they would get a much lower price for them. These details will give a better picture of the weakness of the Philippines administration than all the arguments of the various travellers; the readers will realise that the Spanish are too weak to protect their subjects and that all the blessings they have brought these people have so far had no other purpose than to ensure their happiness in the next world.

We spent only a few hours in Manila, and as the governor left us after dinner to have his siesta we were free to call on Mr Sébir whose carriages had been placed at our disposal the moment he learned of our arrival and who rendered us the most essential services during our stay in Manila Bay. This French merchant, the most enlightened man of our nation I met in the China Seas, had believed that the New Philippines Company and links with offices in Madrid and Versailles would enable him to extend his speculations which had been cut back following the re-establishment of the Indies Company. Consequently he had wound up his business in Canton and Macao, where he had settled several years previously, and set up a business house in Manila where in addition he had been asked to follow through a very important deal affecting one of his

---

[1] Zamboanga, in western Mindanao, played an important role during the wars waged in the seventeenth century against the Moors. Abandoned in 1662, the port was reopened in 1718 and had become a major Spanish base in the Sulu Sea.

CHAPTER XIV

friends; but he was already coming to the conclusion that prejudices against foreigners and the administration's despotism would present insurmountable obstacles to his plans, and when we arrived he was considering closing down rather than developing his business.

We got back into our boats at six o'clock and were back on board our frigates at eight, but for fear that, while we were busy in Cavite with the repairs to our vessels, the suppliers of biscuit, flour &c might make us endure the usual dilatoriness of their nation's merchants I thought it necessary to order an officer to stay in Manila and pay daily visits to the various suppliers the administrator had referred us to. My choice settled on Mr de Vaujuas, a lieutenant on the *Astrolabe*: but this officer soon advised me that his remaining in Manila was unnecessary, and that Mr de Gonsoles Carvagual,[1] the Philippines' Intendant, was taking so much trouble on our behalf that he went every day to verify personally how the workers dealing with our frigates were progressing and that he was as vigilant as if he had himself been part of the expedition; his attention and his kindness deserve a public acknowledgment. His private natural history collection was available to all our naturalists to whom he showed various specimens belonging to the three kingdoms of nature he had put together, and when I left I received from Mr de Gonsoles the gift of a complete collection, in duplicate, of all the shells that are found in Philippine waters. His courtesy extended to anything which might be of interest to us. A week after our arrival we received a letter from Mr Stokinstron, chief supercargo of the Swedish Company, advising us that he had sold our furs for ten thousand piastres and empowering us to draw this amount on him. I was very anxious to obtain these funds in Manila in order to share them among the crews who, having sailed from Macao without receiving any of this money, were worried that they would never see it. Mr Sebir had no funds to transfer to Macao at this time. We approached Mr Gonsoles who was quite unaccustomed to this kind of transaction, but he used the influence which his friendliness has gained him with the various merchants of Manila to press them to discount our bills of exchange, and the funds this released were divided among the sailors before we sailed.

The intense heat of Manila began to have its effect on the health of our crews. Several sailors were affected by colics which

[1] More properly Ciriaco Gonzalez Carvajal.

fortunately had no serious consequences. But Messrs de Lamanon and Daigremont who had brought from Macao signs of incipient dysentery, presumably caused by dried up perspiration, far from finding any relief on land, saw their condition worsen to the point where Mr Daigremont was in a hopeless condition on the 23rd day after our arrival and died on the 25th. This was the second death occasioned by sickness on board the *Astrolabe*, and such a misfortune had not overtaken the *Boussolle* although our men possibly were generally in a less healthy state than those of the other frigate. But it must be said that the servant who had died during the crossing from Chile to Easter Island[1] was a consumptive when he was signed on and that Mr de Langle had agreed to his master's request that he be taken on in the belief that the sea air and hot climates would cure him. As for Mr Daigremont, in spite of his doctors and unbeknownst to his comrades and friends, he wanted to cure his sickness with mulled brandy, hot spices and other remedies which the strongest person would have been unable to cope with, and fell victim of the over-optimistic opinion he had of his own constitution.

Nevertheless, as early as 28 March our work in Cavite was completed, our boats built, our sails repaired, the rigging inspected, the frigates entirely caulked and our salt provisions stored in barrels. We had not wanted to leave this work to suppliers in Manila; we knew that the galleons' provisions had never kept for three months, and we had a great faith in Captain Cook's method and consequently we gave each salter a copy of Captain Cook's process[2] and carried out ourselves this new type of work; we had salt and vinegar on board and merely bought from the Spanish some pigs, the price of which was very moderate.

Communications between Manila and China are so frequent that every week we received news from Macao, and we learned to our great surprise that the vessel *Resolution* commanded by Mr d'antre Casteaux and the frigate *Suptile* under Mr de la Croix de Castries

---

[1] Jean Foll or Le Fol, Vaujuas's servant, died on 11 August 1786 off the coast of British Columbia, later therefore than La Pérouse states here.

[2] James Cook described his method in his journal on 23 November 1773: 'in the cool of the evenings, the Hogs were killed and dressed, then cut up the bones taken out and the Meat salted while it was yet hot, the next Morning we gave it a second Salting, packed it in a Cask and put to it a sufficient quantity of Strong Pickle, great care is to be taken that the meat be well covered with pickle other wise it will soon spoile.' Cook, *Journals*, II, p. 296.

had arrived in Canton River.¹ These ships, sailing from Batavia at a time when the N.E. monsoon was at his strongest, had gone north to the east of the Philippines, having coasted New Guinea and crossed reef-strewn seas for which they had no charts, and after a 70-day voyage from Batavia had reached the entrance of the Canton River where they dropped anchor the day after I left it. The astronomical observations made during this voyage will be of considerable help to increase our knowledge of these seas which are always open to ships that have missed the monsoon, and it is certainly surprising that our India Company had chosen a captain who had no knowledge of this route to command its vessel *La Reine*.²

I received a letter from Mr D'antre Casteau in Manila, advising me of the purpose of his voyage, and shortly afterwards the frigate *Suptile* herself came to bring me further letters.

Mr de la Croix de Castries gave us all the news from Europe which had been past the Cape of Good Hope with the *Calipso*, but they were dated April 24th and our curiosity had still a gap of a year to mourn over; furthermore our families and our friends had not seized this opportunity to write to us, and in the then peaceful state of Europe public happenings were of minor interest compared with news of what causes fear and hope in the hearts of every individual. So we had yet another opportunity of sending our letters to France, and the *Suptile* was in a good enough condition to allow Mr de la Croix de Castries to make up the losses we had sustained in America; he gave four men and an officer to each of the frigates. Mr Guyet, *enseigne de vaisseau*, was transferred to the *Boussole* and Mr le Gobien, *garde de la marine*, to the *Astrolabe*. These reinforcements were very necessary: we had eight fewer officers than when we left France, including Mr de St Ceran whose health

---

[1] Joseph-Antoine Bruny d'Entrecasteaux (1737–93), commander of the French naval station in the Indian Ocean, sailed to Canton in the *Résolution* and the *Subtile* against the prevailing monsoon, a task generally considered to be impossible; he did so successfully and without loss. He was later sent to search for La Pérouse's lost expedition. Anne-Jean-Jacques-Scipion de la Croix de Vagnas, Vicomte de Castries, (1756–1826), had been in command of the *Ariel*, a ship captured by La Pérouse during the American wars when, in August 1784 he was given command of the *Subtile* with orders to sail to the Indian Ocean. He emigrated during the French Revolution, but returned to France in 1803 and served in the navy until 1817.

[2] The *Reine* left too late from India and was unable to battle her way against the monsoon.

was such that I was forced to send him back to the Isle de France in the *Suptile*, all our surgeons having declared that it was impossible for him to continue on the voyage.

Meanwhile our provisions had been stored on board by the date the Intendant had promised, but Holy Week, which suspends all business in Manila, caused some delays in obtaining our personal supplies, and I was forced to put off my departure until Easter Monday. As the north-east monsoon was still very strong a sacrifice of three or four days could not adversely affect the success of the expedition. We loaded back all our astronomical instruments on April third; Mr D'agelet had not, since our departure from France, found a better place to check the accuracy of No. 19; he verified to his full satisfaction that this chronometer lost 12" a day compared with Paris mean time, a difference close to what had been observed at Macao although the temperature differed by ten degrees. We had set up our observatory in the government's gardens at roughly one hunded and twenty *toises* from our ships. The longitude as determined by a very large number of observations of distances was 118 degrees 50 minutes 40 seconds and the latitude measured with a quadrant with a radius of three feet 14 degrees 29 minutes 9 seconds; if we had wanted the longitude of Cavite in accordance with the daily loss attributed to our No. 19 timekeeper it would have been 118^d 46' 8", i.e. 4' 32" less than the result of our observations of distances. Before we sailed, I felt it our duty to go with Mr de Langle to express our thanks to the Governor General for the speed with which his orders had been carried out, and even more to the Intendant from whom we had received so much attention and so many courtesies. Having carried out these duties we both took advantage of a couple of days spent at Mr Sebir's to visit by carriage or boat all the surroundings of Manila; one comes upon no great houses, parks or gardens, but the countryside is so beautiful that a simple Indian village on a riverbank or a European-style house surrounded by a few trees presents a more picturesque spectacle than any offered by our finest chateaux, and the dullest imagination can conjure up an image of happiness in this charming simplicity. Almost all the Spanish traditionally leave the city after the Easter celebrations and spend the hot season in the country. They have not tried to beautify a country which has no need of embellishments; a house similar to a presbytery built at the water's edge with very convenient baths, although without a drive or a garden but shel-

CHAPTER XIV

tered by a few fruit trees – such is the residence of the wealthiest of them, and it would be one of the most pleasant places to live in anywhere if a milder administration and fewer prejudices allowed more civil freedom to the inhabitants. The city's defences have been added to by the Governor General under the supervision of Mr Sauz,[1] a talented engineer, but the garrison is small indeed: it consists in peacetime of a single infantry regiment of two battalions, each made up of a company of grenadiers and eight of fusiliers, the two battalions making an effective total of thirteen hundred men. This regiment is Mexican, all the soldiers are the colour of mulattoes; people assert that they are in no way inferior in bravery and intelligence to the European troops, two artillery companies of eighty men, each commanded by a lieutenant-colonel and having as officers a captain, a lieutenant, an ensign and a supernumerary, three dragoon companies forming a squadron of 150 horse commanded by the oldest of three captains, and finally a militia battalion of twelve hundred men formerly set up and paid for by a very rich Chinese of mixed blood named Tuasson.[2] All the soldiers belonging to this corps are part-Chinese; they carry out the same duties in the town as the regular troops and now receive the same pay from the King, but they would be of very little assistance in case of war. Finally, should the need arise, a militia of eight thousand men can be called up very quickly, which is divided in provincial battalions commanded by European or creole officers; each battalion has a company of grenadiers; one of these companies has been trained by a sergeant who had retired from the Regiment, living in Manila, and the Spanish who are more inclined to criticise the bravery and merit of the Indians than to overpraise them claim that this company is in no way inferior to the European regiments.

The small garrison at Somboango on the island of Mindanao is not drawn from that of Luzon, and two corps of 150 men each have been created for the Marianas Islands and Mindanao which are invariably attached to these colonies.

---

[1] The name is probably a deformation of Souza.

[2] The Chinese in Manila were in a precarious situation. Seldom made welcome by the authorities and often ill-treated, they had rebelled on several occasions and in 1762 when English forces captured Manila a number supported the invaders. In order to show their loyalty towards the Spanish it was not uncommon for Chinese merchants to adopt a local or a Spanish name.

CHAPTER XV

Departure from Cavite. We go along the island of Luzon towards the north. The monsoon has not yet reversed but the winds are more stubborn in the Formosa Channel than on the Luzon coast where we find changeable winds. We set course to cross the channel, hoping to progress more easily north along the Chinese mainland; we encounter a shoal in the middle of the channel; latitude and longitude of this shoal; we turn back towards Formosa and anchor two leagues from the old fort of Zealand; only one boat dares to come up to the frigate and we can get no information from it about the war we know is in progress between this Chinese colony and the home country. We sail the next day and the same evening come upon the Chinese army in the Piscadorès Channel, making for a large river on Formosa which we cannot approach closer than three leagues on account of the banks found at the rivermouth which extend far out to sea; we are struck by a squall and are forced to weigh anchor before dawn; the Chinese army is scattered, and after making various tacks between Formosa and the Piscadorès we decide to bear away to avoid spending the night in a very narrow channel in such bad weather. We coast the Piscadorès; details on these islands. Rain and thunder herald the monsoonal change. We sight the island of Botol Tabagoxima; it is inhabited; the next day we coast the island of Kumi which forms part of the kingdom of Lixeu; the inhabitants arrive and receive our gifts; they urge us to drop anchor near their village, but we go on our way and during the night come upon two uninhabited islands situated to the east of Formosa's northern headland. We finally enter the Sea of Japan and sail along the coast of China which remains constantly fogbound; we waste a fortnight on this useless navigation during which our horizon never reached half a league, and we finally sail for Quelpaert Island, the first landmark when entering the strait between Korea and Japan. We sail along the coast of Korea and daily make astronomical observations which allow us to rectify errors on the charts. Details on Quelpaert Island, Korea, &c. Discovery of D'agelet Island 20 leagues from Korea where we see shipyards and

11. View of Cavite, Philippines, showing boat with outrigger, by Duché de Vancy. SHM 352:17.

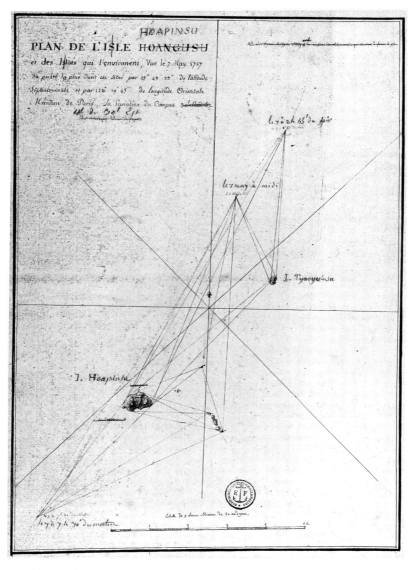

12. Chart of Hoapinsu (Huaping Hsu) and surrounding islets sighted on 7 May 1787. By Blondela. Actual size 34 cm × 47 cm. AN 6 JJ1:39.

CHAPTER XV

boats being built. Latitude and longitude of D'agelet Island. We set our course towards the north-western point of Japan; sighting of Cape Noto and Joolsissima; details on this island, latitude and longitude of this part of Japan. Meeting with several ships belonging to this nation. We return to the coast of Tartary where we land in 42 degrees; we sail along it very close in in spite of the mists, always waiting for the weather to clear so as to leave no gaps on our charts. We put into Ternay Bay; its products, details on the country; we sail after a stay of only three days. Second stay, in Suffren Bay; enormous quantity of cod obtained there. Continuing our northerly route we sight a peak I have named Pic Lamanon; it bore east, and we realise for the first time that we were sailing in a channel we had not noticed because we had hugged the coast of Tartary. We sail towards this island; stay in Baye de Langle; practices and customs of the inhabitants; what they tell us causes us to continue north up to the mouth of the Segalien River; we find the channel between the island and the Tartary mainland obstructed by shoals. Stay in Destiny Bay; salmon in the river still on the newly discovered island, and in Castries Bay on the Tartary coast; practices and customs of the inhabitants who are quite different from the islanders from whom however they are only separated by a very narrow channel; products of the country, shells, birds, fish, &c.

On April ninth according to our style of reckoning and on the tenth according to the Manilans,[1] we sailed with a good N.E. breeze which made us hope that we could round in daylight all the islands of the various passes of Manila Bay. Before sailing, Mr de Langle and I received Mr Bermudès who assured us that the N.E. monsoon would not change for a month and that it would be even later on the coast of Formosa, the Chinese mainland being so to speak the birthplace of the northerly winds that reign for more than nine months of the year over the coasts of that empire, but our impatience did not allow us to listen to the voice of experience; we placed our faith in exceptions: the monsoonal change might occur

---

[1] Having sailed from east to west, La Pérouse had crossed the date line and should have lost a day. By keeping to his own dates, he was following James Cook's practice who only changed his dates when he reached a European port. The route tables of both the *Boussole* and the *Astrolabe* make it clear that La Pérouse sailed from Cavite on 10 April 1787.

at a different time from year to year, and we took our leave. Slight changes in the wind soon enabled us to reach the north of Luzon. I have already mentioned that the monsoon winds are never felt there within a coastal belt of some twelve or fifteen leagues, and one of the advantages of the Manila trade is that one is able to carry on a coastal trade with China at any time of the year. We had hardly turned Cape Bojador[1] when the N.E. winds settled in with a stubbornness that proved only too well how right Mr de Bermudes had been. I had a slight hope that I might find under the lee of Formosa the same varieties as I had met off Luzon, although I could not ignore the fact that the proximity of the Chinese mainland made this unlikely, but in any case all I could do was wait for the monsoon to reverse, as the poor sailing qualities of our frigates which were timber-bottomed and clouted gave us no hope of progressing north against contrary winds. We sighted this island on 21 April. We came upon very fierce tidal currents in the channel separating it from the island of Luzon; they seem to be caused by a normal tide, as our reckoning was no different from our observations of longitude and latitude.[2] On 22 April I saw Lumai Island which is off the S.W. point of Formosa bearing E¼E distant about three leagues.[3] The sea was very rough and the appearance of the coast persuaded me that I would gain more easily the north if I could reach the coast of China. The N.N.E. winds allowed me to sail N.W. and thereby gain in latitude, but in the middle of the channel I noticed that the sea had greatly changed; we were then in 22^d $57'$ of latitude and west of the Cavite meridian, that is to say in 116^d $41'$ of longitude, Paris meridian; the sounding-line gave twenty-five fathoms, sandy ground, and four minutes later only 19 fathoms; this too rapid change in depth convinced me that this was not a depth associated with China from which we were still thirty leagues away, but that of a shoal not indicated on the charts; I continued to take soundings and soon found only twelve fathoms; I changed course towards Formosa Island and the depth continued to be as irregular as before; I then felt it wise to drop anchor and gave the signal to the *Astrolabe*; the night was fine. At daybreak we

[1] Cape Bojeador is the western cape at the north of Luzon.
[2] The tidal rips and currents of the Babuyan Channel, to the east of the route being followed by the two frigates, are particularly violent.
[3] The reference is probably to rocky islets that lie off the south-west cape, Byobi To.

saw no breakers around us; I gave signal to get under way and set sail N.W. for the Chinese mainland, but at nine o'clock in the morning the sounding-line reporting 21 fathoms and a minute later eleven fathoms, rocky bottom, I felt that I ought not to continue such a dangerous investigation, our boats sailing too badly to be able to take soundings ahead of the frigates and indicate the depths; I therefore decided to get away by the same point of the compass as I had arrived and set course for the SE¼E. We thus covered six leagues over an uneven ground of sand and rocks of from 24 to eleven fathoms; the depth then increased and we lost it completely at 10 p.m. approximately eleven leagues from where we had changed course in the morning; this bank whose extent to the north-west we have not determined but from the centre of the line we followed is in 23 of latitude and 116^d 43 of longitude, Paris meridian and its S.E. extremity in 22^d 52 of latitude and 117^d 3 of longitude may not be dangerous as the smallest depth we found was 11 fathoms, but the nature and unevenness of the bottom makes it very suspect, and it is worth noting that these shoals which are very common in the China seas have nearly all tops at sea level which have been responsible for numerous shipwrecks. Our course brought us back to the Formosa coast near the entrance to the bay of the old fort of Zealand where the town of Thaon, the capital of this island, is situated.[1] I had been advised of the rebellion in the Chinese colony and I knew that an army of twenty thousand men, commanded by the Santoq of Canton, had been despatched against it. I would have liked to obtain news of this event, the north-eastern monsoon which was still strong allowing me to waste a few days without endangering the rest of the campaign. I anchored outside this bay in seventeen fathoms, and our boats found 14 fathoms a league and a half from shore, but I was aware

---

[1] The Dutch built the fort of Zeelandia in 1624 at the port of Tainan (La Pérouse's Thaon) which rapidly became an important commercial centre. The Spanish meanwhile were setting up their own settlements in the north of the island: at Chilung (San Salvador) in 1626 and Tanshui (San Domingo) in 1629; these fell to the Chinese in 1642. In 1661 Koxinga (Cheng Cheng-Kung) besieged Tainan with 25,000 men and it fell in February of the following year. From that time, Taiwan (still known to Europeans as Formosa) was administered as a Chinese possession, a dependency of Fukien (Fujian) province. Chinese immigration followed, especially after 1683 when the Chinese conquest can be considered as completed; this led to growing unrest and a number of revolts – no fewer than 23 occurred between 1683 and 1895 when the island was ceded to Japan.

that one could not approach close to the island, that there was only a depth of seven feet in Thaon harbour, and that, when the Dutch owned it, their ships were forced to stay in the Piscadorés Islands where there is a very good harbour which they had fortified.[1] These circumstances made me very unsure about sending a boat ashore, which I could not support with my frigates and which in all likelihood would appear suspect in the warlike state of this Chinese colony; the best I could hope for was that it would be sent back without having been allowed to land; if on the other hand it was held, my position would become very embarrassing and a couple of sampans set on fire would be a poor compensation for such a misfortune. Accordingly I decided to try to attract to the ships Chinese boats passing nearby; I showed them piastres which seemed to me to be the most effective magnet for these people, but apparently any communication with foreigners is forbidden them; it was obvious that we did not inspire any fear in them, since they were passing within range of our weapons, but they refused to come alongside, and only one had the courage to do so; we bought his fish at the price he asked so that he might give us a good name among the other boats if he dared admit he had communicated with us. We could not guess the replies these fishermen were making to the questions we asked them – which undoubtedly they could not understand; not only does the language of these people have no links with European languages, but this kind of pantomime language we think is universal was not any more successful and a head movement which means 'yes' among us may well have a diametrically opposed meaning with them; this little experience persuaded me even more that my curiosity would not be satisfied even if my boat met with the best of welcomes, and I decided to set sail in the morning with the land breeeze. Various fires I saw lit on the coast which I took to be signals made me think that we had caused some alarm, but it was more than probable that the Chinese and the rebel armies were not near Thaiyions where we saw only a very small number of fishing boats which, in a trouble spot, would have made their way elsewhere. What was only a guess became a certainty. The following day, the sea and the land breezes having allowed us to sail ten leagues north, we saw the whole Chinese army at the

---

[1] The port is Penghu where the Dutchman Jan Pieterszoon Coen built a fort, which however had to be abandoned two years later.

CHAPTER XV

mouth of a large river[1] situated in 23 degrees 25 minutes, but whose bar extends 4 or 5 leagues to sea; we sailed athwart this river in 37 fathoms, muddy ground. We could not count all the vessels; some had their sails up, others were at anchor near the coast and a very large number were inside the river. The admiral ship, covered in a large number of flags, was the furthest away; it dropped anchor at the edge of the sandbank one league to the east of our frigates. As soon as night fell, it placed lights on all its masts to serve as a rallying point for several vessels still to windward which, forced to pass close to our frigates to join their commander, took great care to stay at a wide cannon shot, no doubt not knowing whether we were friend or foe. Moonlight allowed us to observe all this until midnight and never have we wished more ardently for fine weather so that we might see the next day's events. The day before, we had sighted the southern Piscadores islands bearing West¼ Northwest. It is likely that the Chinese army having left from Fokien province had gathered on the island of Pombu,[2] the largest of the Piscadores where there is a very good harbour, and had left there that very day to begin its operations, because at 7 p.m. a large part of this fleet was still in the channel; but the weather became so bad that we could not entertain any other plan than setting sail before dawn in order to save our anchor, which we could not have done if we had put off this task for one hour. The sky darkened; at 4 a.m. a wild gale blew up; the horizon did not permit us to see land; nevertheless, at daybreak I saw the admiral ship running before the wind towards the river with a few other sampans I could still discern through the mist; I made for the open sea, all the main sails reefed; the winds were blowing N.N.E. and I hoped to round the Piscadorés while heading N.W.

But to my great astonishment I saw at nine a.m. several rocks forming part of this group of islands bearing N.N.W. and the weather was so hazy that we had been unable to make them out until we were quite close; the breakers surrounding them were indistinguishable from those caused by the breaking waves. Having never seen such a rough sea in my life, I veered back towards

---

[1] This is the mouth of the Pei Chiang River.
[2] There are several large ports in Fujian province from which the expedition could have set out, such as Foochow (Fuzhou), Changchow (Quanzhou) and Amoy (Xiamen). La Pérouse's Pombu is Penghu – the Chinese name of the Pescadores ('Fishermen') Islands is Penghu Liehtao.

Formosa at nine a.m. and at midday the *Astrolabe* which was ahead of us signalled a depth of twelve fathoms as she was changing tack; I at once took soundings and found forty fathoms; therefore in less than a quarter of a league one drops from forty fathoms to twelve, and presumably, soon after, from twelve to two because the *Astrolabe* found only eight while changing course and it was probable that this frigate had not followed this tack for four minutes; which taught us that the channel situated between the islands to the N.E. of the Piscadorès and the Formosa banks is no more than four leagues; this was somewhat narrow to tack at night in frightful weather with an horizon of less than a league, and a sea that was so rough that each time we tacked with a following wind we ran the risk of being swamped by the waves. These various circumstances made me decide to work my way to pass to the east of Formosa; my instructions did not require me to take the channel route, and it was anyhow all too obvious that I could not do so until the monsoon changed, and as this change, which was probably not very far off, is heralded by a very wild gale I felt it wiser to cope with this out at sea, and I set course for the southern Piscadorés Islands which I could see to the west-south-west. Since I was forced to follow this route I wanted at least to survey them as much as the bad weather allowed; we coasted along them at a distance of two leagues and they seem to extend south up to 23^d $12'$ although Mr D'après' map places them 13 minutes further north. We are not quite so certain about their northern limits; the most northerly stretched out to 23 degrees 25 minutes, but we do not know whether there are any beyond that.

These islands are groups of rocks assuming all kinds of shapes; there is one that exactly resembles the tower of Cordouan which is at the entrance to the Bordeaux river[1] and one could have sworn that this rock was cut by humans, but among these islets we counted five islands of a medium height having the appearance of sand dunes; we saw no trees on them; to be frank, the frightful weather endured on this day makes this a very tentative comment, but these islands must be known from the reports of the Dutch who had fortified Pombu port when they were masters of Formosa,

---

[1] This famous lighthouse in the Gironde estuary goes back to the ninth century. It was substantially rebuilt in 1586–1611 and further extended in 1788, plans for this work being under discussion just before La Pérouse sailed from France.

CHAPTER XV

and we know that the Chinese still maintain there a garrison of five or six hundred Tartars who are relieved annually.

As the sea had improved within the lee of these islands, we took several soundings and found such an uneven sandy bottom that the *Astrolabe*, which was within musket range, had forty fathoms while our sounding-line reported only twenty-four and soon we lost the depth altogether. Since night was beginning to fall I set course for the S¼SE and in the morning veered back East S.E. to pass through the channel between Formosa and the Bahee Islands;[1] the next day we were struck by a squall that was as fierce as the day before's, but it only lasted until ten p.m.; it was preceded by heavy rain such as one sees only in the tropics and the sky was afire throughout the night; the fiercest lightning struck every point of the compass, but we heard only one thunder clap. We ran with the wind under the foresail and two topsails, fully reefed, making for the S.E. to round Vella rette[2] which, according to the bearings we had taken before nightfall of the southern point of Formosa, should have been lying four leagues to the east. The winds stayed northwesterly throughout the night but the clouds were being driven very strongly to the S.W. and only a fog which did not extend a hundred *toises* over our heads was affected by the lower winds. I had noticed this for several days, and this observation was of considerable help in deciding me to make for the open sea during this crisis of Nature, which the full moon helped to confirm. We were totally becalmed for the whole of the following day and halfway through the channel between the Bahée Islands and Bottol Tabagoxima[3] the width is sixteen leagues, our observations placing the S.E. headland of Bottol tabagoxima in 21 degrees 57 minutes of latitude and 119 degrees 32 minutes of longitude Paris meridian. The winds allowing me to come within two-thirds of a league of

---

[1] The Bahée Islands are William Dampier's Bashee Islands, so named by him after a local drink which his men greatly enjoyed, 'very much like our English Beer both in colour and Taste.' Dampier, *A New Voyage Round the World*, London, 1703, p. 431. Dampier spent from 6 August to 3 October 1687 among the islands which are the present-day Batan Islands between Formosa and Luzon. The channel itself, through which the French were passing, is still known as the Bashi Channel.

[2] Vela Rete Rocks, formerly known as Shichesi Seki, is the Chhsing Yen group, in latitude 21°44′N and longitude 120°50′E, some eight miles south of Taiwan.

[3] Formerly known as Koto Sho, this island, Hungfou Hsu, lies in latitude 22°3′N and longitude 121°33′E (119°13′E of Paris). The small island close to it is Hsiaohungtou Hsu (known to the Japanese as Shokoto Sho).

this island, I clearly saw three villages on the south coast, and a canoe seemed to be making for us; it would have been rather interesting to visit these settlements which presumably are inhabited by the same people as those of the Bahé Islands whom Dampierre describes as being so good and so hospitable, but the only bay that seemed to promise an anchorage was open to the S.E. winds which seem to blow incessantly because the clouds were being driven strongly from this direction and at around midnight they effectively settled S.E. and enabled me to sail NE.¼N, a trend given to Formosa Island by Mr D'après as far as 23 degrees 30 minutes. We had sounded several times near Botol tabagoxima and half a league from land without finding ground. Every indication is that if there is an anchorage it is very close to the shore. This island, on which no traveller has landed, has a circumference of roughly four leagues; it is separated by a channel half a league across from an islet or very large rock on which one can see a little greenery with a little scrub, but which is neither inhabited nor habitable.

On the other hand, the island seems to contain a fair number of inhabitants as we counted three villages within a distance of one league; it is wooded from a third of the way up from the shore to the peak which seems to be topped with very large trees. The space between these forests and the beach slopes very rapidly still, but it is of the brightest green, cultivated in several places and cut by ravines formed by torrents coming down from the hills. I think that Botol tabagoxima is visible fifteen leagues off when the weather is clear, but this island is often surrounded by fog and it would appear that Admiral Anson only first saw the islet I mentioned which has not half Botol tabago's height.[1] After rounding this island we sailed

---

[1] 'The 3rd of November [1742], about three in the afternoon, we saw an Island, which at first we imagined to be Botel Tobago Xima: But on our nearer approach we found it to be much smaller than is usually represented: and about an hour afterwards we saw another Island, five or six miles farther to the westward. As no chart or journal we had seen took notice of any Island to the eastward of Formosa but Botel Tobago Xima, and as we had no observation of our latitude at noon, we were in some perplexity, apprehending that an extraordinary current had driven us into the neighbourhood of the Bashee Islands. We therefore, when night came on, brought to, and continued in that posture till the next morning, which proving dark and cloudy, for some time prolonged our uncertainty; but it clearing up about nine o'clock we again discerned the two Islands above-mentioned; and having now the day before us, we prest forwards to the westward, and by eleven got a sight of the southern part of the Island of Formosa. This satisfied us that the second Island we saw was Botel Tobago Xima, and the first a small islet or rock, lying five or six

N.N.E., being very careful during the night to look out for any land ahead of us; a strong current bearing north did not allow us to know exactly how far we travelled, but bright moonlight and extreme caution made up for the inconvenience of sailing through an archipelago which is known to geographers only through the letter addressed to Mr de lisle by Father Gaubil,[1] a missionary who had obtained some information about the kingdom of Likeu and its 36 islands from an ambassador of the King of Likeu he met in Peckin.

It will be appreciated that determinations of latitude and longitude calculated in this way are insufficient for navigation, but it is always a great advantage to know that there are reefs and islands in the area where one is. On 4 May at one in the morning we sighted a small island bearing from us N.N.E. We spent the rest of the night tacking under short sails and at dawn I set out to pass half a league to the west of it; we sounded several times without finding ground at this distance but soon learned that the island was inhabited; we saw fires in several places and herds of cattle feeding along the shore. When we had at last rounded its western point, which is the most attractive and most populated side of the island, several canoes came out from the coast to observe us; we seemed to inspire them with extreme fear – their curiosity led them to come within musket-shot range and their mistrust made them immediately hurry away; finally our shouts, our gestures, our signs of peace and the sight of some cloth made two of them decide to come alongside; I had a few medals and a length of Nankin cloth[2] given to each one. It was evident that these islanders had not left the coast

---

miles due East of it.' Walter Richard (compiler), *A Voyage round the World by George Anson*, London, 1776, p. 346. See also G. William (ed.), *A Voyage round the World* ..., London, 1984, pp. 310–11.

[1] Antoine de Gaubil (1689–1759), a Jesuit, went to China in 1721, arriving in Pekin in 1723. He supervised the observatory and became a corresponding member of the Paris Académie des Sciences, keeping in touch with a number of European savants. Some of his letters were published in the widely-read *Lettres édifiantes et curieuses*. His correspondence has now been collected and published, see R. Simon (ed.), *A. Gaubil: Correspondance de Pékin*, Paris, 1970. The 'de lisle' referred to in the text is the geographer and cartographer Joseph-Nicolas Delisle who devoted a great deal of time and energy to organising scientists in various parts of the world to observe the transit of Venus.

[2] Nankin cloth, known as nankeen in Britain, was a yellow cotton material imported from China; the terms *nankin* and nankeen were only then entering current usage in the two languages.

intending to trade because they had nothing to give us in exchange for our gifts and they tied a bucket of drinking water to a rope, indicating by gestures that they did not consider that this was sufficient to repay us and that they were going to fetch provisions on land, which they told us by placing their hands in their mouths. Before coming up to the frigate, they had placed their hands on their chest and raised their arms towards the sky: we imitated these gestures and they then decided to come on board, but always with a mistrust which their expression never ceased to reveal. They nevertheless invited us to go closer to the shore, telling us that we would lack for nothing there. These islanders are neither Chinese nor Japanese but, placed between these two empires, they seem to have something of both; they wore a shirt and trousers made of cotton cloth, their hair tied back over the top of their heads was twisted around a pin which seemed to be made of gold; each one had a dagger, the point and handle of which were also made of gold; their canoes were simply made of a dug-out tree trunk and they were fairly clumsily manoeuvred. It would have been rather interesting to land on this island, but as we had hove to in order to wait for these canoes and the current was running very strongly north we had dropped to leeward and might have struggled in vain to reach it; moreover we had no time to lose and it was very important for us to leave the Sea of Japan before June, a time of storms and hurricanes which make these waters the most dangerous in the world.

Clearly, ships in need would find supplies of food, water and wood in this island and might even manage some kind of trade, but as it is hardly more than three or four leagues in circumference, it is unlikely that the population exceeds four or five hundred people, and a few gold pins are not evidence of wealth. I have kept the name of Kumi Island which I found on Father Gaubil's chart in a latitude and longitude close to what we observed, having placed it in 24^d 33' of latitude and 120^d 56' of longitude Paris meridian; Kumi Island forms part of a group of 7 or 8 islands of which it is the westernmost[1] and it is isolated or at least separated from those

---

[1] La Pérouse was approaching the Sakishima Shoto group east of Taiwan, between the 24th and 25th parallels. The westernmost is Yanogumi Jima in latitude 24°27'N and longitude 120°40'E of Paris, which is separated by some 60 km or 6 leagues from the main group. The island of Kume Shima (his 'Kumi') lies further north. The group is linked by an undersea range to the Ryukyus (Nansei Shoto) which are La Pérouse's 'Likeus'.

CHAPTER XV

which one can suspect lie to the east of it by channels 8 or 10 leagues wide, this being the extent of our horizon at the time and we saw no sign of land. From details provided by Father Gaubil on the large island of Likeu, the capital of all these islands east of Formosa, I tend to believe that Europeans would be accepted there and might be able to trade as favourably as in Japan. At one p.m. I crowded on sails and made for the north without waiting for the islanders who had indicated by gestures that they would soon be back with food supplies, but we were still well provided and a most favourable wind was inviting us not to waste such precious time. I continued on this course all sails high and by sunset we had lost sight of Kumi Island although the sky was clear and our horizon seemed to stretch out for ten leagues. I reduced sail during the night and broached to at two a.m. after covering five leagues because I suspected that the currents had borne us ten to twelve miles beyond our reckoning. At daylight I sighted an island bearing N.N.E. and several rocks and islets further east. I set my course to pass west of this island. It was round and well wooded on its western side; I coasted a third of a league off without finding bottom and saw no trace of any houses; it is so steep that I do not even think it is habitable; it may have a diameter of two-thirds of a league or a circumference of two leagues. When we were athwart of it we saw a second island of a similar size, equally wooded and of much the same shape although a little lower; it bore N.N.E. and between these islands there were five groups of rocks around which a great crowd of birds were flying. I have kept the names of Hoapinsu and of Tyaoynsa for the one further north and to the east, which had been given by the same Father Gaubil to islands east of the northern point of Formosa, located much further south on his chart than the latitude we observed; be that as it may, our determinations place the island of Hoapinsu in 25^d $43'$ $50''$ of latitude and 121^d $14'$ of longitude, and Tyaoynsa in latitude 25 [blank] and longitude [blank] Paris meridian.[1]

We had at last left the Likeu archipelago and were about to enter a wider sea between China and Japan where some geographers claim one can always find a depth.[2] This observation is quite

---

[1] Huaping Hsu, to the north of Taiwan, lies in 25°24′N and 121°58′E of Greenwich (119°38′E of Paris). There are two other islands close by: Mienhua Hsu in 25°29′N and Pengchia Hsu in 25°38′.

[2] The frigates are now entering the East China Sea, an area which was then largely unknown to European navigators.

correct, but it was only really in 24^d $4'$ that the sounding-line gave seventy fathoms and from that time until we were beyond the Japanese channel we sailed continually on the depth; the China coast is even so flat that in 31 degrees we had only 25 fathoms thirty leagues from land. I had planned when I left Manila to survey the entrance to the Yellow Sea north of Nankin if my navigation left me time to spend some weeks on this, but whatever happened it was essential for the success of my later projects to reach the entrance to the Japanese channel before 20 May and I found on the northern coast of China unfavourable winds which allowed me to progress a mere seven or eight leagues a day; the fogs were as thick and as persistent as on the coast of L'abrador: the breeze which was very weak only veered from N.E. to east; we were often becalmed, compelled to anchor and put out signals to remain at anchor because we were unable to see the *Astrolabe* even though she was within hailing distance, and the currents were so strong that we could not hold the lead on the bottom to check that we were not dragging; the tide, however, was running at no more than one league an hour, but its direction was impossible to estimate – it changed every half hour and boxed the entire compass in the space of twelve hours without a single instant of slack water. In a period of ten or twelve days we had only one fine interval which allowed us to see an islet or rock in 30^d $45'$ latt. and 121^d $26'$ of longitude. It was soon hidden by the fog once more and we do not know whether it is adjacent to or separated from the mainland by a wide channel because we never saw the coast and our smallest depth was twenty fathoms.[1]

On 19 May, after a period of calms that had lasted a fortnight with very heavy fogs, the winds settled in the north-west, stormy gale. The weather remained dull and whiteish, but the horizon opened out to several leagues; the sea which had been so fine then became very rough. When this change occurred I was at anchor in 25 fathoms. I gave the signal to leave and without wasting a moment I set course to NE¼E towards Quelpaert Island[2] which was the first landmark of any interest before we entered the Japanese

---

[1] All that can be said is that La Pérouse was sailing past the Shengsiqundao group, a mass of islands and rocks to the south-east of Shanghai, in latitude 30°40' to 45'.

[2] Quelpaert Island is now more commonly known by its Korean name of Cheju-do. It lies at the entrance to the Korea Strait between the East China Sea and the Sea of Japan.

CHAPTER XV

channel. This island which is only known to Europeans as a result of the wreck of the Dutch vessel *Sparrow-hawk* in 1635[1] was at the time ruled by the King of Korea. We sighted it on 21 May in the finest weather imaginable and in most favourable conditions for observations. We determined the latitude of the south point as 33^d 14′ latt. and 124^d 15′ of long. I coasted along the whole south-east shore at a distance of two leagues and we surveyed with the utmost care a length of twelve leagues of which Mr Bernizet drew a plan. One would be hard put to find a more pleasing prospect; a peak of approximately 1000 *toises* and which is visible from 18 or 20 leagues rises in the middle of the island and serves probably as its reservoir[2] from which the houses look as though laid out in an amphitheatre. All the soil seemed to be cultivated up to a considerable height; we could make out with our spyglass the boundaries of the fields; they seem to be very much parcelled out, which suggests a large population, and the various crops which presented a wide range of colours made the appearance of this island even more pleasing. Unhappily, it belongs to people who are forbidden to communicate with strangers and who currently enslave those unfortunate enough to be be shipwrecked on their coast; some of the Dutchmen from the vessel *Sparrow-hawk* found a way, after a captivity lasting eighteen years during which they endured several beatings, to capture a boat and make their way to Japan whence they went to Batavia and finally to Amsterdam. This story, of which we had an account before us, was not of a nature to encourage us to send a boat ashore which we had seen two canoes leave, but they never came within a league of us and it is likely that their only purpose was to observe us and possibly spread a warning all along the coast of Korea. I kept to my course until midnight, sailing North-East a quarter East and hove to in order to await the break of day which was dull but without thick fog. We saw the north-east point of Quelpaert Island bearing

---

[1] The *Sparwer* affair goes back to August 1653 (not 1635 as La Pérouse states) when this Dutch ship was wrecked on Cheju-do while sailing to Nagasaki. The 36 survivors were sent to Seoul in 1654 and forced to join a regiment. Eight of them escaped in 1667 and made their way to Japan. The Dutch had found a compatriot in Cheju-do, a survivor from a 1627 wreck, Jan Welvetree, who had married a Korean woman. The *Sparwer*'s supercargo, Hendrick Hamel, provided in his journal the only information which Europeans were able to obtain for over a century on the closed world of Korea.

[2] Mt Halla San rises to 1,950 metres or 6,398 ft. 1000 *toises* is an impressively accurate estimate.

West, and I steered North North-East to approach Korea. We kept sounding every hour and continually found 60 to 70 fathoms; at dawn we sighted various islands or rocks forming a chain more than fifteen leagues ahead of the Korean mainland; they trended approximately N.E. and S.W. and our observations place the northernmost in 35^d $15'$ of latt. and 127^d $7'$ of long.[1] A heavy mist hid the mainland from us, which was no more than five or six leagues away; we saw it the next day at around eleven a.m.; it was visible behind the islets or rocks which still lined it and shield this peninsula from the fury of the sea. Two leagues south of these islets the sounding-line constantly recorded 30 to 35 fathoms, muddy ground; the sky remained dull and whiteish but the sun broke through the mist and we were able to carry out the best observations of latitude and longitude, which was most important for geography, as no European ship has sailed through these seas which are traced on the world maps only from Japanese or Korean maps that were published by the Jesuits who, to be honest, corrected them according to land routes checked with considerable care and following very good observations of longitude made in Pekin, so that errors are of a minor nature, and it must be agreed that these missionaries have rendered the most signal services to the geography of this part of Asia, which they are the only ones to know and for which they have given us maps that are very close to the truth, in which navigators miss all the details which could not be included as these Jesuits were journeying by land.

On the 25th we passed the Strait of Korea during the night; after sunset we had seen the coast of Japan stretching from East¼NE to East S.E. and that of Korea from N.W. to North. The sea looked very open to the N.E. and fairly high seas rolling in from there confirmed this view. The winds were S.W., light gale; the night was very clear. We ran with the wind under very light sails, making only two-thirds of a league an hour in order to see the previous evening's bearings when daylight came and draw an accurate map of the strait. Our bearings, based on Mr D'agelet's observations, leave nothing to be desired on the accuracy of the chart we have drawn. We sounded at half-hourly intervals and as the coast of Korea seemed to me to be more interesting to follow than that of Japan I came to within two leagues of it and set a course parallel to it.

[1] La Pérouse was now reaching the neighbourhood of Pusan in southern Korea.

PART OF
QUELPAERT ISLAND

13. Chart of part of Quelpaert Island (Cheju-do) sighted on 22 May 1787. By Blondela. Actual size 34 cm × 47 cm. AN 6 JJ1:38.

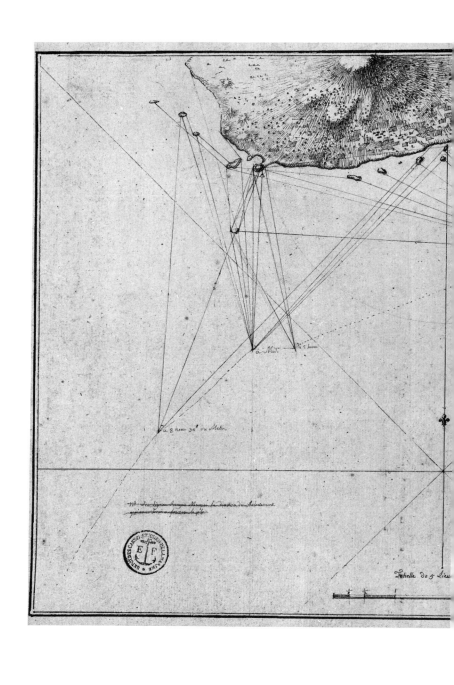

PLAN

DE LA PARTIE DE L'ISLE QUELPAER

Vue Le 22 May 1787 par les frégates françaises La Boussolle et L'Astrolabe. La Presqu'île A est située par 33° 7' 49" de latitude Nord et par 125° 58' 42" de longitude Orientale du Méridien de Paris, la variation du Compas de 1°18' NE.

Voir si les variations sont corrigées et faire les boussolles par le nord corrigé

Marine

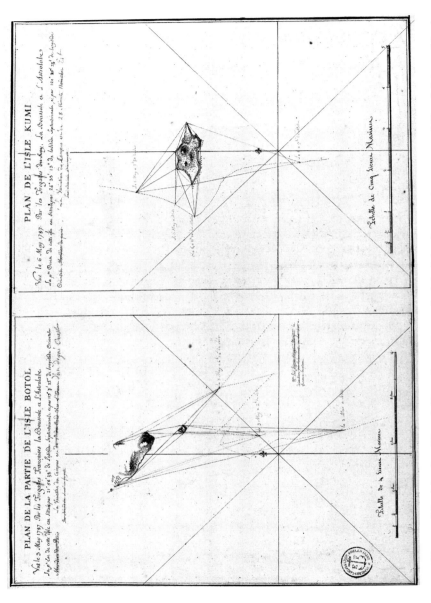

14. Charts of the islands of Botol (Kotosho or Huangfu Su) sighted on 3 May 1787, and Kumi (Mumeshima) sighted on 6 May 1787. By Blondela. AN 6JJ1:37.

CHAPTER XV

The channel separating the coast of the mainland from that of Japan is probably some fifteen leagues wide, but this is reduced to up to ten leagues by rocks which have lined the southern coast of Korea all the way from Quelpaert Island and which finally ended when we rounded the S.E. point of this peninsula, so that we were able to coast the mainland quite near, see the houses and towns along the seashore and identify the openings of bays; we saw fortifications on the hilltops that were absolutely similar to European forts, and it is likely that the main defensive structures of this country are directed at the Japanese. This part of the coast is excellent for navigation as no dangers seem to lurk here and one finds 60 fathoms, muddy bottom, three leagues from shore, but the country is mountainous and looks very arid. The snow had not completely melted away in certain ravines and the soil looked hardly suitable for cultivation; the habitations however were very numerous and we counted a dozen sampans or small boats sailing along the coast; they did not appear to be any different from Chinese ones: their sails were similarly of matting and the sight of our ships seemed to cause them no great fear: it is true that they were close inland and could have reached it before they could be caught up with if our manoeuvres had caused them any such suspicions. I would dearly have liked them to come alongside us, but they went on their way without taking any notice of us and the spectacle we provided, although quite new for these people, did not appear to attract their attention; I did however see at 11 o'clock two boats setting sail to observe us, come up to within a league, follow us for two hours and then return to the harbour whence they had come in the morning, so that it is possible that we caused some alarm on the Korean coast,[1] which is confirmed by fires we saw lit during the afternoon on all the headlands. This day of the 26th was one of the finest of our campaign and of the most interesting because of the bearings we took of a stretch of coast of more than thirty leagues; in spite of this fine weather the barometer dropped to 27 inches 10 tenths and since it had several times given us erroneous indications we continued on our track along the coast

---

[1] King Chongjo (1776–1800) had recently succeeded King Yongjo, the Crown Prince Changhon having been killed in 1763. Korea was going through a particularly difficult period, with endless palace intrigues, an anti-Christian campaign, a Chinese semi-protectorate and the constant fear of a Japanese invasion. The presence of foreigners on its soil would have been most unwelcome.

which we could see in the moonlight, until midnight when the winds veered from south to north without this change being announced by any clouds; the sky was clear and serene, but it darkened greatly and I was forced to get away from the land to avoid being driven towards it by the east winds, but, although no cloud had warned us of this wind change, we nevertheless had received a warning we did not understand and which is not really easy to explain: the lookouts in the tops shouted that they felt burning vapours and like those emanating from an oven door coming in waves every half minute; all the officers climbed the masts and felt the same heat; at the time the temperature on deck was 14 degrees; we sent up a thermometer to the topgallant crosstrees and it rose to 20 degrees; during this time the hot gusts were passing very quickly and in between the temperature did not seem to be any different from that at sea-level. During the night we were struck by a strong wind from the north which lasted only 7 or 8 hours, but the sea was very rough. As the channel between Korea and Japan must be fairly wide in this latitude we were not concerned about the bad weather and the following day I went to within three leagues of the mainland; there was no mist and we could identify the features seen the day before. We had made some progress towards the north in spite of the strong wind and the coast was beginning to trend N.N.W., so we had passed the southernmost part of Korea which was the most interesting point of this peninsula to determine.[1] I felt that after this I should set a course for the S.W. of the island of Nipon whose N.E. point, or Cape Nabo, Captain King had precisely determined.[2] These two points should finally settle the doubts of the geographers who will then only need to let their imagination travel along the outline. On the 27th I gave signal to make for the east but I saw shortly after an island bearing N.N.E. which appeared on none of the charts and seemed to be

---

[1] With the coast now trending towards the north-west, the French would have been sailing into little-known waters with the possibility of meeting superior Korean naval forces.

[2] The expedition of the *Resolution* and *Discovery* had come upon Cape Nabo or Nubo (Nubosaki) around 31 October 1779. George Gilbert records: 'we saw the Land at the distance of about 12 leagues; and a great number of vessels close in shore; this was the SE extremity of the Island, lying in Latitude 35° N, Long: 140° East', Cook, *Journals*, III, p. 710. The weather was unfavourable and Gilbert's estimates are usually approximative: Nabo lies in 35°40′N and 140°52′E. The island of Nipon is Honshu.

CHAPTER XV

about twenty leagues from the coast of Korea; I endeavoured to approach it but it was exactly in the wind's eye; fortunately it changed during the night and at daybreak I sailed to examine this island which I have named D'agelet Island[1] after that astronomer who was its first discoverer. It has only a circumference of three leagues; I coasted along it and almost all around it a third of a league off without finding the bottom; I then decided to lower a boat under the command of Mr Boutin with orders to take soundings up to the land, and he found bottom at twenty fathoms only as he reached the beginning of the waves that were breaking on the shore and at about a hundred *toises* from the coast of this island whose N.E. headland lies in 37^d 25' of latitude and 129 degrees 2 minutes of longitude, Paris meridian. It is some three leagues around and very steep, but covered with the finest trees from the edge of the sea up to the top; a rampart of solid rock, as sheer as a wall, surrounds it with the exception of seven small sandy coves on which it is possible to land, and it is in this cove that we saw being built some boats which were totally in the Chinese manner, but the sight of our ships passing a short cannon shot away presumably frightened the workmen and they had fled into the woods that were no more than fifty paces from their yard. Furthermore all we saw were a few huts, but neither village nor fields, so that it is very likely and almost certain that Korean builders who are only twenty leagues from Dagelet Island spend the summer with their provisions on this island to build boats and sell them on the mainland; this opinion is almost a certainty because, after we rounded the western point, the workmen in the last building yard, who were surprised because they had not seen the vessel hidden by this point, were by their lumber working on their boats and we saw them flee into the forest with the exception of two or three whom we seemed not to frighten. I was very desirous of finding an anchorage to persuade these people by means of gifts that we were not their enemies, but fairly strong currents were bearing us away from the land; night was beginning to fall and I was forced to signal to my boat to return just when Mr Boutin was about to step ashore, for fear of being carried to leeward and unable to join him. I sailed up

---

[1] The name of Dagelet is still currently found on charts. The island is Ullung-do, a possession of South Korea in 37°29'N and 130°52'E or 128°32'E of Paris – the position determined by the French is therefore fairly accurate.

to the *Astrolabe* which was far to the west where the currents had taken her. We spent the night in a calm caused by the height of the mountains of D'agelet Island which intercepted the sea breeze.

At dawn the winds settled in the South-South-East and I set an easterly course for Japan, but I reached that coast very slowly; the winds were continually against us, and time was so precious to us that had it not been for the great store I set on determining one or two features of the western coast of Nipon Island I would have given up this survey and sailed with the wind towards the Tartary coast. On 2 June in 37^d $37'$ of latitude and 132^d $13'$ of longitude, according to our chronometers, we saw two Japanese vessels,[1] one of which passed within hailing distance of our frigates; it had a crew of twenty men all wearing blue cassocks like those of our priests; this vessel of about one hundred tons burden had a single, very tall mast situated amidships, which was no more than, so to speak, a bundle of small masts, joined together by copper rings and woolding; the sail was of cloth; the ties were not sewn but laced along its length; this cloth seemed immense, and two jibs with a spritsail made up the balance of its sails; a small passage-way two or three feet wide protruded on both sides of this vessel, running aft up to two-thirds of its length; green-painted beams protruded over it; the small boat placed across the bow exceeded the vessel's width by seven or eight feet, and moreover the vessel had only a very ordinary sheer, a flat poop with two small windows, and very little in the way of carvings, and was quite unlike a Chinese boat: the only point of similarity was the manner in which the rudder was attached by ropes; but the lateral passage-way which was only two or three feet above the water line, the ends of the boat which must have touched the water when the boat rolled, everything convinced me that these vessels are not meant to leave the coast and that one would be at risk sailing in them in heavy seas during a squall. It is likely that they have ships for the winter that are more able to cope with bad weather. We passed so close to them that we were able to see even the facial features of individuals; they never expressed any fear, not even any surprise. They altered course only when they were a pistol shot away from the *Astrolabe*; they were afraid to

---

[1] The presence of Japanese vessels near Ullung-do was a matter of growing concern to Korean authorities. In the nineteenth century, Japanese fishermen gained the right to approach Korean waters and the inhabitants of both Ullung-do and Cheju-do frequently protested against these threats to their livelihood.

PART OF THE
KOREAN ARCHIPELAGO

15. Chart of part of the Korean archipelago, showing the track of the expedition, May 1787. By Blondela. AN 6 JJ1:36A.

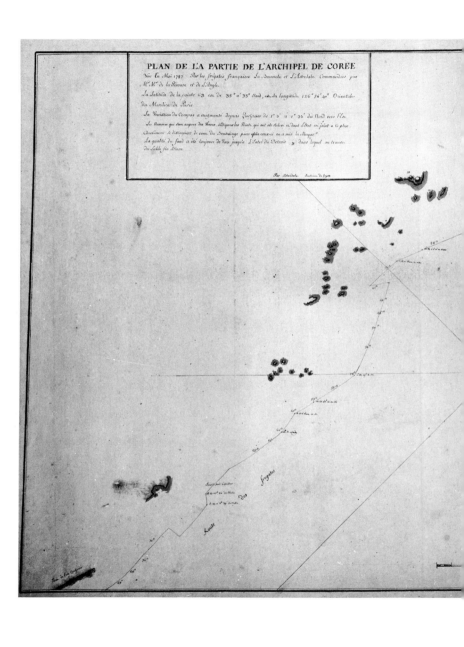

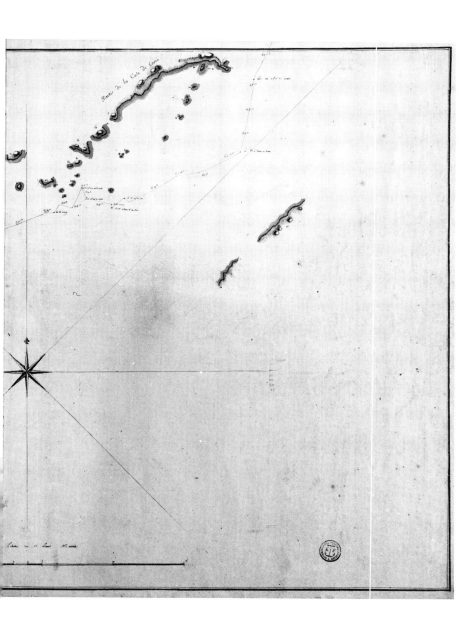

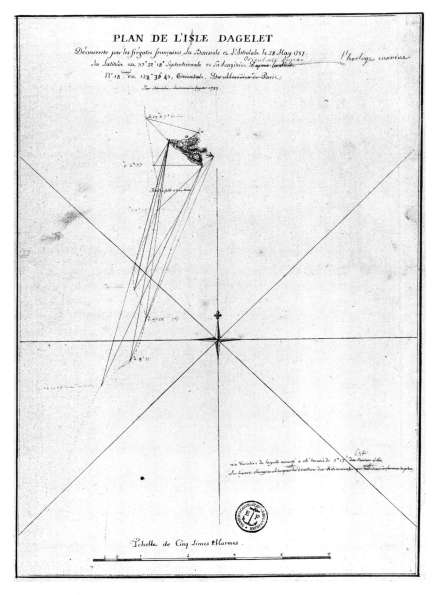

16. Chart of part of Dagelet Island (Ullung-do) sighted on 28 May 1787. By Blondela. AN 6JJ1:40.

collide with this frigate. They had a small white Japanese ensign with words written vertically on it, the boat's name was on a kind of drum placed next to this flagstaff. The *Astrolabe* hailed it as she went by – we no more understood its reply than they understood our question – and it proceeded on its way south quite eager to go and announce its meeting with two European vessels in seas where none had gone before us. On the morning of the 4th in longitude 133^d $17'$ and latitude 27^d $13'$ we thought we saw land but the weather was very misty and soon our horizon did not extend to a quarter of a league; there was a strong southerly gale, the barometer had dropped ten lines since twelve o'clock. I intended at first to heave to in the hope that the sky would clear, but the wind freshened further in the afternoon and the mizzen topsail was carried off; we furled the topsails and hove to with a foresail. At various times during the day we saw seven Chinese vessels masted as earlier described but without the lateral passage-way and, although smaller, built in a style more appropriate to cope with bad weather; they were absolutely like the one seen by Captain King on Cook's third voyage[1] with the three black bands in the concave part of the sail, of a burden also of 30 to 40 tons with a crew of eight men; while the wind was at its strongest we saw one high and dry, its mast was like those of coastal fishing smacks,[2] being merely held by two stays and one forward because they have no bowsprit but only a small mast eight or ten feet high vertically erected to which they tied a small foresail as on a ship's boat. All these vessels were sailing close to the wind on the port tack making for the S.S.W., and it is probable that they were quite close to the land from which they no doubt never go far since they sail only along their shores. The next day was very foggy; we saw two more Japanese vessels, and it was only on the sixth that we finally saw Cape Noto and the island of Jaolsissima[3] which is separated from it

---

[1] The surgeon David Samwell gives the following description: 'She seemed to be about the Size of a small sloop of 50 or 60 tons burden, had one Mast and a large square Sail, the upper part of which had 3 patches upon it of a brown colour.', Cook, *Journals*, III, p. 1286.

[2] The term used by La Pérouse is *cache marée*; a more usual term is *chasse-marée* which describes a small vessel specifically designed to sail to fishing grounds in order to collect the catch from larger fishing boats and bring it back in as fresh a condition as possible to local markets. See J. Boudriot's description in his *Le Vaisseau de 74 canons*, Grenoble, 1977, IV, p. 351.

[3] The most likely identification of Jaolsissima, which La Pérouse reports as lying five leagues from Cape Noto, is Nanatsu-shima which lies in latitude 37°36'N and

by a channel of some five leagues. The weather was clear and the horizon very extensive; although we were six leagues from land we could see its details, the trees, the ravines, the landslides; some small islands or rocks we sailed past two leagues off, and which were linked together by a chain of rocks level with the sea, prevented us from coming closer to the shore; at this distance the sounding-line recorded sixty fathoms, rock and coral bottom. At two o'clock we saw Jaolsissima Island bearing N.E. I set a course to coast along its western side and soon we were forced to hug the wind in order to round the breakers that stretch out about two leagues from the coast and which are very dangerous in the fog which almost constantly envelops the northern coast of Japan during this time of year. A league and a half from these breakers the sounding-line also gave sixty fathoms, rocky bottom, and one could only envisage anchoring here in a situation of extreme necessity. This island is small, flat but well wooded and of a pleasant appearance; it looked as though it had a large population; we saw houses and even some quite substantial buildings and, close to a kind of castle situated on the south-west headland, we saw a gibbet or at least some pillars with a beam across and above them which looked exactly like our European feudal gallows;[1] these pillars may have a quite different purpose, and it would be rather strange if Japanese customs that are so different from ours were the same in this instance.

We had hardly rounded Jaolsissima Island when in a trice we were surrounded by the thickest fog; fortunately we had had time to take excellent bearings of the coast of Japan south of Cape Noto up to this same cape beyond which we could see nothing.

We were quite satisfied with our observations of latitude and longitude, our chronometer No. 19 having operated perfectly since our departure from Manila; therefore Cape Notto on the coast of Japan is a point on which geographers can depend, which with Cape Nabo on the east coast determined by Captain King will give

---

longitude 134°33′E of Paris and approximately 20 miles or 6 leagues from the island of Honshu. An alternative is Hekura-jima in latitude is 37°51′N which is the precise latitude La Pérouse determined for Joalsissima; however, it lies much further from Cape Noto than La Pérouse reported and is a more stormbound and less attractive island.

[1] La Pérouse saw a *torii*, a gateway leading to a temple or shrine, rather than a gibbet.

CHAPTER XV

the width of this empire in its northern part but (what is of far greater importance to geography) will make known the width of the Sea of Tartary towards which I then decided to set course, as the coast of Japan was trending beyond Cape Noto 60 leagues to the east and the constant fog which surrounds these islands might have required us to spend the remainder of the season coasting and surveying Nipon Island up to Cape Sangor;[1] we had a far greater field of discovery before us on the Tartary coast, in the Strait of Tessoy[2] & I felt that I ought not to waste a moment so as to reach it promptly, having had no other aim in my work on the coast of Japan than to assign its true north-to-south limits to the Sea of Tartary. Our observations place Cape Noto in 37^d $36'$ of latitude and 135^d $34'$ of longitude Paris meridian, Jaolsissima Island in 37^d $51'$ of latitude and 135^d $20'$ of longitude, an islet or rock west of Cape Noto in 37^d $36'$ and 135^d $14'$ and the southernmost point which was in sight on the island of Nipon in $37^d 18'$ and 135^d $5'$. These brief observations which will seem very dry and arid to most of our readers cost us ten days of a quite laborious navigation in the midst of fogs, and we think that geographers will consider them to have been put to good use and will no doubt regret that the vast programme of our voyage did not allow us to survey and determine a greater number of features on this coast and more especially towards the south-west in order to ascertain the precise configuration of the strait which separates this empire from Korea. We surveyed the coast of that peninsula with the greatest accuracy up to the point where it no longer trends N.E. and veers towards the west, which compelled us to make for 37 degrees North; the most constant and stubborn of southerly winds were set against the plan I had formulated of sighting and determining the southernmost and westernmost point of the island of Nipon; these same winds followed us until we were within sight of the coast of Tartary which we saw on the 11th. The weather had cleared, the previous day the barometer had dropped to 27 inches 7 tenths and it stayed thus; it was thanks to this that we enjoyed the two finest days of our voyage; since the time of our departure from Manila this instrument

---

[1] Cape Sangar or Sungaar appears on old maps at the north-western end of Nipon (Honshu). It can be identified as Tsugaru, the north-western peninsula of Honshu.

[2] The Strait of Tessoy does not exist, but appears on old maps of the area, about which very little was known at the time, to the north of Honshu and along a presumed extension of the coast of China.

had so often given us good warnings that we owed it some indulgences for its divergences, but the conclusion is that there are certain conditions of the atmosphere which cause a substantial drop in the barometer without leading to rain or wind; the *Astrolabe*'s was at the same level as ours, and I think that a lengthy set of observations is still needed to understand perfectly the language (if I may put it this way) of this instrument which can be in general most useful to a safe navigation, and the advantages offered by Mr Nairne's[1] with his ingenious suspension cannot be compared to any other. Our landfall on the coast was precisely the point that separates Korea from Manchu Tartary;[2] it is a very high land we saw from twenty leagues away. On 11 June it stretched from the N.N.W. to NE¼N and appeared in different tiers, and the mountains, without being as high as those of the N.W. coast of America, were at least six or seven hundred *toises* in height. We began to find ground only four leagues from land in 180 fathoms, muddy sand, and one league from the shore there was still 84 fathoms. I went up to the coast at this distance. The next day, 12 June, it was still very steep, but covered with trees and greenery; we could see snow on the tops of the highest mountains but in small quantities; moreover we saw no sign of cultivation or habitations and so thought that the Manchu Tartars who are a nomadic and pastoral people preferred plains and small valleys where their herds could find a more abundant feed than these forests and mountains. Along this length of coast of more than forty leagues we found no rivermouth; I would nevertheless have greatly liked to stop so that our botanists and our lithologists could observe the nature of this land and its productions, but the coast is straight and since there is a depth of 84 fathoms a league off it is likely that we would have had to come up to within two or three cablelengths from the shore to find a depth of twenty fathoms and we would not have been in a position to get under way with the sea breeze. I was hopeful of finding a more convenient location and went on my way in the finest weather,

---

[1] Edward Nairne (1726–1806) was a celebrated London instrument maker. He contributed to the *Philosophical Transactions* from 1771 and was elected a Fellow of the Royal Society in 1776.

[2] La Pérouse had reached the mountainous region north of Chongjin where the present-day boundaries of Korea, Siberia and China meet. He sailed past the Bay of Peter the Great, now the site of Vladisvostok, during the night of June 12th and fog eliminated any further chance of effecting a landing which the naturalists on board would have appreciated.

CHAPTER XV

with the sky clearer than we had come upon since our departure from Europe.[1] We carried out our observations with the same success on the 12, 13 and 14th, coasting along the shore three short leagues off, but on this final day we were surrounded by mist at 6 p.m. and were becalmed, a light S.E. breeze hardly enabling us to steer.

Up to this point the coast had trended NE¼N. We had already reached the 44th degree of latitude and the one geographers assign to the so-called Strait of Tessoy, but five degrees further west which had to be deducted from Tartary and added to the channel which separates it from the islands to the north of Japan and which would create a kind of Mediterranean Sea from the Gulf of Okots to Korea if Yesso and Locu-yesso were imaginary lands and the voyage of Captain Vries on the *Kastricom*[2] was a tale, which scarcely seemed credible to us.

The 15th and 16th were very foggy days; we did not go very far from the Tartary coast and sighted it during breaks, but the second of these days will remain famous in our journal for the most complete illusion I have ever witnessed in all my sailing days.

At 4 p.m. the clearest of skies succeeded to the thickest of fogs; we saw the mainland stretching from W¼SW to N¼NE and shortly after a great land to the south stretching out to join Tartary in the west, leaving not even an opening of fifteen degrees; we

---

[1] La Pérouse shows awareness of the impatience felt by his naturalists unable to land on unknown shores. This passage no doubt echoes lively or even bitter discussions on board both ships.

[2] The *Castricum* was one of the few European ships to have sailed in the area north of Japan. Maarten Gerritszoon Vries (or de Vries, sometimes found as Fries or even Uries) was a captain of the Dutch East India company who sailed in February 1643 from Batavia with two ships, his own *Castricum* and the *Breskens* which soon lost contact with him and was detained by the Japanese on Honshu. Searching for islands reputed to be rich in precious metals, Vries sighted the north of Hokkaido and the Kuril Islands. In June he discovered Iturup, which he named Staten Eylandt and Urup which he named Compagnies Landt. He veered west towards the Sea of Okhotsk and to Sakhalin where he landed. He returned to the Pacific by Vries Strait, caught up with the *Breskens* which the Japanese had released and returned to Batavia in mid-December. His discoveries were not without importance, but based on vague and inaccurate premises they tended to confuse cartographers. An anonymous report was published in Amsterdam in 1646 and a second one, by Philip de Bakker, in 1692. The journal of the master, Cornelis Janszoon Coen, was edited by P. A. Leupe as *Reize van Maarten Gerrisz Vries en 1643 naar Japan*, Amsterdam, 1858; there is a more comprehensive and more recent edition of documents relating to the voyage by Willem C. R. Robert: *Voyage to Cathay, Tartary and the Gold- and Silver-rich Islands East of Japan, 1643* (Amsterdam, 1975).

could see the mountains, the gullies, in a word all the details of the terrain and we could not imagine where we could have entered a strait which could only be that of Tessoy which we had already rejected; in such a situation I thought that I should hug the wind and steer S.S.E. for the northern point of Yesso Island; but soon these hills and gullies disappeared; the most extraordinary fogbank I have ever seen had caused this error; we saw it dissolve; its shapes, its shades rose up into the clouds and we fortunately had enough daylight left to ensure that no doubt remained about the non-existence of this fantastic land. I sailed all night over the sea space it seemed to have occupied and at daybreak nothing could be seen. The horizon nevertheless was so extensive that we could make out quite clearly the coast of Tartary fifteen leagues away; I set course to approach it, but at eight o'clock the fog surrounded us; fortunately we had had time to take good bearings and to identify the points seen the day before, so that there is no gap in our map of Tartary from our landfall in 42 degrees up to the Strait of Segalien.

The fog was very thick on the 17, 18 and 19th, but we made no progress and remained tacking variously in order to find, the moment there was a break, the headlands we had seen earlier and recorded on our charts; we were only three leagues from land and we saw the coast stretching from the W.S.W. up to the N.N.E. for more than twenty leagues; every shape was clearly distinguishable and the clear atmosphere enabled us to make out all the shades, but we saw no sign of any bay, and a sounding-line of two hundred fathoms gave no depth at four leagues. Soon the fog forced us to veer back to the open sea and we only saw the land at midday; we were quite close to it and have never been in a better position to take bearings; at the time our latitude was 44^d $45'$ and we could see a headland at NE¼N at least fifteen leagues from us and consequently in 45^d $22'$. I ordered the *Astrolabe* to proceed slowly ahead and look for an anchorage. Mr de Langle lowered his boat and sent his first officer, Mr de Monty, to sound a bay we could see ahead and which seemed to offer some shelter. We had 140 fathoms two leagues from land and 200 fathoms a league further off, but the change in the depth seemed to be gradual and it was likely that a quarter of a league from the shore we would have found forty or fifty fathoms, which would have been considerable, but after all one does anchor regularly in such depths. We pursued our route towards the land, but a dense fog rose up from it, brought along by a light northerly breeze, and Mr de Langle was forced to signal his boat to return before reaching the bay where Mr de Monty had been told to take soundings; this officer reached the frigate at the moment the densest fog began to surround us, forcing us to make for the sea. There was another break lasting a few minutes at sunset; and the next day at around eight o'clock we had made a mere three leagues East¼NE in the twenty-four hours and saw only the points we had already laid down on our charts. There was a mountain top that was absolutely like a tabletop: I gave it this name so that navigators could recognise it. During the time we have coasted this land we have seen no sign of habitation, not one canoe has left the shore and this country, although covered with very fine trees which indicate a fertile soil, is ignored by the Tartars and the Japanese who could establish flourishing colonies there, but the policy of the latter on the contrary is to ban all forms of emigration and communication with foreigners, and the Chinese as well as the Europeans are included in this category. On the 21st and 22nd the fog was very heavy, but we were so close to the shore that we could see it as soon as there was the slightest break, and there were some almost every day until sunset. The depth decreased when we reached the forty-fifth parallel and we found 57 fathoms, muddy ground, one league from land. On the 23rd the winds settled in the N.E. I decided to make for a bay I could see in the W.N.W. and where it was likely we could find a good anchorage. We dropped anchor there at 6 p.m. in 25 fathoms, sandy bottom, half a league from the shore. I named this bay Ternai Bay after a general officer of the marine who died during the war at Rodisland where he was in command of the King's squadron, and who had provided an ideal in my youth, had acted like a father to me and was always my protector and my best friend.[1] It is situated in 45^d

---

[1] Charles-Henry-Louis d'Arsac de Ternay (1723–80) joined the navy in 1738 and served in the West Indies and America before joining the Knights of Malta in 1749. Back in active service, he sailed to Canada in the *Actif* in 1755 and took part in numerous campaigns during the Seven Years War, after which he undertook several hydrographic and diplomatic missions. He was governor of the Ile de France in 1772–6 and, war breaking out once more, he sailed for America in command of a squadron which included La Pérouse's *Amazone*; after reaching the Cheasapeake, he sailed up to Rhode Island where he landed the soldiers of General de Rochambeau who later went on to Yorktown. Ternay died at Rhode Island on 15 December 1780, probably of typhoid fever. See M. Linyer de la Barbée, *Le Chevalier de Ternay*, 2 vols, Grenoble, 1972. Ternay had been La Pérouse's mentor and protector and the

11′ of latitude and 135ᵈ 12′ of longitude, Paris meridian. Although it is open to the easterlies, I am led to believe that they never blow towards the coast but along it; the ground is sandy; it diminishes gradually to six fathoms at a cable's length from the shore where the tide rises five feet: this occurs at eight fifteen on days when the moon is full or new but the ebb and flow does not affect the direction of the current half a league from the shore – the current we felt at the anchorage never varied except from south-west to south-east and its maximum speed was a league an hour. Having left Manila 75 days earlier we had in effect sailed along the coasts of Quelpaert Island, Korea and Japan, but these countries being inhabited by people who act in a barbaric way towards strangers we had never been able to consider anchoring there. We knew that the Tartars were more hospitable and anyhow our strength would impress the small populations we expected to meet along the shore; we were all anxious to examine this land which had excited our imagination since our departure from Europe: it was the part of the world which had escaped the untiring endeavours of Captain Cook, and we possibly owed to the tragic event which had put an end to his life the minor advantage of being the first to land here; it was more than evident to us that the *Castricum* had never sailed along the Tartary coast and we were hopeful of finding incontrovertible proof of this during our campaign.

The geographers who, following Father des Anges's report and a few Japanese maps, had drawn the Strait of Tessoy and the boundaries of Yesso[1] and of Company and States Lands had so disfigured the geography of this part of Asia that it was absolutely necessary to bring to a close all these old arguments by facts that were

---

homage which the explorer paid him on this occasion has proved a lasting one, as the name 'Ternej' remains for this bay and the port which has developed on its shore. The entire reference to Ternay, apart from the name itself, was omitted from the Milet-Mureau printed account of the voyage.

[1] Yesso (Jesso, Ezo) was the name given by geographers to land north of Japan and was an amalgam of Hokkaido, part of the Kurils and the south of Sakhalin. States Land is Vries' Staten Eylandt and corresponds to Iturup, one of the Kuril Islands, and Company Land is Urup, north of the former. Father des Anges is Girolamo d'Angelis (1567–1623) who landed in Japan in 1602. When the missionaries were expelled from the country in 1614, his superiors allowed him to stay behind; he spent nine years travelling through northern Japan and is reputed to have converted over ten thousand Japanese; he was finally caught and put to death with 90 of his converts. He is the author of *Relazione del regno di Yezo*, Rome, 1625.

CHAPTER XV

unchallengeable. The latitude of Ternay Bay is exactly the same as that of Acqueis Harbour where the Dutch put in – and the reader will find that its description is quite different;[1] five little coves made up this roadstead, rather like a regular polygon; they were divided from each other by hills covered with trees up to their tops.

The freshest French springtime has never produced such a range of vivid greens, and although we had seen no canoes and not a single fire since we began coasting along this shore we could not believe that such a fertile country so close to China was uninhabited; our spyglasses were turned towards the shore before our boats had been lowered, but we saw only deer and bears feeding peacefully by the sea; seeing this everyone became increasingly impatient to go ashore, weapons were readied as speedily as if we had to defend ourselves against enemies, and while this was being done, sailor-fishermen had already caught twelve or fifteen cods on their lines; Paris residents would have difficulty understanding the feelings of navigators faced by such abundance, but fresh food is a basic need for all men, and the less appetising is still much better for health than the best preserved of salt meats; I promptly ordered the salted food to be locked up, to keep it for less happy circumstances. I had barrels prepared to fill them with clear fresh water from a stream running in each cove, and some members of the crew were detailed to collect vegetables in the fields where they found an enormous quantity of small onions, celery and sorrel. The ground was covered with the same plants as we find in our regions, but greener and more vigorous; most were in flower, and at every turn one found roses, yellow and red lilies,[8] lilies of the valley and

---

[1] The anonymous account of 1646 of the island of Yeso, *Korte beschrijvinghe van het Eylandt bij de Iapanders Eso genaemt*, states: 'Coming in the North Latitude of 45 degrees 10 minutes in the place called by the Inhabitants Acqueis, situated in a large bay up to two Miles deep and wide 1½ Miles, the Land and mountains mainly clayish earth was not cultivated but produces naturally fine fruits...' W.C.H. Robert, *Voyage to Cathay*, (1975), pp. 244–5. However, Coen's journal mentions a village of 'Ackys' in latitude 43°02' and a 'beautiful bay' which he named Bay of Good Hope; these would seem to be the bay and town of Akkeshi in Hokkaido, see ibid., pp. 175–7 and 279.

[2] The saranna lily is the *Fritillaria camschatcensis* (L.) Ker which has edible bulbs and was almost a staple food at the time; its flowers however are usually a darkish purple; other edible lilies are the *Lilium avenaceum* Fisch., the L. pulchellum Fisch. and the L. tenuifolium Fisch., both with red or reddish flowers; a lily with a clear yellow flower is the *Lilium distichum* Nakai.

generally all our meadow flowers; pine trees stood on the mountaintops, the oak trees did not start until half-way and got thinner and weaker as they came closer to the shore; the banks of the rivers and streams were lined with willows, birches and maples, and along the edge of the great forests we could see apple trees and medlars in flower, with clumps of hazel-bushes whose fruits were starting to form.[1] Our surprise kept on increasing as we reflected that a surplus population burdens the vast Chinese empire to the point that laws do not punish fathers who are cruel enough to drown and kill their children, and yet this nation so organised and so praised does not dare cross its wall to spread out and make its living in a land whose vegetation needs to be kept in check rather than encouraged. To be fair we did find signs of humans everywhere, traces of their destructiveness and trees cut by tools; ravages caused by fire were visible in twenty places and we noticed a few shelters put up in corners of different woods by hunters; we also saw some small birch bark baskets sewn with thread in every respect similar to those of Canadian Indians, and some snow rackets; in brief everything confirmed our opinion that some Tartars come down to the seashore for the hunting and fishing season, and we then thought that they collected in groups along the rivers and that the bulk of the population lived inland on a soil that was possibly more suitable for the multiplying of their immense herds.

Three boats from both frigates with officers and passengers landed at six thirty in the cove where the bears had been seen and by seven o'clock they had already fired several shots at various beasts which fled very promptly into the woods. Alone three young fawns fell, victim of their inexperience – the loud merriment of all our newly landed men should have warned them that a couple of hundred paces would have been enough to bring them to inaccessible forests, for these meadows, so attractive to the eye, could hardly be crossed; the grass was thick and three or four feet high so that one was so to speak drowned in it and unable to find one's way; there was in addition the danger of being bitten by snakes

---

[1] Birches and willows are common in the region; the maple is less common, but the variety cannot be identified. The presence of medlars or azaroles seems unlikely in such a cold latitude: the Japanese medlar or loquat, *Eriobotrya japonica*, would find it hard to survive and the German medlar *Mespilus germanica* would fare no better. Hazel bushes would have a better chance of survival, the *Corylus heterophylla* Fisch and the *C. manshurica* Maxim, possibly the former which is noted for its nuts.

CHAPTER XV

which we found in large numbers along the banks of streams, but we had no knowledge of the deadliness of their venom. So this land was no more than a magnificent landscape for us: only the sandy beaches along the shore were practicable and incredible exertions were required to advance through the smallest of spaces; nevertheless his passion for hunting made Mr de Langle and several other officers or naturalists attempt to go through, but with no success, and we were all convinced that a positive result could only come with extreme patience, in very great silence, and by lying in ambush along the tracks of bears and deer which could be identified. This new plan was agreed upon for the morrow, but it was difficult to carry out: one hardly sails for ten thousand leagues in order to go and stand like a post in the middle of a marsh filled with sandflies; nevertheless we tried it on the 25th after running about pointlessly all day, but after everyone had taken up his position at nine o'clock in the evening it was felt by ten o'clock that the bears should have arrived, and we had to agree that fishing is more appropriate for sailors than hunting; and indeed we were more fortunate in that respect; each of the five coves which formed Ternai Bay offered a suitable spot to lower the seine; we each had our stream by which we had set up our cooking arrangements and the fish had only make one leap from the shore into our cooking pots; we caught cod, gurnet, trout, salmon, herring and plaice; our crews had an ample supply of these at each meal, and this fish with the various herbs filling their cooking pot during the three days of our stay guaranteed their good health against attacks of scurvy, of which in fact no member of the crew had yet shown any symptom in spite of the cold damp caused by the almost constant fogs which we had countered by means of braziers placed under the sailors' hammocks when the weather did not allow us to clear the decks.

It was after one of these fishing expeditions that we discovered on the banks of a stream a Tartar tomb next to a collapsed hut which was almost buried in the grass. Our curiosity led us to open it and we saw two people placed side by side, heads covered with a taffeta cap and the bodies wrapped in a bear skin with a belt made of the same skin from which hung small Chinese coins and various other copper valuables; blue beads were scattered as though they had been sowed in this tomb where we also found ten or twelve silver bracelets (I have since learned that they were gold earrings) weighing two *gros* each,[1] an iron axe, a knife made of the same metal, a wooden spoon, a comb and a small blue nankeen bag filled with rice. Nothing had yet decayed and this monument, of a design which seemed inferior to the tombs of Baye des Français, was hardly more than a year old; it consisted simply of a small stack of tree trunks between which a gap had been left to place the two bodies in, with a cover of birch bark, which we were careful to cover up again, after religiously putting back everything in its place, and we took away only a very small number of the various items this tomb contained so that we would have some evidence of our find. Before we made this discovery we had been left in little doubt that the hunting Tartars came on frequent occasions to this bay, but a canoe left near this monument showed us that they came by sea, presumably from the mouth of some rivers we had not yet sighted.

The Chinese coins, the blue nankeen, the taffeta, the caps prove beyond doubt that these people communicate and trade on a regular basis with the Chinese, and it is likely that they are also their subjects.[2]

The rice enclosed in the small nankeen bag is due to a Chinese custom based on the belief that one's needs continue in the next life; and the axe, the knife, the bearskin tunic, the canoe, these various practices are closely similar to those of the American Indians, and since these people have certainly never communicated with each other they prove that men who have reached the same level of civilisation adopt almost the same practices and if they were in exactly the same circumstances they would perhaps no more differ from each other than do the wolves of Canada and those of Europe.

But the pleasant prospect offered by this part of Tartary had however nothing interesting to offer our botanists and lithologists. The plants were exactly the same as those of France and the soil did not contain any different substances either; schist, quartz, jasper, purple porphyry, small smoothed pebbles of rock crystal, such are

---

[1] A *gros* is an eighth of an ounce, or 3.824 grams.

[2] The whole region had long since fallen under Chinese domination. The Treaty of Nerchinsk (1689) between China and Russia had confirmed the Amur River as the official boundary between the two powers. Trade was carried out by sea along the coast of Tartary from Peter the Great Bay to the mouth of the Amur where the Chinese kept a small military garrison. See on this area J.J. Stephan, *Sakhalin*, Oxford, 1971.

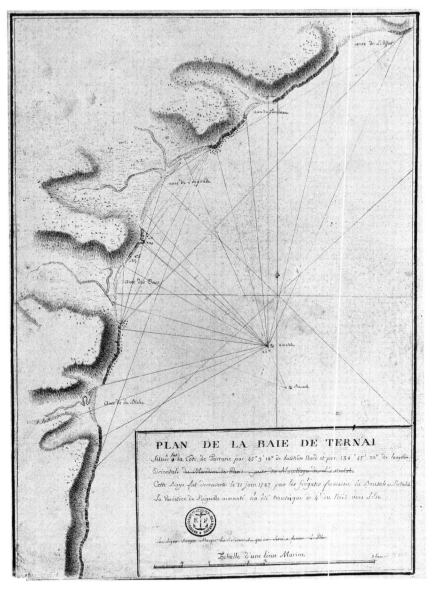

17. Chart of Ternay Bay, Siberia. Bearings taken from the *Astrolabe*. Names, from north to south: Stalking Cove, Tomb Cove, Watering Cove, Bear Cove, Deer Cove. By Blondela (?). AN 6 JJ1:41A.

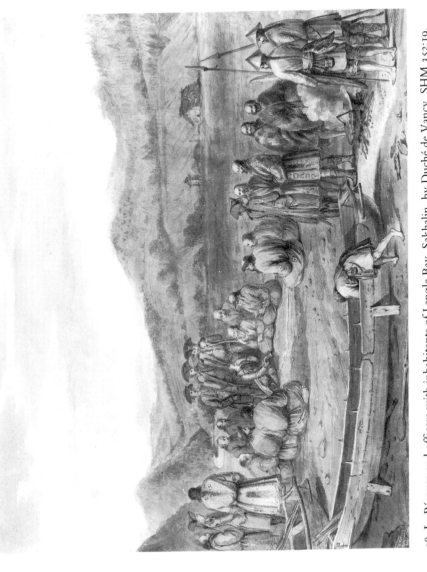

18. Le Pérouse and officers with inhabitants of Langle Bay, Sakhalin, by Duché de Vancy. SHM 3 52:19.

CHAPTER XV

the samples the rivers offered us; iron ore which is so widespread across the globe was visible only in a decomposed form, colouring certain stones like a varnish that has not entered into the substance. Land and sea birds were also very rare, although we did see some crows, turtle doves, quails, wagtails, swallows and fly-catchers, as well as albatrosses, gulls, puffins, bitterns and ducks,[1] but nature was not enlivened by the innumerable flights one finds in other regions: birds in Ternai Bay were hermits, and the deepest silence reigned in the woods. Shells were similarly scarce, and we found on the sand only remnants of mussels, barnacles, winkles and purpura.

Finally, on the morning of the 27th, having deposited ashore a number of medals with a bottle and an inscription showing the date of our arrival, I set sail, the winds having veered south, and hauled along the coast two-thirds of a league off, continuing along a depth of 40 fathoms, mud and sand, and close enough to make out the mouth of the smallest of streams. We covered in this way fifty leagues in the finest weather any navigator could hope for. We thought we could see fires on land during the night of the 27th to the 28th, but they lasted only a few minutes and left us in a state of uncertainty because they were not unlike moonlight beams striking hard shapes. The winds which shifted to the north at 11 p.m. on the 29th forced me to steer east and so leave the land. At the time we were in 46^d $50'$ of latitude, but we came back the next day, although it was foggy, the horizon stretching out to three leagues, and we saw the same coastline we had seen the day before to the north now bearing west; it was lower, more broken up by hillocks, and we only found thirty fathoms two leagues off, rocky bottom; we stayed becalmed on this kind of bank and caught more than 80 cod. A little southerly breeze allowed us to get away during the night and we saw land again at daybreak 4 leagues away; it seemed

---

[1] The albatross was probably the *Diomedea albatrus* which is now quite scarce; among gulls found in this region, one can choose between the *Larus schistisagus*, the *L. argentatus*, the *L. cachinnans*, the *L. glaucescens*, the *L. canus kamschatschensis* and the *Pagophila eburnea*, without thereby excluding less common visitors. The puffin is the *Anas arctica* or the *A. arctica cirrata*. The bittern is the *Botaurus Stellaris* or the smaller *Ixobrychus eurythmus*. There are still numerous land birds in the area, with the exception of the quail (*Coturnix coturnix*), the crow was the *Corvus corone corone* or the *C. corone corvix*, the turtle dove the *Turtur orientalis*, the wagtail the *Motacilla viridis*, the swallow the *Hirundo rustica* and among fly-catchers there are the *Muscicapa latirostris* and the *Ficedula parva*.

to stretch out as far as the N.N.W. but fog hid the more northerly features. We continued quite close along the shore which was trending north a quarter north-east, and on 1 July, a thick fog surrounding us while we were so close to the shore that we could hear the waves breaking on it, I gave orders to drop anchor in thirty fathoms, mud and broken shells. So fogbound were we that we could take no bearings nor send our boats ashore, but we caught more than eight hundred cod. I had those that were in excess of our immediate needs salted and stored in barrels. Our drag-net also brought up a fairly large number of pectinated oysters with such a fine lustre that we thought it was quite possible that they contained pearls, although we only found a couple half-formed near the hinge. This find adds credibility to the report by Jesuits[1] who informed Europe that there was a pearl fishery at the mouth of several rivers of eastern Tartary, but it must be assumed that this occurs towards the south near Korea, because further north the country has too few inhabitants for such an undertaking, because after coasting along for two hundred leagues, often within a cannon shot of the shore and also close in we saw neither houses nor canoes, and when we went ashore all we found were traces of a few hunters who did not seem to settle in the places we were visiting. At 3 p.m. on the 4th there was a fine break in the fog, we saw land as far as the NE¼N and athwartship we saw two miles away in the W.N.W. a wide with a river fifteen to twenty *toises* wide; one boat from each frigate was sent out to examine it; they were commanded by Messrs Vaujuas and Darbaud, and Messrs de Moneron, de la Martinière, Mongès, Father Receveur, Rollin, Bernizet and Colignon went with them; landing was easy and the depth was gradual up to the shore; the appearance of the countryside was fairly similar to Ternai Bay and although three degrees further north, the products of the soil and its composition were also much the same.

Traces of habitations were much more recent here; we could see branches of trees which had been cut with a sharp instrument, with green leaves still adhering to them; two elk skins very artistically

---

[1] Du Halde states that Emperor Kang Hsi issued orders for a survey of the Amur River in 1709 and that three Jesuit fathers took part in the expedition. See Jean-Baptiste du Halde, *Description de l'Empire de la Chine*, 4 vols (Paris, 1735, with revised ed. The Hague, 1736), IV, pp. 12–14. However, the despatch of pearls from this region to Peking does not imply that pearl fishing was occurring on a regular commercial basis at the mouth of the Amur at the time.

stretched over small pieces of wood had also been abandoned next to a hut which was too small for a family but adequate as a shelter for two or three hunters, and I am inclined to think that there might have been a few there whom fear drove into the forest; whatever the truth of this, Mr de Vaujuas felt that he should take one of these skins with him and left in exchange axes and other iron items a hundred times the value of the skin he sent me; the report made by this officer and the various naturalists did not make me wish to prolong my stay in this bay which I named Suffrin Bay.[1] I sailed with a light N.E. breeze which enabled me to get away from the coast; according to our observations it is in 47^d $46'$ of latitude and 137^d $22'$ of longitude.[2] As we left we several times threw the dragseine and brought up the same kind of oysters to which were attached small bivalves of a type commonly found in petrified form in Europe and only recently found in Provence, and some large whelks, common sea-urchins and a faily large number of starfish and holothurians with some very small and quite pretty pieces of coral.[3] The fog and lack of wind forced us to anchor a league further on in 44 fathoms, ooze and sand. We continued to catch cods, but this was a meagre consolation for lost time in a season that was passing too quickly in the light of our wish to carry out a complete exploration of these waters. During a break in the weather which lasted less than ten minutes, we saw from our anchorage eight or ten leagues of coastline stretching to the NE¼N, so that we could sail without problem for seven or eight leagues to the NE¼E and I set course in accordance, taking half-hourly soundings, the horizon ranging less than two musket-shots. Thus we sailed over a depth of fifty fathoms until nightfall; the

---

[1] Pierre-André de Suffren Saint-Tropez (1729–88) joined the navy as a *Garde de la marine* in 1743 and took part in his first sea battle in the *Solide* in 1744. A colourful personality and a sailor renowned for his dash and bravery, he fought in the West Indies, in the Bay of Biscay in 1747 when he was taken prisoner, in the Mediterranean, in the Atlantic at Lagos where he was again captured in 1757, and in India. He rose to the rank of vice-admiral and was appointed *Bailli* of the Order of Malta.

[2] The latitude of Bukhta Syufren – for the name has survived to the present time – is 47°21′N. The longitude is 136°40′E of Paris. The bay is situated south of Zolotoy Point. The river is the Adimi.

[3] The holothurian is the sea cucumber or *bêche-de-mer* (*Holothuria edilis*). More common in warmer waters, it is found with the sea urchin and starfish on the coasts of Japan and Tartary on account of the Tsushima Current which travels from Japan up to this latitude where it meets the cold current descending from the north. This conjunction is in part responsible for the fog which so exasperated La Pérouse.

winds then veered to the north-east, strong gale, heavy rain. The barometer dropped to 27 inches 6 tenths; we struggled against contrary winds throughout the sixth of July; we observed a latitude of 48 degrees and our longitude according to the chronometers was 138^d 20′. The weather cleared at midday and we sighted some mountain peaks stretching to the north, but fog hid the lower part of the coast and we could not make out any headlands even though we were only three leagues away. The following night was very fine; we followed the coastline in moonlight; it was trending N.E., then N.N.E. We hauled along it at dawn and were hopeful that by nightfall we would reach the 50th degree of latitude, this being the limit I had set myself for our navigation on the Tartary coast and from where I would then return to Yesso and Locu Yesso if they existed, being quite confident that we would at least come upon the Kurils by sailing east, but at eight a.m. we sighted an island which seemed to be quite extensive and formed an opening of 30 degrees with Tartary.[1] We could not see any headlands and could only make out mountain peaks stretching towards the south-east, which showed we had already progressed a fair distance into the channel which separates it from the mainland. At that moment, our latitude was 48^d 35′, the *Astrolabe*'s, which had dragged two leagues ahead of us, 48^d 40′. My first thought was that this was Segalien Island whose southern point had been placed two degrees too far north by the geographers,[2] and I thought that if I entered this channel I would have to follow it right up to the point where it emerges into the Sea of Okots, on account of the stubbornness of the southerly winds which blow constantly here during this season, which would

---

[1] La Pérouse had reached Sakhalin, the great island north of Hokkaido, variously referred to as Segalien or Oku Yesso (Inner or Landward Yesso). Japanese exploration of these northern regions dates from 1635, roughly the period when the Russians themselves were reaching the Amur River estuary. Disparate information reaching cartographers then created the impression that there were two, or even three great islands. A map by J.B. D'Anville published in Fr du Halde's *Description de l'Empire de la Chine* indicated the presence of 'Saghalien anga hata' – or the cliffs at the end of the Black River – whereas Japanese charts showed an island north of Ezo (Hokkaido) with another, further north, named Karafuto. The Japanese term *ezo* in fact meant an area inhabited by the Ainu. La Pérouse's work in this region proved of great importance in clarifying this confused geographical picture.

[2] La Pérouse's first impression is clearly that Sakhalin was an island, as most maps suggested at the time. Then, as navigators – who, like La Pérouse usually entered the Tartar Strait from the south – began to sail further north and found it, not merely narrowing, but increasingly impeded by shoals and rocks, the theory that it was a peninsula attached to the Siberian mainland rapidly gained ground.

CHAPTER XV

present an insurmountable obstacle to my desire to fully explore this sea, and after drawing up a most precise chart of the Tartary coast, all I needed to do to carry out this programme was to haul along on the west the first islands I would come upon up to the 44 degrees, and so I set my course towards the point I had sighted to the south-east.

The appearance of this land was quite different from that of Tartary. All one could see was arid rock with snow lingering in the hollows, but we were too far away to see the low lands which, like those of the mainland, might be covered in trees and greenery. I gave the tallest of these mountains which looked like the ventilator of a furnace the name of Pic Lamanon[1] because its shape looked volcanic and this physicist has made a particular study of the various materials that are melted by the heat of volcanoes.

The southerly forced me to tack all sails out to round the southern point of this new land the end of which we could not see; as yet we have only briefly sighted a few peaks, as a thick fog had come up around us, but the sounding-line had been working for three or four leagues from the Tartary coast towards the west and, running east, I altered course when we found 48 fathoms. Not knowing how far this depth was situating us from the island we had just discovered and in the middle of darkness, we nevertheless obtained a latitude on 9 July with an horizon of less than half a league; it gave 48^d $15'$. The winds stayed obstinately southerly during the 9th and 10th and they were accompanied by such heavy fog that our horizon scarcely extended to a musket shot; we fumbled along this channel, quite certain that we had land from S.S.E. to east and north up to the S.W., and the new reflexions this sighting of land to the S.S.E. induced made me fairly sure that we were not in the Segalien Channel, to which no geographer has ever assigned such a southerly position, but to the west of the land of Jesso, the east of which the Dutch had presumably travelled along, and since we were sailing quite close to the coast of Tartary we had entered without being aware of it the gulf which it forms with this part of Asia. All that remained for us to do was to find out whether Yesso is an island or a peninsula, making with Chinese Tartary roughly

---

[1] Lamanon's name remains as Cape Lamanon, but it is only 45 m high. It is likely that what was seen on this occasion was Kamabuse Yama, a steep mountain 1,094 m high.

the same shape as Kamchatka does with Russian Tartary. I waited with great impatience for a break in the fog to decide finally what I should do to settle this question; it occurred on the 11th, and it is only in these fogbound regions that one sees – although quite infrequently – such extensive horizons, as though nature wanted by intervals of great clarity to compensate in some way for the almost endless darkness that encompasses these waters. Be that as it may, at two p.m. the curtain rose and we saw land from N¼NE to N¼NW. The opening was now no more than 22 degrees and several people claimed to have seen mountain summits closing it off altogether. This element of uncertainty made me quite unsure of which course I should follow. It was highly inconvenient to make twenty or thirty leagues north if we had really seen the end of the gulf, because the season was advancing and and we could not hope to make these twenty or thirty leagues against the southerly winds in less than eight or ten days since all we had gained by tacking in this channel for the last five days was twelve leagues; on the other hand the aim of our mission would not be fulfilled if we missed the strait separating Yesso from Tartary; and I felt that the best policy to adopt was to anchor and try to obtain some information from the natives. On the 11 and 12th the weather was quite clear as the breeze was quite brisk, and we had to reef the sails. We came up to within less than a league of the island; the coast was trending perfectly north and south; I wanted to find a bay where our ships would be sheltered, but the coast did not reveal the slightest bight and the sea was as rough half a league from shore as it was out in the offing, consequently although we were sailing over a very even sandy bottom which, in the space of six leagues, only changed from eighteen fathoms to thirty, I was compelled to struggle all sails out against the south winds.

The distance I was from the coast when I first saw it had misled me; as I approached I found it as green and as wooded as that of Tartary. Finally, on 12 July in the evening, the southerly breeze having greatly moderated, I came up to the land and dropped anchor in 14 fathoms, mud and sand, two miles from a small cove where there was a river.[1] Mr de Langle who had anchored an hour earlier came at once to see me; he had already lowered his boats and his longboat and suggested that we go ashore before nightfall to

---

[1] Zaliv Delanglya is close to modern-day Tomari. The river is the Ilinka.

examine the land and see whether there was any hope of obtaining any information from the inhabitants. Already with our glasses we could make out a few huts and a couple of islanders who seemed to be fleeing towards the forest. I accepted Mr de Langle's suggestion; I asked him to include in his party Mr Boutin and the Abbé Mongès, and after the frigate had anchored I lowered the Biscay boat, commanded by Mr de Clonard followed by Messrs Duché, Prevot and Colignon and told them to join Mr de Langle who had already landed. They found the only two huts in the bay abandoned, but so recently that the fire was still burning; none of the small furnishings had been removed and a litter of pups whose eyes were not yet opened had remained, the mother could be heard barking in the woods, which made us conclude that the owners of this house were not far away. Mr de Langle ordered that some axes, various iron tools, some beads and anything which in a general way could be useful and attactive to these people be left in the hut, quite sure that when he had gone back to his boat the inhabitants would return and our presents would then prove that we were not their enemies. At the same time he had the drag-seine lowered and caught in two attempts more salmon than the crew could consume in a week. Just as he was about to return to his ship he saw a canoe draw up to the shore with seven men who in no way seemed frightened of our number; they pulled up their little boat on the sand and sat down on mats amidst our sailors with an air of confidence which made us think well of them; among them were two venerables elders with long white beards, dressed in a cloth made of tree bark rather like the loin-cloths of Madagascar, and two out of the seven islanders wore clothes of well-padded nankeen; the style of their dress was much the same as worn by the Chinese; others had only a long robe which was fastened precisely with small buttons and a belt, which meant they did not need to wear trousers; their heads were bare and only two or three wore a bear skin headband; their forelocks and faces were clean-shaven, with all the hair at the back grown eight or ten inches long but differently from the Chinese who leave only a round tuft they call *Pintsec*;[1] they all wore boots made of sea otter skin with a Chinese-

---

[1] The Chinese term is *bianzi*. An alternative is to see La Pérouse's rendering as a corruption of 'pigtail' which had fairly recently entered English usage in connection with this Chinese practice and which he might have believed to be a Chinese word.

style foot that was very artistically tooled; their weapons consisted of bows, pikes and arrows, iron-tipped. The oldest of the seven islanders, the one shown the most respect by the others, had very unhealthy eyes, although he was not blind – he wore a kind of shield on his head to protect them from excessive sunlight. These inhabitants' manners were serious, dignified and very friendly. Mr de Langle gave them everything he had left and made them understand by gestures that the darkness was forcing him to return to his ship, but that he was very anxious to meet them again on the morrow to give them more presents; in their turn they made it perfectly plain that they would sleep in the neighbourhood and would not miss the meeting, and it was generally felt that they were the owners of a store of fish that had been found on the bank of the small river, raised on stakes four or five feet above ground level.

Mr de Langle had inspected it and respected it as he had the abandoned huts, but he had found there some salmon, herrings, dried and smoked, with bladders filled with oil and salmon skins as thin as parchment. This store was too large for the subsistence of one family and he concluded that these people bartered these various items. The boats did not get back to the ships until eleven at night, and the report he made greatly aroused my curiosity. I impatiently waited for daylight and I was on land with the longboat and the large boat before sunrise; the islanders arrived in the cove shortly after me; they came from the north where we considered their village to be situated. They were soon followed by a second canoe, and we saw twenty-one inhabitants; among them were the proprietors of the huts that had been abandoned the day before, whom the presents left by Mr de Langle had reassured, but there was not a single woman and we have reason to believe that they are of a very jealous nature; as we could hear dogs barking in the woods, it is likely that they were with the women; our hunters wanted to enter the forest, but they greatly urged us not to go towards the place where the barks were heard, and as I intended to ask them questions of importance and wanted to inspire confidence in them I gave instructions that they should not be upset in any way.

Mr de Langle arrived shortly after me with all his staff and before we began our talks which were preceded by gifts of every kind; they seemed to value only useful items, so iron and cloth were

CHAPTER XV

prized over everything. They had as good a knowledge of metals as we had: they preferred silver to copper, copper to iron &c, but they were very poor – only three or four had silver earrings with blue beads in every respect similar to those I had found in the Ternai Bay tomb and which I had taken for bracelets. Their other small ornaments were of copper like those of the same tomb I have already described; their tinder-box and pipe were Chinese or Japanese, of perfectly worked pale copper; they made us understand by signs that the blue nankeen some of them wore, the beads and the tinder-boxes came from the country of the Manchus, pointing towards the west and saying the word manchu exactly as we do. Then noticing that we were holding some paper and a pencil to draw up a vocabulary of their language they guessed our intention, anticipated our questions, brought forth by themselves various objects, added the name of the country and were kind enough to repeat it four or five times until they were quite sure we had grasped their pronunciation; the ease with which they had guessed, on seeing our paper and our pencil that we were trying to compile a vocabulary, made me fairly certain that they are not unfamiliar with the art of writing, and one of these islanders who since that occasion drew for us the outline of the country held the pencil absolutely in the way the Chinese hold their brushes. They appeared quite eager to obtain our axes and our cloth, did not even hesitate to ask for them, but they were also careful always to take only what was theirs and what we had given them; it was clear that their notion of theft is no different from ours, and I would not have hesitated to ask them to guard our belongings. Their scrupulousness in this connection included even not picking up a salmon on the sand, although they were laid out in their thousands because our fishing endeavours had produced as great an abundance as one the previous day and we had to ask them several times to take anything they wanted.

We finally got them to understand that we wanted them to outline the shape of their country and that of the Manchus; then one of the old men rose and with the end of his pike drew the coast of Tartary on the west, trending roughly north and south, on the east facing it and in approximately the same direction he drew his island and placing several times his hand on his chest he made us understand that he had just drawn his own country; he had left a strait between Tartary and his island and turning towards our ships which could be seen from the shore he showed by a line that it was passable. To the south of his island he had shown another place and left a strait between them, indicating that it was a route for our vessels. His ability to guess all our questions was very great, but not as considerable as that of a second islander of about thirty years of age who, seeing that the outlines drawn on the sand were disappearing, took one of our pencils with some paper; he drew his island, which he called Tapschoka[1] and, by means of a line, the small river on the banks of which we were standing, which he showed at two-thirds of its length from north to south;[2] he then drew the land of the Manchus, leaving as the old man had done a strait at the end of the funnel and to our great surprise he added the Segalien River[3] pronouncing the name as we do; he placed the mouth of this river a little to the south of the northernmost point of his island and showed by means of lines the number of days which a canoe would take to go from the spot where we were to one facing the mouth of the Segalien – he drew seven, and since these people's canoes never go more than a pistol shot away from the land and follow the shore of small coves we concluded that they hardly covered nine leagues a day in a straight line, as landings are possible everywhere along the coast, and they go ashore to cook and eat their food and probably take frequent rests; thus we thought that we were only sixty-three leagues from the end of the island. This same islander repeated what had already been said to us, that they obtained nankeen and other articles by communicating with the people who live along the Segalien River, and he also showed by means of dashes how many days were needed for the canoes to go up this river to the places where trade was carried out. All the other islanders were present during this conversation and endorsed their compatriot's comments by their gestures. We then wanted to know if this strait was very wide: we endeavoured to make him understand this by means of signs. He guessed our meaning and placing his two hands horizontally and two or three

---

[1] This term, if local, is hard to identify. The Kurils were called 'Tchoupka' by the Ainu, meaning the place where the sun rises; see J. Stephan, *The Kuril Islands*, Oxford, 1974, p.6. It appears likely that the name was also generally used to refer to Sakhalin.

[2] The Ilinka River is situated two-thirds from the southern end of Sakhalin, not from the north point.

[3] This Segalien River is the Amur; both names were in use on maps of the time, as for instance in Robert de Vaugondy's 'Partie orientale de l'empire de Russie en Asie' of 1750.

CHAPTER XV

inches from each other he made us understand that this was his indication of the width of the little river at our watering place; spreading them further out he indicated that this second width was that of the Segalien River, and moving them much further away he made us understand that this was the width of the strait which separates his country from Tartary. We now only needed to know the depth and we led him to the bank of the small river from which we were a mere ten paces away and we dipped a pike into it. He seemed to understand us and placing one hand vertically above the other at a distance of five or six inches we thought he was indicating in this way the depth of his river; he separated them more for the depth of the Segalien River and finally stretched out his arms as far as they could go to represent the depth of the strait; we still needed to know whether he had indicated absolute or relative depths, because in the first case the channel would have been only a fathom deep and these people, whose canoes had never gone near our vessels, might think that three or four feet of water would be enough for us just as three or four inches would be sufficient for their canoes, but it was not possible to obtain any clarification on this matter and in any case Mr de Langle and I believed that it was of the utmost importance to find out whether the island we were sailing along was the one geographers had named Segalien Island without suspecting how far it extended to the south, and I gave orders to make all the necessary preparations on the two frigates to sail the next day.

We spent the rest of the day visiting the country and its inhabitants. We had met none since our departure from France who more aroused our curiosity and our admiration. We knew that the most highly populated and possibly those with the oldest administrative structures live in countries close to these islands, but it does not appear that they have ever conquered them because there is nothing to attract their greed, and we found it quite opposed to our way of thinking to encounter among hunting and fishing people who cultivate nothing and have no herds more politeness, more gentleness, more seriousness and maybe a greater intelligence than in any nation of Europe; I certainly do not believe that the better class of Europeans is not in every respect considerably superior to the twenty-one islanders of Bay de Langle – so named after Mr le Vicomte de Langle who discovered it and was the first to land there – but nobility of manners, knowledge &c belong in Europe to a small number of individuals, and our peasants, our artisans, our sailors, do not have the leisure to concern themselves with such things; they were therefore far more widespread among these islanders who all seemed to have received the same education. They showed none of the stupidity of the Indians of Baye des Français; they were surprised by our crafts and cloths, they looked at them carefully, turned them over in every direction, discussed them among themselves and tried to find out how they had been made; the shuttle was not unknown to them; I brought back in the King's collection a loom absolutely similar to ours on which they make their linen, but the thread is made with the bark of a willow that is very common in their island and which seemed to be much the same as ours. Although they do not till the soil they use its wild products with the utmost skill. We found in the hut a great many roots from some kind of lily which our botanists identified as the yellow or Kamchatka lily;[1] they dry it and it becomes their winter supply. There was also a great deal of garlic and angelica;[2] these plants are found along the edge of the forest. Our brief stay did not allow us to discover whether they have some form of government and all we could do in this matter would be to hazard a few guesses – we shall leave these to our readers – but there is no doubt that they show extreme consideration towards elderly men, that their customs are very gentle, and that if they were a pastoral people and had large herds I would certainly use no other model to conjure up the practices and customs of the patriarchs. They are generally well built, strong, with fairly pleasing features, very hairy in a distinctive fashion; they are short: I saw none who reached a height of five feet five inches and several were less than five feet tall. They allowed our artists to draw them, but constantly declined the requests of Mr Rollin, our surgeon, who wanted to take measurements of their bodies: and they may have seen this as a magic trick, because it is known from travellers' accounts that the concept of magic powers is widespread in China and in Tartary, and our

---

[1] The Russian scientist and historian Stepan Petrovich Krasheninnikov had described this plant (*Lilium flore atro-rubente*) and its uses. He was the author of a *History of Kamtschatka and the Kurilsi Islands*, published in St Petersburg in 1757 and in a shortened English version in London in 1764, the work of James Grieve. See also E.A.P. Crownhart-Vaughan, *Explorations of Kamchatka: 1735–1741*, Portland, 1972, pp. 111–12.

[2] Garlic (*Allium sativum sibericum*) was used as an antiscorbutic. The angelica, which is also mentioned in Cook's voyages, was probably the *Angelica archangelica*.

CHAPTER XV

missionaries have often been dragged before the courts and accused of being magicians because they used to lay hands on children when they baptised them. Be that as it may, this mild refusal, with their determination to move away their wives and hide them are the only criticisms we could make of their trust, and we can declare that the inhabitants of this island are a very well governed people, but are so poor that they will not for many years attract the cupidity of conquerors or traders; a little oil and some dried fish are really small items to export; we bought during that day only two marten pelts.[1] We saw a few bear and sea-otter skins cut into pieces and made in clothes, but not many, and the pelts obtainable in this island would be quite unimportant for traders. All the silver ornaments of these 21 islanders did not weight two ounces, and a medal with a silver chain I placed around the neck of an old man who seemed to be the head of this gathering struck them as being priceless. We found lumps of coal along the shore but not a single pebble with gold, iron or copper; I am very inclined to believe that they have no mines in their hills. All wore in their thumb a heavy ring rather like a *gimbelete*;[2] these rings were made of ivory, horn or lead. Their nails were very long like the Chinese. They also had their form of greeting which consists in kneeling down and touching their foreheads on the ground. Their way of sitting on mats was the same; they ate like them with little sticks. If they share a common origin with these people and the Tartars their separation goes back a long way because the prints of nature have been obliterated and their features have nothing in common.

Our Chinese did not understand a single word of their language,[3] but they understood quite well the two Manchu Tartars who had crossed over from the mainland to this island two or three weeks previously, possibly to trade for fish.

We met these only in the afternoon. The conversation was carried out not by gestures but by speaking through one of our Chinese who knew Tartar very well. They gave him exactly the

---

[1] La Pérouse obtained skins of the *Martes zibellina*, which had become scarce at the time. The Russians, who supplied the Chinese market, had been forced for a number of years to launch expeditions towards the north and well down the coast of Alaska to obtain adequate quantities, but the sable, which these were, remained elusive.

[2] The *gimblette* was a small hard ring-shaped biscuit. The name comes from the Provençal word *gimbeleto*.

[3] These are Chinese sailors taken at Macao.

same geographical details about the country, changing the names because presumably each language has its own. The dress of these two Manchu Tartars was of grey nankeen, similar to that of the coolies or dock-hands of Macao. Their hat was pointed and made of bark; they wore the *pentsec* or Chinese hair tuft, and their manners and features were quite inferior to those of the islanders. They said that they lived eight days' travel up the Segalien River. All these reports together with what we saw on the Tartary coast, which we hauled along so close in, convince me that the seashore in this part of Asia is hardly inhabited from the 42 degrees or the Korean frontier up to the Segalien River, that mountains which may be impassable separate this maritime region from the rest of Tartary[1] and that access from the sea is only possible by way of river mouths, although we saw none of any size.[2] These islanders' huts are cleverly built; every precaution is taken against the cold; they are made of timber covered with birch bark and topped by a framing covered with dried straw just as our peasant houses are covered with thatch; the door is very low and set in the gable-end; the hearth is in the middle under an opening made in the roof to let the smoke out; small benches or planks eight or ten inches high are built in all around and the floor is covered with mats. The hut I have just described was situated in the middle of a real forest of rose-trees a hundred paces from the seashore; they were flowering and the scent was delightful, but it nevertheless could not make up for the stench of fish and oil which would have triumphed over all the perfumes of Arabia. We carried out an experiment that showed that the pleasure derived from the sense of smell is, like taste, ruled by habit: I gave one of these elderly men a flask filled with a very pleasant perfume; he put it to his nose and made it clear that he felt for this water the same revulsion as we did for his oil. They continually had a pipe in their mouths; their tobacco was of a good quality, with large leaves. I thought that they got it from Tartary, but they clearly explained that their pipes came from the island

---

[1] In spite of the paucity of information available to him La Pérouse's comments on the region are accurate. The pigtail worn by the mainlanders illustrates the tighter hold which China had on the coastal areas compared with Sakhalin which was only a tributary of Peking.

[2] La Pérouse adds the following marginal note: 'These islanders never gave any indication that they carried out any trade with the Tartary coast which they knew quite well since they drew it, but with the people who live an eight days' journey up the Segalien River.'

CHAPTER XV

situated to the south and undoubtedly from Japan. But our example never induced them to try snuff, and it would not have rendered them a good service to get them accustomed to a new necessity. To these details I append a vocabulary that will be found at the end of this volume. It was with some surprise that I found the word *chip*[1] for a vessel and *thou tri* for two and three. Had these English terms escaped the confusion created by the tower of Babel or rather were they not evidence that a similarity between a few words does not indicate a common origin?

On the 14th at daybreak I gave the signal to weigh anchor with a southerly breeze and in misty weather which soon turned to heavy fog, and until 19 July there was not the slightest bright interval. I set course for the N.W. towards the Tartary coast and when, according to our reckoning we were at the point from where we had sighted Pic Lamanon, we hugged the winds and tacked under light sails in this channel awaiting the end of this darkness which I felt was like nothing encountered in any other sea. Finally the fog disappeared for a while. On the morning of the 19th we saw the island from NE¼N to ESE, but it was still so shrouded in mist that we could make out none of the points we had seen earlier. I made to approach it but we soon lost sight of it; however, guided by the sounding-line, we continued to coast along it until two p.m. when we dropped anchor off a very good bay in 20 fathoms, in small gravel, two miles from the shore. The curtain rose at four o'clock and we saw the land behind us to the North¼NE. I named this bay, the best we had anchored in since we left Manila, d'Estin Bay.[2] It is

---

[1] It should be noted at this point that in French *ch* is pronounced *sh*, and that *ou* is pronounced as *oo*.

[2] Henri-Charles-Hector-Théodat d'Estaing (1729–94) was born at Ravel, in Auvergne, of an old noble family. He first served with the musketeers and rose to lieutenant-general during the Seven Years War. He fought with Lally in India and was taken prisoner in 1757 and again in 1760, spending some time in England. Appointed governor of the French Leeward Islands and lieutenant-general of the naval forces, he rose to vice-admiral in 1777, soon taking command of the French naval forces operating in America and later of the combined French-Spanish forces. He captured Newport, Rhode Island, in 1778 but bad weather forced him to leave for Boston; in 1779 he captured St Vincent and Grenada in the West Indies, but unsuccessful at Savannah in Georgia he returned to France in 1780. Named governor of the province of Touraine in 1785, he took up a liberal position during the French Revolution and was appointed full admiral in 1792. He was however eventually arrested as a suspect, accused of siding with the royal family and executed on 28 April 1794. The bay which was given his name is probably Bukhta Lesovkogo, close to the modern port of Lesogorsk which is in latitude 49° 27'N.

295

situated in 48^d $55'$ of latitude and 140^d $27'$ of longitude, Paris meridian; our boats landed at 4 p.m. by ten or twelve huts arranged in no particular order a fair distance from each other and about a hundred paces from the seashore. They were a little larger than those I described earlier; however, they had been contructed out of the same materials; but they were divided into two rooms: the inner apartment contained all the small items of the household, the hearth and the benches around the sides, but the antichamber was quite bare and seemed intended for visitors, strangers presumably not being allowed in the presence of the women – some officers met two who had fled and hidden in the long grass; when our canoes landed in the cove they uttered cries giving the appearance of being as frightened as if they were afraid of being eaten; they were nevertheless under the protection of an islander who was bringing them home and seemed to be assuring them that we were not so much devils as white men. Mr Blondelas had time to sketch them and his drawing gives a very good impression of their features; these are somewhat unusual but quite attractive: their eyes are small, their lips thick, the upper lip being painted blue or tattoed – we were not able to verify this. Their legs were bare; they were wrapped in a long cloth dressing gown and as they had become wet in the dew covering the grass this dressing gown was stuck in folds to their bodies and allowed the artist to depict their shape which is not very elegant; they wore their hair to its full length and their forehead was not shaven as was the men's.

Mr de L'angle who was the first to land found all the islanders gathered around four canoes loaded with smoked fish; they were helping to push them into the water and he learned that the 24 men forming their crews were Manchus who had come from the banks of the Segalien River to buy this fish; he had a lengthy conversation with them through the intermediary of our Chinese whom they had warmly welcomed. Like our first geographers they told us the land we were following was an island; they gave it the same name, adding that we were still five days by canoe from its extremity, but that with a favourable wind one could cover this distance in two days and sleep every night on land; thus everything we had been told at Baye de L'angle was confirmed during this second stop, but less intelligently expressed by the Chinese who was acting as our interpreter and did not know French. Mr de Langle also came across in a corner a circle of fifteen or twenty stakes each topped with a bear's head; theses animals' bones were scattered in the neighbourhood. Since these people fight bears hand-to-hand, do not have firearms and their arrows can only wound them, we concluded that this circle was intended to preserve the souvenir of their exploits, and that the 20 heads on display were to recall the victories they had had over ten years – the state of decomposition of the majority indicated that they went back this far.[1] The products and composition of the soil of Baye d'Estin were much the same as those of Baye de L'angle. Salmon was as plentiful and each hut had its store, but we discovered that these people only eat the head, the tail and the spine, and that they smoke and dry out, in order to sell them to the Manchus, the two sides of this fish, keeping for themselves only the smell that infects their house, their furniture, their clothes and even the grass surrounding their villages. Finally at eight p.m. our boats left, after showering gifts on the Tartars and the islanders. They were back at eight fifteen and I ordered everything to be readied in order to sail the next day.

It was a very fine day. We made excellent observations of latitude and lunar distances, which enabled us to correct the reckoning of the previous six days since our departure from Bay de Langle situated in 47^d $49'$ of latitude and 140^d $27'$ of longitude which was only $5'$ from that of Baye d'Estin, the trend of the west coast of this island from the point where we saw Baye de L'angle in 47^d $39'$ up to 52 being absolutely north and south.[2] We hauled along it a short league away, and at seven p.m. a thick fog having come up around us we dropped anchor in 37 fathoms, mud and small pebbles. The coast was much more mountainous and steep than in the southern part; the shore was also much sheerer; we saw neither habitations nor fire, and as night was falling I did not send any boat ashore, but for the first time since leaving Tartary we caught eight or ten cods, which seemed to be an indication of the closeness of the mainland which we had lost sight of back in 49 degrees of latitude, but compelled to follow one or the other coastline I had given the

---

[1] La Pérouse's conjecture is correct: Krasheninnikov reported that the people of Kamchatka were accustomed to display over their houses the heads of bears they had killed, 'like trophies' and as proof of their courage. See Crownhart-Vaughan (1972), p. 125.
[2] This passage seems more confused than it is. La Pérouse's latitude for Bay de Langle was 47°49′ (a chart drawn at the time gives 47°48′36″). He merely adds here that when he first sighted it he was in 47°39′.

preference to the island so as not to miss the strait if there was one leading to the east, a task which required extreme care on account of the fog which broke only very briefly; so I stuck to it, so to speak, and never went more than two leagues away, from Baye de Langle up to the end of the gulf in 52^d 10′ of latitude, and my conjectures on the closeness of the Tartary coast were so sound that as soon as our horizon opened up a little we were able to discern it quite clearly. The channel began to narrow around the 50th degrees and was no more than twelve or thirteen leagues wide. On the evening of the 22nd I anchored a league from land in 37 fathoms, ooze, athwart a small river, three leagues north of a very remarkable peak the base of which is situated on the shore and whose summit seen from every point of the compass retains a perfectly regular shape; it is covered in trees and greenery up to the top. I called it Lamartiniere Peak because it offers great opportunities for botanical research which has been this savant's main occupation.[1]

As I had seen no habitations as I sailed along the coast from Baye d'Estin I wanted to dispel my uncertainty in this respect and I had four boats readied under Mr de Clonard to go and survey the cove into which the small river, the ravine of which we could see, was running. He was back at eight p.m. and to my great surprise brought back all his boats filled with salmon although the crews had no lines or nets. This officer told me that he had landed at the mouth of a stream whose width did not exceed four *toises* and the depth one foot, that he had found it so full of salmon as to be in a way paved with them and our sailors had killed twelve hundred in the space of an hour by clubbing them. Moreover all they had found were two or three abandoned shelters which they assumed had been built by Manchu Tartars when they crossed over from the mainland to trade in the southern part of the island. The vegetation was even more vigorous than in the bays where we had already landed, the trees were larger, celery and cress grew in the greatest abundance along the banks of this river; it was the first time we had come across this latter plant since our departure from Manila. They could also have collected several bags of juniper-berries,[2] but we gave preference to herbs and fish. Our botanists made a good

---

[1] The mountains forming part of the Zapadny chain come close to the shore in this part of the coast; it is not possible, from the brief mention given here, to identify Pic Lamartiniere with certainty.

[2] The *juniperus communis*.

collection of fairly rare plants and our lithologists brought back numerous spar crystals and other silicate stones which were fairly interesting, but found no marcasite, no pyrites, nothing in a word that would indicate the presence of mines of any type of metal in this country;[1] pines and willows greatly outnumbered the oaks, maples, birches and medlars, and if other travellers had landed a month after us on the banks of this river they would have been able to pick quantities of gooseberries, strawberries and raspberries which were still in flower.

While our crews were obtaining this abundant harvest on land, we were catching a great many cod from the ships and anchoring in this bay for a few hours gave us with enough fresh food for a week. I named this river Salmon Stream, and set sail at daybreak to continue hauling along very close in to this island which was not coming to an end in the north although every headland that I saw protruding a little raised my hopes. On 23 July we observed $50^d\ 54'$ and our longitude had practically not changed since Baye de L'angle; in this same latitude we saw a very good bay, the only one since we had begun coasting along this island that offered a safe shelter for ships against the winds of the channel; there were a few houses scattered on the shore close to a gully which indicated the presence of a river somewhat larger than those we had already visited; I did not consider it necessary to examine more closely this bay which I named Baye de La Jonquière[2] whose width I nevertheless sailed along; a league from shore the ground was mud, 35 fathoms. But I was in such a hurry and clear weather was so rare and so precious for us that I felt I ought to take advantage of it only to progress north. Since reaching the 50 degrees of latitude I had

---

[1] In fact, Sakhalin has valuable resources of coal, petrol, iron, gold, copper, silver and other minerals. The Russian authors of a relatively recent book on the island called it 'a real treasure'. See Gor and Leshkevich, *Sakhalin*, Moscow, 1949, p. 3.

[2] As he sails along the coast, La Pérouse is continuing to honour those who helped him during his career. Clément Taffanel de la Jonquière, born in 1706, was a distant cousin of the Galaup family and a friend of Ternay; he acted as La Pérouse's protector in his early years, taking him with him as a *garde de la marine* in the *Célèbre* in May 1757. See H. Manavit, 'Ce que Lapérouse doit à La Jonquière', *Revue du Tarn*, III (1969), pp. 163–72. La Jonquière rose to brigadier of the naval forces in 1765, commodore in 1771 and lieutenant-general in 1780. However, La Pérouse's tribute did not last: the name his disappeared. The coast forms a very wide bight in this latitude, with smaller bights and rivermouths. The port of Aleksandrovsk-Sachalinskij is situated in $50°54'$ and the smaller town of Mgaci in $51°05'$; either of these localities could be what La Pérouse saw on this occasion.

gone back to my earlier opinion: I could no longer doubt that the island we had been following from the 47th degree and which according to the report of the natives seemed to extend much further south was Segalien Island whose northern extremity has been fixed by the Russians at 54 degrees and which is one of the world's longest islands from north to south; consequently the so-called Strait of Tessoy is no more than the one separating Segalien Island from Tartary, in approximately 52 degrees. I had gone too far not to want to explore this strait and ascertain whether it was passable; I was beginning to form an unfavourable opinion of it because the depth was lessening very quickly as one went north and the coast of Segalien Island was nothing more than drowned sand dunes almost level with the sea, like sand banks. On the evening of the 23rd I dropped anchor in 24 fathoms, muddy bottom, three leagues from land; I had found the same depth two leagues further east, three miles from the shore, and from sunset until the time we anchored I had covered a distance of two leagues westward at a right angle from the line of this coast to find out whether, if we got away from Segalien Island, the depth would increase, but it remained the same and I was beginning to suspect that the slope ran south to north along the length of the channel rather like a river whose depth diminishes as one nears its source. At daybreak on the 24th we set sail, making for the N.W. The depth diminished down to 18 fathoms in three hours; I set a westerly course and it remained exactly the same. In the end I decided to cross this channel twice, east and west, in order to verify that there was not a greater depth and thus find the fairway through the strait, if there was one. These procedures were the only reasonable ones in our circumstances because the depth was falling so rapidly when we advanced towards the north that for every league in that direction the depth reduced by three fathoms. Consequently, assuming a gradual levelling up, we were no more than 6 leagues from the end of this gulf in which we could not sense any current, which seemed to prove that there was no fairway and explained the perfect evenness of the slope.[1] We anchored on the 26th in the evening on the coast of Tartary in 21 fathoms and the next day at midday, the fog having cleared at last, I decided to sail N.N.E. towards the middle

---

[1] The depth of the strait near and beyond this point fluctuates between 6 and 10 fathoms.

CHAPTER XV

of the channel in order to complete our study of this aspect of geography which had cost us such exertions. We sailed in this way with a perfect view of both coasts and, as I expected, the depth fell by three fathoms per league and after covering four leagues we dropped anchor in 9 fathoms, sandy bottom. The southerly winds were so constant that they had not altered by 20 degrees in a month and we ran the risk in thus sailing before the wind of becoming embayed at the end of this gulf and of having to wait for the turn of the monsoon in order to get out of it; but this was not the worst inconvenience – not being able to hold at anchor with a sea that was as rough as any on a European coast lacking shelter was far more serious. The southerly winds whose origin, if I can put it that way, lies in the China Sea always reach the back of the Segalien Island gulf without meeting any obstacle; they affect the sea quite as strongly as the westerlies do around Cape Finistere and are more regular than the trade winds between the tropics; but we had gone so far that I wanted to see, or touch, their final point. Unhappily the weather had become very uncertain and the sea was worsening. Nevertheless we lowered our boats to take soundings all around us. Mr Boutin was instructed to go towards the south-east and Mr de Vaujuas to take soundings towards the north, with express orders to take no risks that might jeopardise his return. These orders could only be entrusted to an officer known for his extreme caution because the worsening condition of the sea and the strengthening wind might compel us to raise anchor in order to save our vessels, and I instructed this officer to do nothing, under any circumstances, which might endanger either our ships if we had to wait for his longboat, or his own longboat should circumstances become so serious that we found ourselves compelled to leave it behind.

My orders were most scrupulously carried out. Mr de Vaujuas travelled a league north and found no more than six fathoms.[1] He reached the furthest point where his circumstances allowed him to

[1] La Pérouse had reached Bukhta Nevelsokogo, opposite Cape Uangi on Sakhalin. At this point the depth is 6 fathoms; a little further on 8 and 10 fathoms may be found, but the coast of Tartary advances here to form Cape Lazarev, while Uangi juts out on the western side, and consequently the main Tartary Strait gives the impression of ending here. If the weather had been more favourable and he had been less pressed for time, Vaujuas could have gone a little further north and discovered the passage through, even though it is narrow and sailing through it would have involved the two frigates in running considerable risks.

sound; having left at seven p.m. he was not back before midnight; the sea was already rough and after our misfortunes at Baye des Français I was beginning to feel extremely concerned; his return seemed a meagre compensation for the very hazardous situation in which our vessels found themselves, because at dawn we were forced to sail and the sea was so bad that it took us four hours to raise our anchor. The messenger of the capstan and its chain broke, the capstan itself was damaged, three men were fairly seriously injured and, although there was a strong gale blowing we had to let out all the canvas our masts and yards could carry. In this way we made two crossings east and west of the channel. Fortunately, a few minor wind changes from the south to S.S.W. and S.S.E. were in our favour and in 24 hours we covered five leagues. In the evening of the 27th, the fog having lifted, we found ourselves on the coast of Tartary outside a bay which seemed to be very deep and to offer a safe and convenient anchorage. We had no wood left whatsoever and our supply of water was greatly reduced. I decided to put in here, signalling the *Astrolabe* to go ahead and take soundings. We dropped anchor off the northern point of this bay at 5 p.m. in eleven fathoms, ooze, and Mr de L'angle, at once lowering his boat, went himself to sound the back of this roadstead and reported that it offered a most satisfactory shelter behind four islands which protected it from the off winds. He had landed at a Tartar village where he had received an excellent welcome; he had also discovered a watering place where the clearest of water could cascade right into our longboats, and the islands from which the good anchoring place was no more than three cablelengths away were covered with trees. Following Mr de Langle's report I gave orders to take all necessary steps to enter the bay[1] at dawn and we dropped anchor to the west of Observatory Island at eight a.m. in six fathoms, muddy bottom.

---

[1] Castries Bay – as La Pérouse will call it – is now known as Zeliv Chikhacheva, but the French name has not disappeared: the town and port of De-Kastri is situated inside the bay which does indeed provide excellent shelter in this inhospitable region.

CHAPTER XVI

Stay in Castries Bay on the coast of Tartary. Description of this bay and of the Tartar village; customs and practices of the inhabitants, their respect for tombs and for property; extreme confidence they inspire in us. Their tenderness towards their children; their unity between them. They show some reserve towards us, taking us for sorcerers. Meeting with four canoes, outsiders in this bay, belonging to a neighbouring people; the crews of these canoes tell us that they came from the Segalien River and were returning home with a load of grain and nankeen; they draw on our charts a sandbank with seaweed, joining Tartary to Segalien Island and indicate by gestures that they dragged their canoe over this bank. Products of Castries Bay; its fish, shells, quadrupeds, birds, stones, plants; departure from Castries Bay after a storm. The northerly winds blow for two days, enabling us to reach the southern point of Chotka Island. Discovery of Castries Strait separating Yesso from Japanese Locuyesso or Chotka Island from the island of Tssissa. Call at Crillon Bay on Chotka Island. The inhabitants come aboard. Description of these inhabitants and their village to which we send an exploring party. We cross the strait and identify all the land formerly discovered by the Dutch of the Castricum; Staten Island, Vries Strait, Company Land. This navigation which some geographers were beginning to reject may be the most accurate of all the ancient expeditions. Four Brothers Island, Marikan Island. We pass through the Kurils and give the pass the name of Boussole Channel. Thick fog even more constant than on the coast of Tartary. We give up the plan of exploring the northernmost Kurils and make for Kamchatka.

Accepting that it was impossible to emerge to the north of Segalien Island had presented us with a new problem; it was very doubtful whether this year we could reach Kamchatka.

Castries Bay in which we had just anchored is situated at the back of the gulf and two hundred leagues from the Strait of Sangar, the only exit we were sure of to get out of the Sea of Japan. The southerly winds were more constant, more settled, more stubborn than in the China seas where they came from, because they were squeezed between two coasts. They did not vary for more than two compass points east or west; whenever the breeze freshened, the sea became alarmingly rough, worrying for the state of our masts, and finally our ships were not good enough sailers for us to hope to make two hundred leagues against the wind in the fine season in such a narrow channel where almost constant fog makes tacking so difficult. Nevertheless this was the only option open to us, unless we waited for the northern monsoon which might be delayed until November, and I did not contemplate this possibility for one minute. On the contrary, I felt that we should make every effort to see to our needs in wood and water as speedily as possible and I announced that our stay would not exceed 5 days. As soon as we had moored, Mr de Langle and I gave our boats and longboats instructions as to their particular destination which remained unchanged during the whole of our stay; the sloop collected our water, the large boat our wood, the smaller boats were given to Messrs Blondelas, Belle garde, Mouton, Bernizet and young Prevôt to chart the bay; our yawls which had a small draught were sent fishing for salmon in a small river[1] which was full of them, and our Biscay boats were used by Mr de Langle and I to go and supervise our various tasks and take us with our naturalists to the Tartar village, to the various islands and generally to all points that seemed worth inspecting. First and most important was the operation of our chronometers, and our sails were hardly furled than Messrs Dagelet, Lauriston and Darbaud had set up their observatory on the island of that name three cablelengths from our ships; this island was also to supply our carpenters with the wood we almost entirely lacked; a graduated pole was placed in the water at the foot of the observatory to ascertain the height of the tide; the quadrant and the regulator clock were set in place with an energy worthy of better results; the astronomical work was going on without stop. The brief stay I had announced did not allow for a moment's rest, the morning and the afternoon were devoted to measuring relative altitudes, the night to taking the distance of stars, the comparison of the workings of our chronometers had

---

[1] The river is the Bolshoy Somon.

CASTRIES BAY

19. Chart of Castries Bay, Siberia. Four islets are named: Basalt, Observatory, Oyster, South. The northern headland is Assas Point, the southern one Clostercamp. By Bernizet, July 1787. AN 6 JJ1:41B.

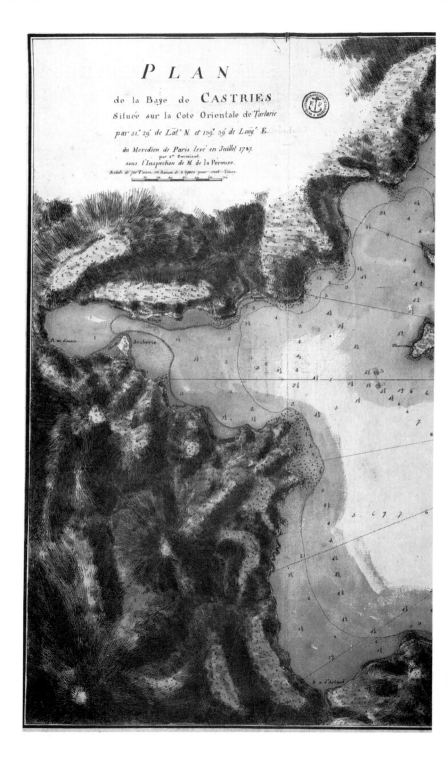

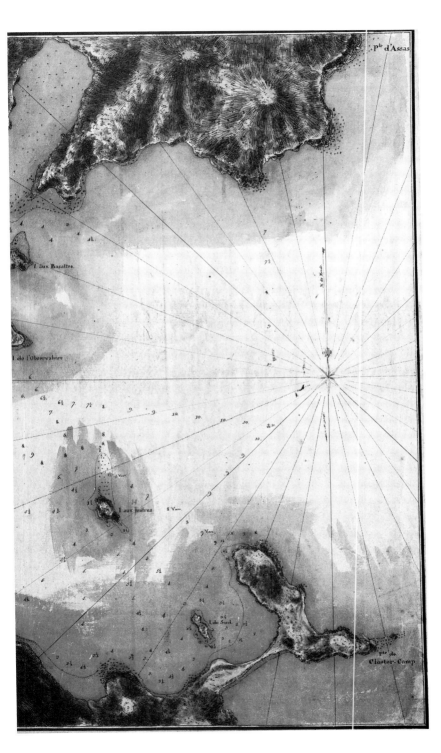

20. French officers among local tombs, Castries Bay, Siberia, by Duché de Vancy. SHM 352:20.

CHAPTER XVI

already begun. Our No. 19 left little uncertainty because its results compared with lunar distances had always been the same, or at least had not gone beyond the margin of error which can arise with such observations; the same could not be said of No. 18 which was on the *Astrolabe* – its variations were not regular and Messrs de Langle and l'Auriston did not know what daily error to allocate to it. The clumsiness of a carpenter destroyed all our hopes: he cut down a tree near the astronomers' tent which in falling broke the sight on the quadrant, put the regulator clock out of order and practically wiped out the work of the two previous days, whose net output (if one can put it that way) was reduced to the latitude of our anchorage in $51^d\ 29'$ and $139^d\ 41'$ of longitude, Paris meridian, according to our No. 19 and taking into account the daily difference of $12'$ which had been worked out for it at Cavitte.[1] High tide at the new and full moons was calculated to be ten o'clock, its maximum height at the same periods to be 5 feet 8 inches and the speed of the current less than half a knot. The astronomers, reduced by this misfortune to casual observations, accompanied us on our various travels during the final two days. Baye de Castries, from the name of the minister who gave orders for this campaign and proposed me to the King as its leader,[2] it offers a safe shelter to vessels in bad weather and it would be possible to winter there; the bottom is of mud, rising gradually from 12 fathoms to 5 as one nears the shore, the rocks of which extend three cable lengths out to sea, so that it is quite difficult to land a boat when the tide is low – one then has to struggle against weeds between which there remain only two or three feet of water and which put up an invincible resistance against the boatmen.

---

[1] The latitude determined by the French expedition is correct. The longitude of De-Kastri is $140°47'$E of Greenwich or $138°27'$E of Paris. The ships were anchored east of the present town site, but there still remains an error of approximately one degree in the longitude recorded in the journal.

[2] Charles-Eugène-Gabriel de Castries (1727–1801) was a military man who had fought at Dettingen, Fontenoy and Lawfeldt; he rose to brigadier in 1756, led an expedition to Corsica and in 1757 commanded an army corps under Marshal Charles de Soubise in Germany. He was appointed governor of the province of Lyonnais in 1766 and in 1780 became Minister of Marine; he carried out a number of major reforms during his term of office, improved the teaching of hydrography, and established a naval code and rules governing recruitment and promotion within the service. He resigned in 1787. His part in planning for La Pérouse's expedition and in selecting La Pérouse to lead it was of major importance, but La Pérouse's acknowledgement of his role was omitted from this section when the 1797 account was published, no doubt because by then Castries had emigrated and was a declared opponent of the Revolution. He died in exile at Wolfenbüttel (Brunswick) in 1801.

No sea is more prolific in various types of fucus,[1] and the vegetation of our finest meadows is not any greener or richer. A very large bight on the edge of which the Tartar village was situated, and which at first we thought was deep enough to receive our vessels because the tide was high when we dropped anchor in the offing turned out to be nothing more than a vast prairie of seaweed among which we could see salmon leaping, coming from a river whose waters lost themselves in this weed, and where we caught more than two thousand in two days.[2]

The inhabitants for whom this fish represented a most abundant and dependable food supply watched our successful fishing expeditions quite unconcerned because they were certain that the quantity was inexhaustible. We landed at the foot of their village the day after we arrived in the bay. Mr de Langle had gone there first and his gifts had won us some friends.

Nowhere in the world could one meet better people. The chief or elder came to welcome us on the beach with a few other inhabitants; he prostrated himself in the Chinese manner and then led us to his hut where were his wife, his daughters-in-law, his children and grandchildren. He stretched out a clean mat on which he bid us sit down, and an iron cooking-pot filled with a small grain we could not identify[3] was set in place over the fire with some salmon, to be offered to us. This grain was their most precious food; they told us that it came from the country of the Manchus and we verified that they give this name only to the people who live seven or eight days' journey upriver on the Segalien and have direct contact with the Chinese. These people made us understand by gestures that they belonged to the Orotchy nation and, showing us four canoes belonging to outsiders, which we had seen arriving the same day in the bay and which had stopped in front of their village, they named their crews Bichys,[4] a name indicating that these people came from further south, but maybe less than seven or eight leagues from the Orotchys's village, because these people like those

---

[1] Fucus (*Fucus vesiculosus*) is used here to refer to seaweed or rock-weed. The word has now a more restricted use, but in the eighteenth century it was a more generalised term.

[2] Marginal note: 'This fucus or sea weed is exactly the same as is used in Marseilles to pack the various cases of oil or liquor.'

[3] This is probably a Siberian variety of millet, *Setaria italica*.

[4] The Oroki were Tunguses living in the Amur region. The word *bit'chy* meant a southern village or clan.

CHAPTER XVI

of Canada have a different name and language in each settlement. These strangers, about whom I shall have more to say later, had lit a fire on the sand on the shore near the Orotchy village; they were cooking their grain and salmon in a cooking-pot hanging from a tripod made of three sticks tied together from the top of which hung an iron hook. They had come from the Segalien River and were bringing back to their country some nankeen and grain which no doubt they had bartered for oil, dried fish, and possibly skins of bears or elk, the only quadrupeds with squirrels and dogs whose remains we had seen.

The Orotchys's village consisted only of four huts, solidly built with full lengths of pine trunks with proper notched joints at the corners. A fairly well made structure held up the roof which consisted of tree bark; a bench, similar to the one found in huts on Segalien Island, went round the apartment, and the hearth was placed in the same way in the centre with an opening large enough to let the smoke out. We have reason to believe that these four houses belonged to four different families who lived in the greatest unity and trust; we saw one of these families leave, presumably for a long voyage as we did not see it return during the 5 days we spent in this bay. The owners placed a few planks in front of their door to keep out dogs and were thus able to leave it full of their belongings. We soon became so convinced of the extreme trustworthiness of these people and the quasi-religious respect they have for property that we left in the middle of their huts and under the protection of their honesty, our sacks filled with cloth, beads, iron tools and generally with everything we used for barter, without their ever taking advantage of our confidence and we left this bay believing that theft was a crime they did not even suspect existed.

Each hut was surrounded by a salmon drying yard; they were hung from poles in the full sun, having previously been smoked for three or four days around the fire in the centre of their hut from where it was the women's task to carry them out into the open air where they acquired the hardness of wood.

They fished in the same river as we did, with nets or spears and we saw them eating with disgusting avidity the mouth, the gills, the small bones and sometimes the whole skin of the salmon they were cleaning out very skilfully and they, so to speak, sucked those mucilaginous parts just as we swallow an oyster. Thus the largest proportion of their fish reach the homes after being cleaned out, except when there has been a great haul in which case the women who are required to prepare this fish, dry it and slice it in two to expose it to the smoke, would seek out with the same avidity the fish that were whole and devoured in as disgusting a manner these mucilaginous parts which seemed to them to be the most exquisite of dishes.[1] It was in Castries Bay that we learned for the first time the use of the lead or bone ring these people and those of Segalien Island wear on their thumb: it provides them with a point of purchase when cutting and cleaning the salmon with a sharp knife they all wear on their belt.

Their village was built on a low, marshy tongue of land facing north, and which struck us as uninhabitable in winter, but facing it on the other side of the gulf on higher ground, facing south and at the entrance of the forest stood a second village consisting of eight huts, larger and better built than the first ones; below and quite near we visited three *jourtes* or underground houses similar in every respect to those of the Kamtchatdales described in the fourth volume of Cook's last voyage and large enough to contain the inhabitants of the eight huts during the worst of the cold season.[2] Finally at one end of this settlement were four tombs, better made and as large as the houses; each contained three, four or five coffins, properly made, decorated with China cloth some of which was in gold or silver brocade; bows, arrows, nets and generally speaking the most precious of these people's belongings were hanging inside these monuments whose door, made of carefully worked planks, closed by means of a cross-piece held up by two supports at each end.

---

[1] Other travellers do not refer to this practice. The point was that fish was the villagers' main food and they needed to smoke enough for the winter; fresh fish and more especially salmon was only available during the better season and mucilaginous parts were indeed sought after as delicacies obtainable only during a few months of the year.

[2] Captain King described the village of Koratchin on 8 May 1779: 'This Ostrog consist'd of 3 log houses, [three jourts or houses made underground] & 19 Ballagans or Conic shap'd huts, support'd by posts 10 feet high. The houses are made very hot by means of a large oven or stove, which being well eated warms the room and keeps it so.... These log houses are almost all built after the same fashion & consist of a square room with broad benches along one or two of the sides, two small Windows, but for want of Glass they have [mica]. On one side is a door, entering into the Kitchen about half the breadth of the room & half its length, the other half being taken up by the [ ] building of the stove or oven; The door of the Principal room [opens] from a broad passage or entry running the breadth of the house, in which passage are their Sledges & many other domestic things....', Cook, *Journals*, III, pp. 661–2.

CHAPTER XVI

Their houses were as filled with material as the tombs; nothing they use in winter had been removed; the clothes, furs, snowshoes, bows, arrows, pikes, all had remained in this totally deserted township in which they reside only in the bad season. They spend the summer on the other side of the gulf from where they could see us land, enter the huts, even the tombs, without ever joining us or expressing the slightest concern that we might remove their belongings which they knew we greatly desired since we had already bartered with them for some on several occasions. Our crews had sensed as clearly as the officers the price to be paid for this worthy trust, and shame and contempt would have been visited upon any man cowardly enough to carry out the smallest theft.

It was clear that we had visited the Orotchys only in their country homes where they were harvesting their salmon which, like wheat in Europe, is their basic food. I saw so few elk skins among them that I am led to believe that the results of their hunting are meagre. I also reckon that a few wild tubers of the yellow lily or *sarana* which the women pull up along the edge of the forest and put to dry near their fires represent quite a small part of their diet.

One might have thought that such a large number of tombs (for we found them on all the islands and in all the coves &c) was an indication of a recent epidemic which had worked its ravages throughout these parts and reduced the current generation to a very small number, but I tend to believe that the various families which make up this nation were dispersed in the neighbouring bays to fish for salmon and dry it, and that like our city folk they only come together in winter; each one brings his own supply of fish to last until the sun returns, and I thought it more likely that the religious respect these people have for the tombs of their ancestors lead them to maintain them, to repair them and to delay in this way, maybe for several centuries, the inevitable consequences of the passing years. I noticed no external difference between all these inhabitants, but the same cannot be said of those whose ashes are resting in a style whose splendour varies according to their wealth; it is fairly probable that the toil of a long life is hardly sufficient to meet the costs of one of these sumptuous burial places, which however are only relatively sumptuous for one would gain a false impression if one equated them with the edifices put up by civilised people.[1] The poorest people are exposed in the open air in a coffin placed on a stand held up by poles 4 feet high; they all have their bows, arrows, nets and a few lengths of cloth near their monument, and it would presumably be sacrilegious to take them away.

We believe that these people, like those of Ségalien Island have no leaders and are the subjects of no government.[2] The mildness of their manners, their respect for elders remove any inconvenience this anarchy might have. We never witnessed the slightest quarrel. Their affection towards each other, their tenderness towards their children made a touching picture, while at the same time our senses were repelled by the horrible stench of this salmon which filled their houses and their surroundings. The bones were scattered about, the blood was lying around the hearth, and voracious although fairly gentle and friendly dogs were licking and devouring these remnants. These people are moreover revoltingly dirty and smelly; none probably are less strongly built or have features further from those we associate with the ideal of beauty. Their average height is less than four feet ten inches, their limbs are thin, their voices weak and high-pitched like those of children, the cheekbones prominent, their eyes small, rheumy and diagonally slit, their mouth wide, with a snub nose, a narrow chin, almost no beard and a complexion that is olive and covered with a varnish of oil and smoke. They let their hair grow long and plait it rather as we do; the women's falls loose over their shoulders, and the broad image I have painted is as appropriate for their appearance as for the men from whom it would be hard to distinguish them were it not for a slight difference in dress and if their bosom which is not kept in with a belt did not reveal their gender; they are however not compelled to hard work which could have, as with the American Indians, affected any attraction their features might have; their tasks are limited to sewing, cutting out clothes, laying the fish out to be dried and attending to their children whom they suckle up to the age of three or four, and I was greatly surprised to see one of that

---

[1] Chinese customs had strongly influenced the funeral practices reported here. The natives of Castries Bay were only a hundred kilometres or so from the Manchu outposts on the lower Amur River.

[2] La Pérouse gained a false impression from his meeting with a small group of informants who possibly all belonged to the same family: each village had its chief whose position was formally recognised by the Chinese authorities.

CHAPTER XVI

age who after stringing a little bow, firing an arrow with fair accuracy and hitting a dog with a stick, threw himself on his mother's breast and took the place of a child aged five or six months who had fallen asleep on her knees.

The women seem to be fairly well respected among them; they never concluded a deal with us without their agreement, and all their silver earrings and other copper ornaments sewn on their clothes in a fringe are entirely reserved for women and young girls; the men and young boys wear a nankeen jacket, or one of dogskin or fish skin, cut like a carter's vest; if it reaches below the knee, they wear no breeches; if it does not they wear some in the Chinese style that come down to their calf; they all have boots made of sea otter fur, but keep them for winter, and at all times and every age, even children still at the breast, they wear a leather belt to which are tied a knife in a sheath, a flint and a small bag for their pipe and tobacco.

The women's dress differs slightly. They are wrapped in a wide robe made of nankeen or salmon skin which they know how to tan to perfection and make as supple as the skin of our gloves; this dress comes down to their heels and is sometimes fringed with items of copper that make a small tinkling sound. The salmon used for this clothing are not caught in summer; they weigh thirty or forty pounds. Those we caught in July only weighed three or four pounds, but their number and their delicate taste make up for this disadvantage and we all feel we have never eaten any that were better. We cannot speak of their religion as we saw neither temples nor priests but only a few roughly carved idols hanging from the frame of their huts; they represented children, arms, hands, legs and were very much like the *ex votos* of several country chapels. It could be that these images we may have wrongly identified as idols are only meant to remind them of a child eaten by bears or of hunters killed by these animals; however, it is hardly likely that such simple people are free from superstitions, and we suspected that they sometimes took us to be sorcerers; they replied with anxiety, although politely, to our various questions and when we picked up some paper and a pencil to write them down, they seemed to take the movement of our hands to be magical signs and then refused to answer our questions, letting us understand that this was evil, and it was only with the utmost difficulty and the greatest patience that Mr Lavau, the *Astrolabe*'s surgeon, succeeded in drawing up the vocabulary of the Bitchy and Orotchys people that will conclude this chapter; on this point our gifts could not overcome their preconceptions, they even accepted them only reluctantly and often stubbornly refused to take them at all. I felt that they would have liked more delicacy in the manner we offered them, and to test this suspicion I sat down in one of their huts and bringing a couple of small children aged three or four close to me I offered them, after a little petting, a length of pink nankeen I had in my pocket; I saw a great satisfaction reflected in the eyes of the whole family and yet I am sure they would have refused my gift if I had given it directly to them. The husband went out of his hut and soon returned with his best dog which he begged me to accept; I declined and tried to make him understand that it would be more useful to him than to me, but he insisted and seeing that he was not having any success he got the two children who had received the nankeen to come up and placing their little hands on the dog's back made me understand that I should not refuse it from the children; this delicacy of manners can only exist among well-controlled people; I believe that a nation with neither herds nor agriculture cannot attain a greater degree of civilisation. I must point out that dogs are their most precious belongings; they harness them to small sledges, very light and well made, similar in every respect to those of Kamchatka; these dogs, of the German shepherd breed, are strong although of medium size, very docile and gentle and seem to have their master's temperament, whereas those of Baye des Français, much smaller but of the same breed, were wild and ferocious. The one we took was less than a fortnight old and which we kept on board for several months wallowed in their blood when we killed a sheep or an oxen, ran towards the hens like a fox and generally behaved more like a wolf than a domesticated dog. He fell into the sea one night when the ship gave a sudden strong roll, possibly with the assistance of a sailor whose ration he had stolen.

But the travellers whose four canoes were beached in front of the village had aroused our curiosity as much as the inhabitants; they belonged to the Bitchy people to the south of Castries Bay; they gave us to understand by highly expressive gestures that they came from the Segalien River and were on their way back to their village with a load of rough nankeen cloth and grain.

We used all our skill to question them on the country's geography; we drew the coast of Tartary with a pencil on some paper, with the Segalien River and the island of the same name facing it

CHAPTER XVI

which they also call Chotka, and we left a pass between them; they took the pencil from our hands and joined by a line the island and the mainland, then pushing their canoe over the sand they explained that, after leaving the river, they had pushed their boat in this manner on the sandbank which joins the island to the mainland they had just drawn; after which, pulling up some weeds from the bottom of the sea, which as I explained earlier this gulf is full of, they planted it on the sand to make it clear that the sandbank they had crossed is also covered with seaweed. This report given on the spot by travellers coming from the river, and which was so much in accord with what we had just seen since we had only stopped when we found six fathoms, left us in no further doubt; but for it to agree with what the people of Baye de Langle had told us, it is enough that there remain, at high tide, somewhere along the bank, openings with three or four feet of water which is more than adequate for their canoes. As however it was a question of some interest and it had been determined only in the presence of Mr de Langle, I went ashore the next day and we had the same conversation in sign language; finally Mr de Langle and I asked Mr Lavau (who is particularly gifted when it comes to speaking and understanding foreign languages) to make further enquiries – he found the Bitchys unwavering in their report[1] and I then gave up the plan I had made when I arrived of sending my longboat to the end of the gulf which could not be more than ten or twelve leagues from the Baye de Castries. But this plan had serious disadvantages. The slightest southerly breeze causes the sea to swell to the point that an undecked vessel runs the risk of being swamped by the waves which often break as on a bar; moreover the almost endless fog and

---

[1] The explanations La Pérouse obtained at this point reinforced the opinion he had now formed that the strait was impassable for ships. However, the Bit'chy travellers who were returning from an expedition to the Amur River may simply have dragged their canoes over low water and sandbanks from the river delta to the strait and, since they were going back south, they may not have crossed over to the island at all. Be that as it may, the conversation confirmed La Pérouse's decision to sail back and prevented him from solving the enigma. This was achieved by the Japanese Mamiya Rinzo in 1809 and the Russian Gennadii Nevelskoi in 1849, but their findings were not publicised so that in 1855 a British squadron chased Russian ships as far as Castries Bay and, in the belief that they were blockaded in a gulf, waited patiently for several weeks for them to surrender – whereas the Russians had sailed through the strait to the safety of Nicolaiesk-na-Amur. See on this episode J.J. Stephan, 'The Crimean War in the Far East', *Modern Asian Studies*, III (1969), pp. 268–74.

the constancy of the south winds also made the longboat's return quite uncertain and we did not have a moment to lose. Therefore, instead of sending the longboat to settle an issue of geography about which I could no longer harbour any doubts, I decided to speed up our departure and at last get out of the gulf in which we had been sailing for three months, which we had explored to the very end, crossed thirty times in every direction and sounded more than two thousand times, less in order to leave no element of uncertainty on our chart than for our own safety, as only the sounding-line could guide us in the midst of the fog which surrounded us for so long, but this did not exhaust our patience and we did not leave a single feature of the two coasts unsurveyed. Only one point of interest remained to be cleared up, which was the southern extremity of Segalien Island which we knew only as far as Baye de Langle in 47^d 49', and I must admit that I might have left that task to others if I had been able to disembogue because the season was advancing and I was not unaware of the great difficulty involved in sailing for two hundred leagues against the wind in such a narrow channel, with so much fog and where the southerly winds had never varied by more than two points east or west. I knew from the account of the *Kastricum* that the Dutch found northerly winds during August, but they had sailed on the east coast of their so-called Yesso and we were engulfed between two lands whose extremity was situated in seas dominated by the monsoon that blows on the coasts of China and Korea until the month of October.

Nothing, it seemed to us, could deflect the winds from the first force which had driven them; these thoughts made me only more anxious to speed our departure and I decided that it would definitively take place on August second. Each day until then was spent surveying some part of the bay and the various islands of which it was formed; our naturalists made a number of excursions all along the coast that should have satisfied our curiosity. Mr de Lamanon himself, whose convalescence was making very slow progress, wanted to go with us; the lava and other volcanic matter which he learned these islands were made of did not permit him to spare a thought for his weakness; he found with the Abbé Mongez and Father Receveur that most of the substances around the bay and on the islands that form the entrance to it were compact or porous red lava, grey basalt in layers or dikes, and

CHAPTER XVI

finally trapps[1] that seemed not to have been affected by heat but had supplied the material for the lava and the basalt that had been molten in the furnace. Various forms of crystallisation were found amongst this volcanic material whose eruption seemed very ancient; they could not discover the craters of the volcanos and a stay of several weeks would have been needed to study and follow their traces.

Mr de la Martiniere travelled with his usual energy along the ravines and the riverbeds in order to seek out new species, but he only saw the same ones as in Ternai or Suffrin bays, a little less advanced, and the vegetation was roughly the same as around Paris on 15 May. Strawberries and raspberries were still in flower, the fruit on the currant bushes were beginning to redden, and celery and cress were quite scarce. Our conchologists were more fortunate – they found some extremely fine lamellar oysters wine- or black- coloured, but so tightly fixed to the rocks that a great deal of skill was required to get them off, and their lammellae were so thin that it was very difficult for us to keep them in one piece. We also netted a few whelks of a very fine colour, some scallops and small mussels of the most common variety as well as various clams.

Our hunters caught several grouse, some cormorants, guillemot, black and white wagtail and a small flycatcher, sky blue in colour, which we have not found described by any ornithologist, but none of these species were widespread.[2] Nature, as far as living beings are concerned, is as it were torpid in these frozen climes. The cormorant, the sea gull that gather in flocks under happier skies live here like hermits on tops of rocks. A sorrowful and dark sense of mourning seems to reign along the shore and in the forests where only the cawing of a few crows can be heard and where white-headed eagles and other birds of prey seek their retreat; alone the

---

[1] The terminology used by La Pérouse makes it clear that he was summarising reports handed to him by the naturalists. This is shown by his use of the term trapp (volcanic rock of a columnar structure), a word of Swedish origin (*trappa*) which had only recently entered the French language. See F. Brunot, *Histoire de la langue française des origines à nos jours*, Paris, 1933–53, VI, 1, p. 619.

[2] The grouse or hazel-hen was probably the *Falcipennis falcipennis*, the cormorants the *Phalacrocorax capillatus*, the *P. urile* and the *P. pelagicus*, the guillemot the *Uria aalge inorata*, the *U. lomvia*, or one of the *Cepphus*, the wagtail the *Motacillina lugubris* or (if the French did not see a black and a white bird, but a black and white one) the pied wagtail *M. alba yarelli* which however is rare in this region; the flycatcher was probably a migrating *Erannornis*.

swift or the sand martin seem really to be at home: we saw their nests and their flight wherever rocks along the sea formed a cave, and I believe that the most widespread animal on earth is the sand martin or the common swallow, having found one or the other in every country I have landed on.[1]

Although I did not have the soil dug, I believe it to stay frozen below a very slight depth because the water at our watering-place was only a degree and a half above freezing point and the temperature of all the running water we tested with a thermometer never exceeded four degrees even though the mercury remained at 15 degrees in the open air, but this temporary heat does not penetrate; it merely hastens the growth of vegetation which must emerge and die in less than three months and brings to life a multitude of flies, mosquitoes and sandflies who balance by their importunities the benefits brought about by the sun coming closer.

The natives cultivate no plants; however they do seem fond of vegetables and the grain provided by the Manchus, which we took to be a small hulled millet, they found a delight; they carefully harvested various wild roots which they dried and kept for the winter, including the yellow lily or *sarana* which is a real onion. Quite inferior in physique and skills to the inhabitants of Segalien Island, they lack the latter's knowledge of the shuttle and wear only the most common of Chinese material and the skins of a few land beasts or of seals. We killed one with sticks, which Mr Colignon had found sleeping by the sea; it differed in no way from those of the Labrador coast or Hudson's Bay. This encounter was followed by a great misfortune: being caught in a shower while sowing European seeds in the forest, he wanted to light a fire to dry himself and unwisely used gunpowder to start it; the fire spread to the powder-flask he was holding, the explosion broke the bone in his thumb and he was so seriously wounded that his arm was saved only by the skill of our senior surgeon, Mr Rollin, whose attention was more usually turned towards those who seemed to be in the best of health: he had noticed among several of them signs of incipient scurvy such as a swelling of the gums and legs; these had developed on land and would have disappeared if we had stayed a

---

[1] The swift or martlet was probably a *Riparia riparia*. The comment on the swallows is valid: the *hirundo rustica* and the *hirundo daurica* are both very widespread.

CHAPTER XVI

fortnight; we could not spent this long in Baye de Castries, and we felt confident that wort, spruce beer, and an infusion of quinine mixed in with the crew's drinking water would wipe out these minor symptoms and give us time to wait for a port of call where we could stay longer.

On the second of August, as announced, we sailed with a light westerly breeze that blew only in the bay itself. The southerlies were waiting for us at Clostercam Point.[1] They were quite moderate and unaccompanied by clouds. We tacked fairly successfully and the tacks were helpful. I was particularly eager to examine the coast of Tartary we had lost sight of from the 49th to the 50th degrees because we sailed too close to Segalien Island. Consequently on the way back I hugged the mainland up to the point where we had last taken bearings within sight of Pic Lamanon. On the sixth of August the weather which had been quite fine deteriorated. We were struck by a southerly squall that was alarming on account less of its violence than of the effect it had on the sea. We were forced to load our vessels with all the canvas the masts and the beams of the frigates could bear to avoid drifting too much and losing in one day everything we had gained in three. The barometer dropped to 27 inches 5 tenths; the rain, the fog, the wind, our situation in a channel where the coast was hidden by mist, everything made our position at the very least very wearying, but these squalls we were complaining about were the forerunners of northerly winds on which we had not counted. They declared themselves on the eighth after a storm &c and enabled us to reach on the evening of the 9th the latitude of Baye de Langle which we had left on 12 July. After the misfortune that had befallen our astronomical tent in Baye de Castries, it was very important to find this point which had been perfectly determined in latitude and longitude during our first call, in order to compare it with the longitudes that our timekeepers would give us 27 days later, and we found that our No. 19 had a variation of 34' of longitude too far east, which spread over 27 days would give a daily error of 5 seconds to be added to the 12 seconds observed at Cavitte, but Mr D'agelet had noticed the precise moment when this slight change had occurred and he

---

[1] Castries had played an important part in the Battle of Klosterkamp, on 16 October 1760, fought against the army of Ferdinand of Brunswick. The feature which La Pérouse selected to commemorate this French victory is Mys Orlova, a long peninsula in the south of Castries Bay.

worked it out perfectly by increasing the daily loss to 8 seconds, making a total of 20 seconds. The distances observed then retained the same difference and we can assume that all the west of Ségalien Island as well as the eastern coast of Tartary which is the second side of the channel are placed on our charts with an error of less than 20′. A bank, the depth of which is very even and which does not represent a danger, stretches ten leagues from north to south in front of Baye de Langle and roughly eight leagues west; we passed it during the night, sailing south, and I hove to at ten o'clock in the evening until dawn in order to avoid missing the slightest opening; the next day we continued to follow the coast at a distance of two leagues and we saw to the S.S.W. a small flat island which formed a channel of about six leagues with Segalien Island. I called it Moneron Island after the engineer who was serving on this expedition[1] and we sailed between the two islands, where we never found less than 50 fathoms of water. Shortly after, we sighted a peak a thousand or twelve hundred *toises* high.[2] It consisted merely of bare rock with snow in the fissures; we saw neither trees nor greenery. At the same time, we could see more land, lower, and the coast of Segalien Island ending in a headland – no double mountains were now visible. Everything told us that we were reaching its southern end and that the land and the peak were on the island of Tssissa. We dropped anchor in the evening with this feeling of hope that was transformed into a certainty the next day, 10th August, when the calms forced us to anchor off the southern tip of Segalien Island, which I named Cape Crillon after the Comte de Crillon; it is situated in 45^d $57′$ and longitude 140^d $34′$.[3] It forms the end of this island, one of the longest from north to south known on earth, separated from Tartary by a channel of which we had explored both

[1] The name remains as Ostrov Moneron in $47°17′$N and $141°15′$E.
[2] La Pérouse named it Langle Peak. It is Rishiri-zan (1,719 m) situated on an island east of the low lands in northern Hokkaido.
[3] La Pérouse was entering the strait which now bears his own name; 45 km wide, it separates Sakhalin from the northern tip of Japan. Point Crillon, the western extremity of Aniva Bay, latitude $45°53′$N and longitude $142°05′$E, has retained its French name under the Russianised form of Mys Kriljon. La Pérouse had already given Crillon's name to a mountain on the Northwest Coast of America. During the Japanese occupation of Sakhalin (1905–45) these European names disappeared and La Pérouse Strait became Soya Kaikyo. According to Tadao Kobayashi, 'Lapérouse dans les mers proches du Japon', *Colloque Lapérouse Albi 1985*, pp. 398–9, the name La Pérouse Strait was first used by Milet-Mureau in the printed version of the voyage.

CHAPTER XVI

sides and which ends in shoals between which there is no pass for vessels, but it is likely that there is a way through for canoes between these weeds. This island is Locu Yesso and the island of Tssissa athwart our ships, separated from the island of Segalien by a channel of twelve leagues which I have named Canal de Castries,[1] is Japanese Yesso and extends to the south as far as Sangar Strait. The Kuril chain lies much further east and forms with Yesso and Loku Yesso a second sea communicating with that of Okots and whence one can reach the coast of Tartary only by crossing the strait of Castries in 45^d 40′ or that of Sangar, after passing through between the Kurils. This aspect of geography, the most important of all those which modern travellers left for their successors had cost us many exertions and a great deal of care because fog renders this navigation extremely hazardous and since 10 April, date of our departure from Manila, until 10 August when we crossed the strait, we put in for only three days in Baye de Ternai, one day in Baye de Langle and finally five days in Castries Bay, because I count as nothing the times we anchored off the coast even though we sent boats ashore to survey the land and they got us some fish. It was at Cape Crillon that we were first visited by islanders because on both coasts we had paid them visits without their showing the slightest curiosity or expressing the slightest inclination to see our vessels. On this occasion they gave some signs of mistrust and did not come near until we had spoken a few words of the vocabulary Mr Lavau had drawn up at Baye de L'angle; but if their fear was fairly marked at first their confidence soon developed. They climbed onto our vessels as if they were calling on their closest friends, sat down in a circle on the quarter-deck and smoked their pipes. We overwhelmed them with gifts. I had nankeens, silk cloth, iron tools, beads, tobacco and generally everything I thought might please them given to them, but I soon noticed that they saw brandy and tobacco as the most precious of items, and yet I had these handed out more cautiously because tobacco was necessary for our

---

[1] Canal de Castries, as it also appears on the large summary map of the area drawn up on board, is La Pérouse Strait. The name is appropriate as a tribute to the navigator's work in this area, and two other reasons can also be adduced for the change: Castries had been given his honour a few days earlier; in Milet-Mureau's eyes, perpetuating in this major geographical feature the émigré nobleman's name was a political embarrassment – and not a little dangerous during the upheavals of the Revolution.

crews and I feared the after-effects of brandy. We noticed, more especially in Baye de Crillon, that their features were handsome and very regular; they were strongly built and gave an appearance of vigour; their beards came down to their chests and their arms, neck and back were hairy to a remarkable degree; this aspect was general among them for one would find in Europe only individuals as hairy as these islanders.[1] I believe that they were on average an inch shorter than French people, but this is hardly noticeable because their finely proportioned bodies and their obvious muscularity create an overall impression of very fine men; their complexion is as dark as that of Algerians or other people of the Barbary Coast.

Their demeanour was grave and their thanks expressed by noble gestures, but their requests for new gifts were repeated to the point of being importunate and their gratitude did not go so far as to make them offer anything in return, not even the salmon which filled their canoes and which they took back to land when we refused the excessive price they were asking for it. However they had received purely as gifts some nankeen, silk stuffs, iron instruments, beads, &c. Our joy at finally coming upon a strait other than that of Sangar had made us generous but we could not fail to notice that when it came to gratitude they were very inferior to the Orotchys of Baye de Castries who never sought any gifts, often stubbornly refused to take them and firmly insisted that they should be allowed to pay us back, but if their moral attitude was much lower in this respect than that of the Tartars, their physical appearance and their energy in turn was beyond comparison.

These islanders had woven all the clothes they wore; their houses were far cleaner and more elegant than those of the mainland; their belongings were well finished and nearly all Japanese-made. They had an item of trade that was very important, unknown in the Tartary channel, which provided them with all this wealth. It was whale oil and their manner of extracting it was not the most economical: it consisted in cutting up the flesh of the whale into pieces and letting it rot in the open air on a slope facing the full sun.

---

[1] These islanders were Ainu, the aboriginal inhabitants of this region who are believed to be descended from ancient circumpolar people. They were gradually driven north by the advancing Japanese and are now much reduced in number. Distinctive for their beards and thick, wavy hair, they were often referred to as 'the hairy Ainu'.

CHAPTER XVI

The oil that runs out of it is caught in bark vases or in gourds made of seal skin; it is remarkable that we did not see a single whale on the western coast of the island and that this cetacean abounds on the eastern side. There can be no doubt that this race of men is quite different from the one we observed on the mainland although they are only separated from it by a channel of three or four leagues choked by sandbanks and seaweed; nevertheless their way of life is similar – hunting and more especially fishing provide almost all their subsistence. They leave fallow a most fertile soil and presumably both have neglected cattle raising, herds for which could have come from the upper reaches of the Segalien River or Japan, but the same diet has resulted in quite different constitutions; it is true that the cold in the islands is less harsh than at the same latitude on the mainland but this cannot of its own be the reason for this great difference. One must therefore go back to a different origin. I believe that the Bitchys, the Orotchys and the other Tartars of the seaboard up to the neighbourhood of the northern coast of Segalien share a common origin with the Kamchadals, the Kuriacs and these types of men who like the Laplanders and the Samoyedes are to mankind what their stunted birches and pine trees are to the trees of most southerly forests, but the inhabitants of Segalien Island are on the contrary quite superior by their physique to the Japanese, the Chinese and the Manchu Tartars; their features are more regular and closer to European forms; it is very difficult to dig into and interpret the archives of the world, and to discover in them the origin of peoples, and the travellers must leave the elaboration of theories to those who read their narratives.

Our first questions concerned the geography of the island, part of which we knew better than they did. It seems that they are accustomed to outline a terrain, because right away they traced the part we had visited up to the Segalien River, leaving a relatively narrow pass for their canoes. They marked each resting place and gave it a name; in the end one cannot doubt that although distant from the mouth of that river by more than 150 leagues they all knew it perfectly well and without this river the Bitchys, the Orotchys, the Segalians and generally all the people of these maritime regions would know as little about the Chinese and their goods as the inhabitants of the American coast, but their knowledge failed when it came to drawing the other side of their island.[1] They drew it following the same north and south direction, and did not appear to have noticed that the direction changed, so that they left us in some doubt and for a time we thought that Cape Crillon concealed a deep gulf after which Segalien Island continued to the south; such a view was hardly reasonable; the strong current that came from the east indicated the presence of an opening and we had not met such a one in the 52nd degrees towards the end of the Tartary channel where, on the contrary, the tide did not run at a *toise* an hour, but as we were in a deep calm and caution did not allow us to drift with the current which could have borne us too close to the headland Mr de Langle and I thought it wise to despatch a boat to the shore under the command of Mr de Vaujuas, and we instructed that officer to climb the highest point of Cape Crillon and make a note of all the land he could see beyond it; he was back before nightfall. His report confirmed our first opinion and we remained convinced that one must exercise the greatest care when one wishes to learn about a large country from such questionable information; these people have no notion of a change in trends and draw the back of their island as if it continued along its length, a creek the width of three or four canoes impresses them as an immense harbour, one fathom is an immeasurable depth, their scale of comparison in a word is their canoe with a draught of four inches and only two feet in width.

Before returning to the ship, Mr de Vaujuas visited the village on the headland where he was made most welcome; he carried out a little barter and brought us back a great dal of salmon. He found the houses to be better built and especially wealthier and better furnished than those of Baye de l'Angle and Baye d'Estin. The interior in several cases was decorated with large varnished vases from Japan. Since Segalien Island is only separated from Tssissa Island by Castries Strait which is twelve leagues wide, it is easier for those who live along the shores of the strait to obtain Japanese goods than it is for their northern compatriots who, in their turn, are closer to the Manchu Tartars and the Segalien River, but unlike these here

---

[1] Except in the north, Sakhalin is very mountainous, and the inhabitants of the west coast had little reason to take their canoes into the Sea of Okhotsk to reach a coast which, for miles, was steep and barren and largely uninhabited. Trade was carried out along the west coast and with the Siberian mainland. The situation is much the same today: most of the ports and settlements are strung along the west coast or along the narrow south-western peninsula.

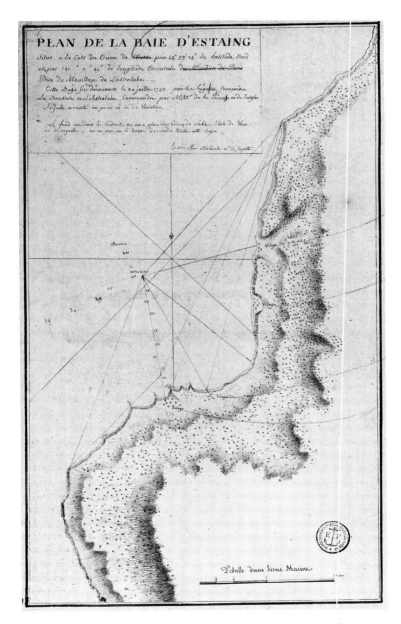

21. Chart of D'Estaing Bay, Sakhalin. By Blondela. Four villages are shown along the coast. Actual size: 30 cm × 48 cm. AN 6 JJ1:43.

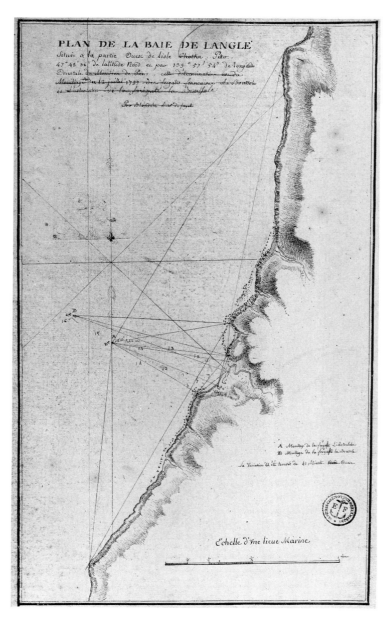

22. Chart of Langle Bay, Sakhalin. By Blondela. A small village is shown by the main river. The *Astrolabe* anchored closer inshore than the *Boussole*. AN 6 JJ1:42.

they do not have the whale oil which seems to form the basis of their trade. Our islanders left before nightfall and explained in sign language that they would be back the next day; they were indeed on board at daybreak with a few salmon they exchanged for axes and knives. They also sold us a sword and a costume made of local material, and seemed to see with regret that we were getting ready to sail; they strongly urged us to round Cape Crillon and put into a cove they sketched out and called *Tabouoro*. It was Aniva Gulf,[1] but a light breeze had come up from the N.E. and I gave the signal to raise anchor, first towards the south-east to round Cape Crillon well out to sea, as it ends in a rock or islet towards which the tide was running very rapidly. As soon as we passed it we saw from the topmasts a second islet appearing four leagues from the point towards the south-east. I named it the Dangerous One because it is level with the waves and it may be hidden at high tide; I set course to pass it to windward and I rounded it one league off. The waves were breaking all around it without my being able to ascertain if this was due to the tide or to surrounding reefs; at this distance the sounding-line constantly recorded 23 fathoms and when we had passed it the depth increased and we soon had 50 fathoms where the current seemed to be moderate. Until that moment we had sailed through tidal rips stronger than those of the Four or Raz near Brest. They are felt only on the coast of Segalien Island or in the southern part of this strait. The south coast towards Tssissa Island is much less exposed, but we were soon affected by a swell from the open sea or the east which made us risk during the entire night to collide with the *Astrolabe* because there was a dead calm and neither frigate answered the helm. The next day we found ourselves to be a little further south than was indicated by our dead reckoning and only ten minutes north of the village of Aqueis[2] as it is called in the Voyage of the Kastricum. We had just crossed the Strait of Castries that separates Yesso and Loku-yesso, quite close to where the Dutch anchored at Aqueis; the fog had hidden it from them and it is quite probable that mountain peaks on both islands gave them

---

[1] Aniva is a large bay or gulf at the southern end of Sakhalin. Maarten Vries who spent several days there in July 1643 called it Aniwa, which seems to be a local name. See W. C. H. Robert, *Voyage to Cathay*, pp. 127–43. Cornelis Coen's journal mentions a village in a valley called 'Tamary' by the natives – this may be La Pérouse's Tabouoro.

[2] This 'Aqueis' is Akkeshi, on Hokkaido. Coen's journal in ibid., pp. 175–7, 279.

the impression that they were linked together by low lands and following this false impresson they drew a continuation of the coast on the very spot where we passed through; apart from this one error, the details of their navigation should not be scoffed at. We sighted Cape Aniva almost in the same compass point as it appeared on the Dutch chart. We also saw Meure Gulf, a name for the area between Cape Crillon and Cape Aniva. Their latitude differed by only ten or twelve minutes and their longitude from Cape Nabo by less than one degree, an amazing precision for the time when that campaign took place and I made it a rigid rule not to alter any of the names they had given when I was able to identify them by comparing their reports, but it is rather noteworthy that the Dutch sailed from Aqueis to the gulf of Aniva without suspecting, when they anchored in Aniva, that they were on another island, such were the similarities in the external appearance, the customs and the way of life of these people.

The next day was very fine but we made little progress to the east. We could still see Cape Aniva to the north-west and its eastern coast trending north towards Cape Patience[1] in latitude 49^d. This was the limit reached by Captain Vries and since his latitudes from Cape Nabo are roughly correct, the Dutch charts, enough features of which we checked to warrant our confidence, give us the width of Segalien Island up the 49 degrees. The weather was very good but the east S.E. winds which had been blowing constantly for four days slowed our advance towards the States and Company islands.[2] Our latitude was observed at 46^d $10'$ and our longitude $143^{d.}$ We could see no land and several times attempted to find a depth in two hundred fathoms; on the 16th and 17th the sky was cloudy and whiteish, and the sun was not visible; the winds veered east and I tacked south in order to approach States Island which I saw quite clearly on the 19th. We saw Cape Tron to the south and Cape Vries[3] to the SE¼E. This was the compass point where we were

---

[1] Vries's Cape Patientie is Mys Terpenija, the headland which marks the eastern limit of the bay of the same name. This was the northernmost limit of the Dutch voyage.

[2] These names appearing on the Dutch charts give the impression that the islands were larger than they really are. Staten Landt (Eylandt) is the island of Iturup, and Compagnies Landt is Urup, both part of the Kurils.

[3] Tron or Trou (Trouw) is Zal Prostor, the headland on north-west Iturup. Cape Vries is the eastern point.

CHAPTER XVI

due to find them according to the Dutch charts and modern navigators have never sailed with greater accuracy.

On the 20th we saw Company Island and recognised Uries Strait[1] in spite of its being shrouded in fog; we coasted three or four leagues off the northern shore of Company Island. It is arid, with neither trees nor greenery and struck us as uninhabited and uninhabitable; we noticed the white patches the Dutch refer to, which we took at first to be snow, but on closer examination we saw large fissures in whiteish rocks like plaster. At 6 p.m. we were athwart the N.E. point of this island which ended in a very steep point which I named Cape Kastricum after the ship that made this discovery. We could see beyond it four small islands or islets[2] and to the north a wide channel that seemed open towards the E.N.E. and separated the Kurils from Company Island, the name of which should be religiously preserved and prevail over those that may have been imposed by the Russians more than a hundred years after Captain Vries' voyage.

The 21st, 22nd and 23rd were such foggy days that we could not pursue our route to the east through the Kurils which we could not see two cablelengths away. We remained tacking variously by the opening of the channel where the sea did not seem to be disturbed by any currents, but our observations of longitudes taken on the 23rd revealed that in two days we had been carried 40 minutes towards the west and we confirmed this observation on the 24th by identifying the same points we had seen on the 21st in exactly the position expected in accordance with the longitude we had observed. Although the weather was very foggy we were able to make a little progress during part of this day because it cleared frequently, and we saw and identified the northernmost of the Four Brothers and the two headlands of Marikan Island [3] which we took

---

[1] Uries or Vries Strait (Proliv Friza) separates Urup from Iturup. The *Korte Beschrijvinghe van het Eylandt bij de Iapanders Eso genaemt* records that 'They have landed on said Compagnies Landt, and visited one of the Mountains of the same near a Snow-water draining place, found there shining earth as it were full of Silver being white like Fuller's earth mixed with white Sand, not hard, when they put the same in water it liquefied.' Robert, *Voyage to Cathay*, p. 244. These islands are of volcanic origin and contain rich deposits of sulphur.

[2] These islands are Brat, Chirpoev, Chipoij and Broutona. The names given by Vries have been replaced by native names; those given by the Russians, such as Morten, Petrovich, Spanberg in 1738–9, have not been retained.

[3] The Four Brothers are those mentioned in the previous note. Marikan is now known as Sumusir.

at that time to be two islands – the southernmost bore East 15^{d}South. We had only made four leagues towards the north-east in three days, and with the fog becoming thicker and continuing without a break on the 24th, 25th and 26th we were forced to go on tacking between these islands, the extent and trend of which we did not know, not having, as had been the case on the coast of Tartary and Loku-yesso the advantage of taking soundings to estimate the proximity of the land, beause here one cannot find bottom; this situation, one of the most tiring and wearisome of the campaign, lasted until the 29th; then it cleared and we saw mountain tops to the east. I set out towards them. Soon low-lying land appeared and we recognised Marikan Island which I consider the first of the southern Kurils. Its length from N.E. to S.W. is roughly twelve leagues, a large bluff ends each of its extremities and there is a peak, obviously of volcanic origin, judging from its shape, in the centre;[1] I set course to approach the N.E. point of Marikan Island, planning to leave the Kurils by the pass which I assumed lay to the north of this island; I could see two others to the East N.E. but further off and seemingly leaving a channel of four or five leagues, but at eight p.m. the winds veered north, and quite weak; there was a heavy swell and I had to change course and bear west to get away from the coast because the swell was driving us towards the land and we would not have found bottom one league from the shore with a sounding-line 200 fathoms long. These northerly winds led me to decide to come out through the channel on the south side of Marikan Island and north of the Four Brothers. It struck me as wide, in a southerly direction approximately parallel with Uries Strait, which was taking me away from my route, but the wind left me no other option; clear days were so rare that I felt I should take advantage of the only clear moment we had had in ten days.

We crowded on sails during the night to reach the entrance to this channel; there was little wind and the sea was extremely rough. At dawn we could see the south-west point of Marikan (which I named Cape Rollin after our senior surgeon) approximately two leagues away, and we remained in a dead calm without being able to lower the anchor should we be carried towards the land as the sounding-line could not find the bottom. Fortunately the current was taking us slowly towards the middle of the channel, and we

---

[1] This peak is Mt Zavaritskii, an active volcano.

CHAPTER XVI

made about 5 leagues towards the E.S.E. without there being sufficient wind to steer. In the south-west we could see the Four Brothers Islands and as we had some very good readings of longitude for determining their position as well as that of Cape Rollin on Marikan Island we are sure that the channel is 15 leagues wide. The night was very fine, the winds settled in the E.N.E. and we entered the pass in moonlight. I have named it *Boussole* Channel[1] and I believe it to be the finest one can find between the Kurils. We did well to take advantage of this particular moment – the sky clouded over at midnight and the thickest fog surrounded us at dawn the next morning before we knew for certain that we had passed through the channel. I continued on the south-east tack in the midst of this fog, intending to approach the northern islands with the first break and to survey them if possible as far as Point Lopatka[2] but the fog was worse here than on the coast of Tartary; for ten days we had had only 24 hours clear when there was almost a dead calm and we were lucky to have had half a fine night to get through the pass.

At six p.m. I tacked north towards the land which I believed to be twelve leagues away; the fog was still as bad; at around midnight the winds veered to the west and I sailed east, waiting for daylight to get closer to the coast. Daylight came without the fog clearing; however, the sun broke through a couple of times during the morning and opened up our horizon for a few minutes only to one or two leagues; we seized the opportunity to calculate some altitudes and check the time to determine our longitude. These observations left us in some uncertainty because the horizon was not clear; they did tell us however that we had been carried some ten leagues south-east which was quite in accordance with the results of various bearings taken the day before during the calms; the fog then returned with the same obstinacy; it was as heavy the next day and I decided to forego the exploration of the northern Kurils and make for Kamchatka. We had determined the position of the southernmost Kurils, those which had caused some uncertainty among geographers; the position of Marikan Island was well determined as was that of Point Lopatka, it seemed to me that there could not be any significant error left in the direction of the islands that lie

---

[1] The name remains under its Russian form of Proliv Bussol. The channel is 60 km wide; its average depth is 2,200 m.
[2] Lopatka is the southern cape of the Kamchatkan peninsula.

between these two features and I did not believe that I should sacrifice for a search that was practically useless the health of the crews who were beginning to need some rest and whom the continual fogs were keeping in a state of dampness that was very unhealthy in spite of all the precautions we took to protect them; consequently I set sail for the east-north-east and gave up my intention of anchoring at one of the Kurils to examine the nature of the soil, and the customs of these people who, I am assured, are the same as those of Chotka or Tssissa, judging from the accounts of the Russians who have drawn up a vocabulary of these islanders' language which is quite similar to the one we had established in Baye de Langle, and the only difference lies in the way we heard and transcribed their pronunciation which cannot have had the same effect on Russian ears as on French ears; moreover the appearance of the southern islands, along which we had sailed at a short distance from shore, is quite horrible, and I believe that Compagnies Land, the Four Brothers, Marikan Island &c are uninhabitable. An arid rock lacking greenery and even any loam can only serve as a refuge for shipwrecked mariners who would have no better choice before them than to make promptly for the islands of Tssissa or Chotka by crossing the channels which separate them.

The fog was as persistent until 5 September, but since we were out in the open sea we crowded on sail in the middle of this gloom and at six that evening a break allowed us to see the coast of Kamchatka; it extended from the W¼NW to the N¼NW and the mountains we saw in that part of the compass were precisely those of the volcano[1] to the north of St Peter and St Paul from which however we were 35 leagues away since our latitude was only 51° 30′. The whole of this coast appeared frightful; it was painful and almost frightening to behold these enormous rock formations covered with snow in early September and on which no vegetation had ever grown. We sailed north; during the night the winds veered north-west; the next day the weather stayed clear. We had come nearer to the land; it was much more attractive to see close to, and the base of these enormous mountain tops covered in perpetual ice was carpeted with the finest greenery. In its midst one saw various clumps of trees; the total picture this presented was a thousand times more sublime than that of our monotonous European plains

---

[1] The volcano is Avachinskaya, 2,718 m.

CHAPTER XVI

where a listless yellowy green constantly reminds one of the farmer's exertions.

On the evening of the 6th we saw the entrance to Awatka Bay. The light the Russians have set up on the eastern point of this entrance was not lit during the night; the governor told us the next day that he had made unsuccessful efforts to keep the fire going. The wind repeatedly put out the wick which was sheltered only by four ill-fitting planks of pine.[1] The reader will realise that this monument, which was in keeping with Kamchatka's standards, was not copied on any of the lighthouses of Ancient Greece or Italy, but one might also have to go back to the heroic times before the Siege of Troy to find such a friendly hospitality; the governor came to meet us in his canoe over a distance of 5 leagues; he had spent the night on the mountain attending to the wick and still felt guilty at not having managed to keep it lit. He told us that our arrival had long been announced and that he thought the peninsula's governor-general who was expected at St Peter and St Paul within 5 days had some letters for us.

Hardly had we dropped anchor when we saw coming on board the good priest of Paratanka with his wife and all his children. It was then easy for us to see that we would be able to bring back to the stage, in the following chapter, all the actors of Captain Cook's last voyage or at least their story, as several had died and Major Beem[2] had returned to Petersburg.

---

[1] Cook's ships had found the light operating and useful, but the winds were much more moderate on that occasion. '...in the evening saw the lookout house of Awatscha Bay, on which a good light was shewn us after dark,' wrote James Burney. (Cook, *Journals*, III, p. 701). James King recorded in the log of the *Resolution*: 'the light house NWbW 3 miles, there was a good light kept on it.. [the next day] at 9 weight and came to Sail turning up the bay with light Airs.' (ibid.)
[2] For Behm and others, see the notes to ch. XVIII.

CHAPTER XVII

Supplement to the previous chapter. The eastern navigation on the coast of Tartary was absolutely new for Europeans; the Emperor of Japan at all times forbade navigation to the north of his empire and the Jesuits who had drawn the map of China had never had permission to approach the coast; new details on the coast of Tartary; impossibility of finding a river up which the smallest boats could sail; the Chinese and the Manchu Tartars have only ever penetrated this coastal area at the Segalien River; pearl-oysters; doubts expressed on the so-called pearl fisheries mentioned by the Jesuits because there are no men to employ there; the shores of Northwest America are more populated than those of Eastern Tartary. The winds always blow along the channel and never allow the islanders and the people of the mainland to undertake longer travels in their canoes.

Very marked difference in the physique of the islanders and of the mainlanders, although their customs, way of life and religion are practically the same.

Poverty of the country, impossibility of carrying out any useful commerce there; whales are very common on the eastern shore of Ocu yesso, but sea otters are unknown. Father de Angelis and several other missionaries knew about Yesso without suspecting the northern limits and their accounts created the utmost confusion in the geography of this area; several geographers were already rejecting the Dutch account which is the most accurate possible, and the Russians were wrong in thinking the Sea of Okots was contiguous with that of Korea. The discovery of the Strait of Castries will open this route to their ships; landmarks in this strait; the Peak de Langle: it is visible 40 leagues away; need to retain the local names, at least those that go back furthest in time – this rule has invariably been followed in the charts drawn up to illustrate the narrative of our voyage.

Our route from Manila to Quelpaert Island on the southern coast of Korea was new only to us because the Dutch still carry on today

CHAPTER XVII

their trade with Japan and each year send one or two ships to Nangasacki[1] but I do not know whether they sail by way of the Formosa channel or whether they pass east of that island. I was assured that the captains, before leaving Batavia, swear to keep the details of their navigation secret and to allow no one to copy their charts which are handed to them in manuscript form. Do such precautions indicate that other Europeans than themselves would be accepted in Japan and might trade there in competition with them, or is this oath nothing but an ancient custom which they have omitted to change?

Be that as it may, we think that we can say the time has come when all the veils concealing private routes are about to be lifted. The art of navigation has made sufficient progress in recent times not to be impeded by such obstacles; soon geography will only be a science, because the discussions and criticisms will be pointless when all the main points have been accurately determined as to their latitude and longitude, and everyone will be in a position to know the extent of the lands they inhabit and the surrounding seas. Although the Tartary seas we have explored form the limits of the most ancient inhabited continent, they were as little known to Europeans as the Strait of Ainan or the archipelago of St Lazarus &c.[2] The Jesuits whose accounts have given us such a good knowledge of China could not provide any information on the eastern reaches of that vast empire. Those who had travelled to Tartary were not allowed to come near the coast, and the prohibition, applied throughout all time by the Emperor of Japan banning sailing to the north of his possessions, was a new reason to believe that this part of Asia concealed riches which Japanese and Chinese policies were concerned not to let Europeans discover, but the

---

[1] The Japanese period of isolation began in the 1620s. The chief aim was the extirpation of Christianity which the shoguns saw as a likely prelude to European conquest. Missionaries were prevented from landing and those already in the country were expelled or put to death, the great majority of them being Roman Catholics. The Dutch, who belonged to the Reformed religion, were tolerated in small numbers but purely as traders and were restricted to the island of Deshima in Nagasaki harbour. The Japanese themselves were forbidden to travel abroad, in order to avoid their becoming corrupted by foreign ideas, and the building of large sea-going ships was banned.

[2] The Strait of Anian is the fabulous North-West Passage linking the Atlantic and Pacific oceans. The St Lazarus (San Lazaro) archipelago was a group of islands situated near the Pacific opening of a passage into the American continent believed to have been discovered by Admiral de Fonte.

details contained in the previous chapter must have proved to readers that the coast of eastern Tartary is less inhabited than that of North America; separated in a way from the mainland by the Segalien River whose course runs almost parallel to its trend and by inaccessible mountains, it has only ever been visited by the Chinese and Japanese from the sea and the very small number of inhabitants one does meet originate from people living in the north of Asia and they have nothing in common with the Manchu Tartars and even less with the inhabitants of Oku yesso, Yesso and the Kurils. One feels that such a country backing onto mountains less than twenty leagues away from the shore cannot have rivers of any size; the Segalien River collects all the water on the western side; those that run east split into streams in every valley and no country is better watered or more plesantly fresh in the fine season. I reckon that the total population of the little communities living between the point where we landed in the 42 degrees and Baye de Castries at the estuary of the Segalien River does not reach three thousand people, and without this river down which the Manchu Tartars came in their canoes as far as the sea, from which they spread out along the northern and southern shores, the Chinese empire would be as little known to the Bitchys and Orotchys as is that of Constantinople. But the route opened up by this river is now very frequented; there is not one individual living on the mainland or the islands of Yesso or Oku yesso who is unaware of the Segalien, just as the inhabitants of Egypt and Judea knew about the Nile; but trade is only carried out eight or ten days' journey up the river; it appears that its estuary like that of the Ganges is uninhabited, and this is no doubt due to the sterility of the country which is drowned and covered in marshes where the herds that represent the main wealth of the Tartars cannot find a wholesome subsistence. I know that the Jesuits informed Europe that there was a pearl fishery on this coast, and we indeed found oysters with pearls in them. I must admit that I do not know where to place this fishery, unless it is on the Korean border or at the mouth of the Segalien; in that case I suppose that this fishery cannot in any way be compared with those of Bassora or the Gulf of Monaar[1] which provide work for five or six

---

[1] Bassorah, now Basra in present-day Iraq, was once a major pearl trading centre in the Persian Gulf. Monaar is the Gulf of Mannar in north-west Sri Lanka, a pearling centre whose origins go back to antiquity.

thousand people, but it is possible that some families of fishermen gather to look for pearls which they barter for nankeen or other cheap Chinese trade goods; nevertheless I did show some excellent imitation pearls to the Bitchys and the islanders of Oku yesso and I did not notice that they were any more impressed by them than by the most common types of beads. We sailed quite close along the coast, often within a gunshot without noticing any village. At Baye de Ternai we saw some bears, hinds and fawns grazing like domestic animals and raising their heads in astonishment when our ships entered the bay. Two tombs and some burned trees were the only indications that this country had any other inhabitants. The Baye de Suffrin was equally deserted, and 25 to 30 people made up the population of Baye de Castries which was spacious enough for ten thousand. Our naturalists found on the shore and at the rivermouths neither pyrites, nor drift-ore or gold particles in the sand, nothing in a word that indicates a country rich in metals. We came upon some silex, chalcedony, crystalline spar, porphyry and a great deal of volcanic matter with very little schorle but quite a number of fairly fine crystals, zeolites and other incrustations often found in the lava of extinct volcanoes. The coast of Oku yesso which is the east side of the Tartary channel is even more fertile than that of the mainland; it seemed to me that the vegetation was stronger, but the islanders do not work the soil to any greater extent. The animal kingdom supplies almost all their subsistence for I do not include the few lily bulbs and the garlic the women collect along the edge of the forest and dry out. I am even led to believe that hunting is more a form of relaxation than work for these people, and that fish, fresh or dried, represents their staple diet, as wheat is for the French. The two dogs I was given at Baye de Castries refused at first to eat any meat and threw themselves at the fish with a voracity that can only be compared to that of starving wolves. Necessity alone compelled them gradually to accept other food.

However a few bear and moose skins these people wore left me in no doubt that they hunt these animals in winter, but the mainlanders especially who are so weak do not dare attack them with their arrows. They indicated by gestures that they laid out traps by attaching some bait to a tightly bent bow; the animal when eating this bait caused the release of an arrow directed at the bait. The islanders, largely because they are stronger and better built, seemed to derive some pride from a number of scars they would willingly display, making us understand that they had fought bears, using stakes after wounding them with arrows.[1]

The canoes are made out of a dug-out pine tree and can take seven or eight people. They manoeuvre them by means of very light oars and undertake in these frail craft voyages of two hundred leagues from the southern extremity of Oku yesso and Yesso in the 42 degrees right up to the Segalien River in the 53 degrees, but never go further from the land than one pistol shot, except when they cross from one island to the other and wait for an absolutely flat sea. The winds which always follow the direction of the channel never drive the waves towards the shore, so that it is possible to land in every cove as one might in the most sheltered anchorages. With them they take pieces of birch bark with which, with a few pine branches, they erect a hut in a trice. The streams which are lined with salmon offer a guaranteed sustenance. Every canoe leader has his own cooking-pot, tripod, tinder-box and amadou; wherever they land their hut is up, their fish caught and the cooking completed in the hour, and this navigation is as safe as on the Languedoc canal;[2] they arrive within a specified time and stop every night in the same bays and near the same streams. They marked on our chart the number of nights required between Cape Crillon and the Segalien River, and from this it appears that they make eleven leagues a day although they have neither masts nor yards on their canoes; sometimes they attach a shirt to two oars fixed crosswise and thus sail, which is less tiring than rowing; one sees near the villages small canoes for one or two men only, which they do not use for long journeys, but which are meant to be rowed into streams where they fish; they are so light that when they reach an area where the depth is twelve or fifteen inches they use small crutches instead of poles and, remaining seated, they push against the bottom and propel their craft at a considerable speed; when the water is deeper they handle these small craft with paddles; the customs of the two populations differ only slightly – the same way

---

[1] La Pérouse's comments on the importance of fish in the local diet, the bear hunt and its symbolic importance in local culture are confirmed by Krasheninnikov who adds that, not only did dogs principally feed on fish, but themselves caught salmon in the rivers.

[2] The Languedoc Canal is known as the Canal du Midi or Canal des Deux-Mers, linking the Atlantic Ocean with the Mediterranean Sea. Begun in the seventeenth century, it was carrying up to 1600 barges and boats a day past the central inland port of Toulouse at the time La Pérouse was writing.

CHAPTER XVII

of life, the same sea and land architecture, the same respect for elders, but in this comparison I rather tend to opine that the Tartars would have the advantage in respect of moral behaviour and the islanders in respect of skills and certainly of character and other qualities related to a belief in one's own ability. We thought we could discern on Oku yesso a class distinction that was not apparent in Tartary: there was in each canoe one man who did not mix with the others, did not eat with them and seemed to be totally subordinate to the rest. We suspected that he was a slave – this is a mere conjecture, but he was at least of a much lower status.

The people of Yesso and Okuyesso also possess an item of trade which the Bitchys and Orotchys lack entirely; it is whale oil. These cetaceans abound on the eastern coast of their islands where we sighted as many as in the Strait of Le Maire, but we did not see a single one in the Tartary channel – as though they had not yet discovered the strait of Castries or of Sangar from which no doubt the strength of the current repels them. A more direct intercourse between the islanders and Japan also gives the belongings found in their huts an appearance of opulence not found on the mainland except in the tombs for which they reserve all their wealth, and we found no monument of this type on the island, but we did notice that they had the same idols hanging from the ceilings of their huts, and the leader of one of the two canoes of Baye de Crillon to whom I gave a bottle of brandy threw a few drops into the sea before leaving, explaining that this was an offering to the Supreme Being, but it seems that the sky is His temple's vaults and that the heads of families are his priests.

It is easy to conclude from this account that no commercial considerations could attract Europeans into these seas; a little whale oil and dried or smoked fish with just a few bear or moose skins are very minor items with which to meet the costs of such a long voyage; I should even add as a general axiom that one can only hope to carry out any significant trade with a major nation, and that even if dried fish and oil were items of some importance one would not manage to fill a vessel of 300 tons along these various coasts that extend for more than a thousand leagues; although the quality of the dried salmon of Baye de Castries quite impressed us and it was quite easy to buy, I must admit that I felt some reluctance in so doing because I was afraid that these poor people might sell us their winter stock and die of hunger during the bad season.

We did not come upon any sea otters; we showed them some samples and we got the impression that they did not know these furs; they did not seem to prize them any more than that of the seals which they use to make boots; it is likely that this amphibian is found only along the eastern seaboard of the northern Kurils, which would indicate that its real native land lies east of Asia, towards the coast of America, where as I have already stated it is found in great numbers from the tip of Onoloska as far as St Diego on the western coast of California. Reading the various accounts that have given many erroneous impressions of the vast country we have just explored, one finds here and there some extremely accurate comments which, in truth, were not easy to disentangle from the rest. Fr des Anges had undoubtedly known these people and the description he gives of them is accurate, but he places them at the southern extremity of Yesso, facing Japan; he had neither understood nor dared to suspect the vast extent of this country, and the strait of Tessoy he mentions, which the islanders said was obstructed by seaweed and so close to the mainland that one can see with the naked eye a horse grazing on the other side, is nothing more than the end of the gulf into which we sailed and from which we saw Point Boutin on the island of Oku yesso jutting out towards the mainland and ending on the seaward side like a sandbank a *toise* or two in elevation. The accounts of Kempher,[1] and the letters of Fr Gaubil also contained some truths, but neither were reports of what the Japanese or the Tartars had told them and they had spoken with men who were too ignorant for their reports to be accurate. Finally, the Russians had found it more convenient to deny that these two islands, larger than the British Isles, existed; they had mixed them up with the Kurils and assumed that there was no land between these

---

[1] Engelbrecht Kaempfer (1651–1716), a German doctor, travelled in Russia and down to Persia as part of a Swedish diplomatic mission. He then entered the service of the Dutch East India Company and sailed as a surgeon to India and Batavia; in 1690 he sailed for Japan where he spent two years. He returned to Germany and on his death his papers were purchased by Sir Hans Sloane and sent to England where his *History of Japan* was translated by J.G. Scheuchzer and published in two volumes in 1727. Fr Antoine de Gaubil (1689–1759) was a French Jesuit who travelled to China in 1721 and reached Pekin in 1723; a number of his letters were printed in France in his lifetime, but they have now been published in a more complete collection: R. Simon (ed.), *A Gaubil: Correspondance de Pékin*, Paris, 1970.

CHAPTER XVII

islands and the Asian continent,[1] which would have opened the seas of Japan and Korea to their ships from Okots. But this assumption eliminated the Dutch voyage of 1634, and we are bold enough to claim that Captain Vries's navigation is the most accurate that could have been made at a time when the methods of taking readings were so undeveloped. It seems that the Dutch tried to make up for this disadvantage by paying the most scrupulous attention to their dead reckoning and to the precision of their bearings, and if the strait of Castries escaped their investigations it will be no surprise to sailors who know these fogbound latitudes. The longitude and the latitude of this strait have been so precisely determined during the voyage that there is now no problem in sailing through this passage into the Korean Sea; De Langle Peak, which rises more than twelve hundred *toises* above sea level and which can be seen on a clear day from a distance of forty leagues, is an excellent landmark for the southern coast of this channel, which is to be preferred to the northern one because the currents are more moderate. If the strains of this campaign give France and the other nations of Europe nothing more than the precise knowledge of the geography of this part of our continent they will at least be of more immediate value to the Russians who may one day develop major shipping enterprises in Okots and will cause the arts of sciences of Europe to flourish in these countries which today are inhabited only by a few bands of roaming Tartars and to a greater extent by bears and other forest animals.

I shall not try to explain how the Yesso, Oku yesso and all the Kurils are inhabited by a race of men which is different from the Japanese, the Chinese, the Kamchatdales and the Tartars from whom they are separated by a strait less than a league wide and at most five or six feet deep. As a traveller I report facts and note differences; there are enough other writers who would develop these data into philosophical systems. Although I did not land on the Kurils, I am sure from the Russian accounts and the similarity

[1] La Pérouse is unaware of the political and commercial background: the Russians did not deny the existence of the various islands, but the Treaty of Nerchinsk (1689) barred them from the regions to the south of the Amur River. Korea and Japan had closed their doors to foreign traders, while China restricted foreign trade to Canton in the south and Kiakhta (now Ka'chta) in the north. The seas around Japan and Korea consequently offered few openings for trade and were highly dangerous for would-be fur traders; the Aleutian Islands and Alaska, however, were much more open and presented far greater opportunities for a lucrative trade.

of the language (which will be evident from the vocabularies appended at the end of the present work) that the inhabitants of the Kurils and those of Yesso and Oku yesso share a common origin. Their customs, their way of life also hardly differ from of the mainlanders, but Nature has given the physique of these two people such a marked difference that this impression, better than any medal or other monument, proves beyond doubt that these islands were not peopled from the continent and that it is a colony which may even be unconnected with Asia;[1] although Oku yesso lies west of the Kurils by more than 150 leagues and it is impossible to cross this space with boats as frail as their canoes, nevertheless they were able to communicate among themselves with no difficulty because all these islands form a kind of circle and are separated by channels of various widths and none I estimate to be more than fifteen leagues; it should therefore be possible to go by canoe from Kamchatka to the mouth of the Segalien River by following this island chain as far as Marikan Island, and pass from Marikan to the 4 Brothers, Compagnie, States, Yesso and finally Loku yesso, and so reach the frontiers of Russian Tartary, but one would utter in vain among these islanders the names of Yesso and Oku-yesso which presumably are Japanese: neither the Tartars nor the so-called Oku yessians would have any knowledge of them. They call their island Chotka and Yesso Tssissa; this confusion of names is very harmful to the advance of geography or at least it unnecessarily strains one's memory and I believe that when local names are known they should be religiously preserved, and failing them those given by the earliest navigators. I have had this rule followed invariably in the charts drawn during this voyage, and if we did deviate from it, it was out of ignorance and not for the pointless and ridiculous honour of bestowing a new name.[2]

[1] La Pérouse identified the distinctive Ainu people in southern Sakhalin. They had been forced to migrate towards the north by advancing Japanese settlements. The Tungus had settled in maritime Siberia, while the Kamchadals occupied the northern Kurils. The population mix in northern Sakhalin was however more complex than the French realised, with a mixture of Palaeo-Siberian Chukchi and some immigrants from Alaska.
[2] The names which geographers allocated to Japan over the centuries were a source of constant confusion. Yesso, or Yezo, sometimes Jedso, is Hokkaido, while Niphon often stood for Honshu or Japan itself. Oku (oki) Yesso can be read merely as 'the islands of Hokkaido'. Ezo na Chishima was the Japanese name for the chain of islands north of Hokkaido. Chishima is still the Japanese name for the Kurils.

CHAPTER XVII

VOCABULARY

of the bearded inhabitants of Chiocka, found at Baye de Langle

A few words from this extensive language are spoken gutturally, but the pronunciation should be soft and like that of people who lightly roll their r's. I have shown this by ch; the qs found at the beginning of a few other words is meant to express a kind of hissing which has to be sounded before uttering the syllables that follow.[1]

Names of the main parts of the body

CHIOCKA	FRENCH
Chy	eye, eyes
tara	eyebrows
quechetau	forehead
étou	nose
notamekann	cheeks
tsara	mouth
yma	teeth
aou	tongue
mochtchiri	chin
qs-chara	ears
tapine chime	shoulder
taks sonq	arm
tay	forearm
tayha	wrist
taypompé	hand and the fingers in general
tchouaipompé	thumb
tchame	chest (front and upper)
toho	female breasts
haûne	stomach
tsiga	male genitals

---

[1] It must be remembered that the native words were recorded by the French and that these spellings reflect the French phonetic system; in some case there may even be touches of a local French accent, although there is nothing in records to indicate whether La Pérouse himself still retained traces of his own native Languedoc in his speech.

CHIOCKA	FRENCH
chipouille	female genitals
ambe	thighs
aouchi	knees
aïmaitsi	legs
acouponai	malleolus, ankle
paraourai	feet (top of)
kaima pompéame	big toe
tassoupompéame	index finger
tassouhapompéame	middle finger
tassouame	the third and the little finger
ochetourou	nape of neck
saitourou	back
assoroka	buttocks
tcheai	bend or back of the knee
oatchika	calf of the leg
otocoukaïon	heels
téhé	beard
chapa	hair

Names of all kinds of objects[1]

choumau	stone, generic term
ni	tree trunk, wood in general
hocatoûrou	canoe
tacôme	canoe hauling-line
oukannessi	oars or paddles
kochkoûme	small square vase made of birch bark with a handle-piece. It is used for drinking and for bailing canoes.
manchous	People of Tartary close to the Amur River and of Chiocka where they live, they showed these people in the north-west and indicated that ships could pass through the strait between them.
tchoiza	sea
ouachekakai	a kind of spade used to get water out of the canoes
soïtta	canoe thwart

[1] Marginal note by La Pérouse: 'All items followed by this sign (M) are supplied by the Manchu Tartars with whom they trade.'

CHAPTER XVII

CHIOCKA	FRENCH
taratte	very long and strong strap 6 to 8 *lignes* in length; it is used mainly to moor the canoes.
moncara	iron axe (M)
ho	great iron decorated lance
couhou	bow
haï	arrows, ordinary in iron like a snake tongue, some barbed, others smooth (M)
tassehaï	double-forked arrows also of iron (M)
étanto	wooden heavy-headed arrows
tassîro	great cutlass (M)
matsiranitsi and makiri	small knife with sheath. It hangs from the leather belt which keeps their tunics overlapped (M)
matsirai	name they give our sheath knives
toche	large rough pieces of birch bark
qs-sichechai	pinewood plank
achka	hat or bonnet
tobéka	sea-calf skin made into a long cloak
achtoussa	tunic in fine birch bark, very skilfully prepared
sératousse	large dogskin tunic or jacket
tetarapé	kind of shirt made of rough cloth edged at the bottom and the neck with blue nankeen
ochsse	skin socks or ankle-boots sewn to shoes
tchirau	Chinese-style shoes with a curled-up pointed end
mirauhau	small leather pouch with four curled corners. It serves as a pocket and hangs from the leather belt
tcharompé	ear pendants, normally consisting of six to eight blue beads (M)
hakame	large ring made of iron, lead, wood and the tooth of a sea-cow. Ornament or distinctive sign, forced onto the left thumb (M)
achtkakaroupai	small fan-shaped sunshade or eye-shade, to protect the eyes of old men from the sunlight
hiérachtchinam	large strong mat on which they sit and sleep
Tama	single blue beads, all native people are definitely attracted by this blue colour
hounechi	fire
camoui	dog
tai-po	a musket

CHIOCKA	FRENCH
niutou	birch bark water bucket of the same shape as ours with its handle
onachka	drinking water
chichepo	sea water
abtka	small rope
otoroutchina	grass in general or meadows
choulaki	moss (plant)
tsiboko	water-parsley or wild celery
mahouni	wild rose tree
taroho	rose flower, commonly called dog rose
mahatsi	kind of tulip
pechkoutou	angelica (plant)
sorompai	large wooden spoon
kaïani or kahani	ship, vessel
chouhou	copper cauldron (M)
nissy	pole or long stick
pouhau	hut or house
nioupouri	huts or village
oho	plain where the huts are built
naye	river flowing in the same plain
tsouhou	sun
hourara	firmament
tébaira	wind
oroa	cold
tébairouha	winter or snow (season)
hourarahaûne	clouds
tchikotampé	our neckties or handkerchiefs
otoumouchi	jacket brass buttons with a round head (small) (M)
tchitsa	name of an island or people they indicate as being to the south of Chiocka
qs-lari	bird feather
tsita	bird in general or birdsong
étouchka	choucas, type of crow
tsikaha	common swallow (small)
mâchi	gull (seashore palmiped)
omoche	common two-winged or dipterous fly
mocomaie	large common clam (bivalve shell)
pipa	large mother-of-pearl shell (bivalve)

CHAPTER XVII

CHIOCKA	FRENCH
otassi	gurnet (type of fish)
toukochiche	salmon
émoé	fish in general or special name of a type of burbot
chauboûne	type of carp or carp-like fish
pauni	bone or vertebral column of fish they grill and set aside in heaps. Is it to feed their dogs in winter? Is it to make a kind of bread like the inhabitants of northern Iceland?
chidarapé	milt, roe and air bladder of fish which they also set aside.
houaka	no, it cannot be, I cannot or I do not want to
ta-sa	who? what? what is it? interrogative pronoun
he & hi	yes
hya	no
tape or tapai	this, that this one, that one, demonstrative pronoun
coukaha	come here
ajbai	to eat (action)
cbuha	to drink
mouaro	to lie down, or snore
étaro	to sleep
chiocka	name of their large island
tanina	other name given to this land, but most gave me chioka
kaïne	sewing needle

Name of Numbers

tchinai	1
tou	2
tchai	3
ŷnai	4
aschnai	5
yhampai	6
araouampé	7
toubi schampai	8
tchinaibai schampai	9
houampai	10

tchinaibi kassma	11
toubi kassma	12
tchailbi kassma	13
ynaibi kassma	14
aschaibi kassma	15
yhambi kassma	16
araouambi kassma	17
toubi schambi kassma	18
tchinaibi schambi kassma	19
houampaibi kassma	20
houampaibi kassma tchinaiho	30
ynai houampai touch-ho	40
aschnai houampai touch-ho	50
tou aschnai houampai touch-ho	100

If in this language there exists a difference between singulars and plurals, it is not expressed in the pronunciation. There are no words for our whistles nor for our pins.

I have not seen these people dance or sing; but they can all make pleasing sounds with the main stalk of a large celery of the euphorbia type open at either end. They blow into the small end; these sounds are fairly similar to the softer tones of a trumpet; the tune they play is indeterminate and with no metre whatever; it is a sequence of high and low sounds ranging over one and a half or two octaves, that is to say twelve or sixteen notes. We did not see any other musical instruments.

CHAPTER XVIII

Anchorage in Awatska Bay. Visit from the Kamchadals and their priest; their generous offer to go hunting for our food requirements. Welcome by Lieutenant Kaborof, governor of the ostroc of St Peter and St Paul; he announces the forthcoming arrival of the governor of Okots at St Peter and tells us that Kamchatka is now no more than a dependency of the Okots administration; services he renders us while awaiting the governor's arrival to whom he sends a courier with our letters.

Arrival of Mr Coslof at the harbour of St Peter and St Paul; he is followed by Mr Schemales and the unfortunate Iwaskin who arouses our utmost interest; extreme courtesy displayed towards us by the governor; his attentions, his civility; he showers gifts on us and totally refuses any payment for the oxen he sends us. Kamchadal ball, details on their dances; arrival of a courier from Okots with our mail from France, the news of my promotion to commodore is celebrated to the sound of the entire artillery of the post.

We discovered the tomb of Mr de la Croyere and fixed to it an inscription engraved on copper and one on Captain Klerk's; the governor plans to have two monuments erected that will be more worthy of these two famous men.

Mr Coslof's new ideas on the administration of Kamchatka; arrival of an English vessel in 1786. The proposals it puts forward for trade, they are sent to the Court at St Petersburg, loss of this ship on Copper Island, false impressions of this expedition.

We obtain permission to send our interpreter to France with our mail. Departure from Awatska Bay; we receive salutes from the batteries of the ostroc and those on the narrows, and return both these salutes.

Hardly had we moored outside the port of St Peter and St Paul when we received a visit from the thayou[7] or village chief and

---

[1] *Toion*, village chief, a term applied by the Russians to local headmen in all their territories and outposts, including eastern Siberia and Alaska.

several other Kamchadals. They all brought us a few gifts of salmon or ray, and offered to go shooting bears or ducks that proliferate around the ponds and rivers; we accepted these kind offers, we lent them some muskets, we gave them lead and powder, and we were never short of game during our stay in Awatska Bay; they asked for no payment in return for their work, but we had been so well supplied in Brest with a multitude of articles that are very precious to the Kamchadals that we insisted on their allowing us to show our appreciation, and our wealth allowed us to tailor our gifts more to their needs than to the presents they brought us. The administration of Kamchatka had completely changed since the departure of the English; it was now no more than a province of Okots.[1] Captain Schemale,[2] the same man who had succeeded Major Béem,[3] was still in the country with the title of special commander of the Kamchadals, and Mr Renikin, the real successor to Major Béem, who had arrived in Kamchatka shortly after the departure of the English, governed the country for only four years and returned to St Petersburg in 1784. We obtained all these details

---

[1] Empress Catherine II ('the Great') reorganised the administrative structure of Siberia during the 1780s, eliminating the former Department of Siberian Affairs based on St Petersburg. Siberia was set up as a *tsartvo*, or separate kingdom, and divided into three provinces, (*namestnichestva*). The province of Irkutsk was subdivided into four regions (*oblasti*), one of which was centred on Okhotsk. See S.D. Watrous (ed.), *John Ledyard's Journey through Russia and Siberia 1787–1788*, Wisconsin, 1966, p. 59.

[2] Schemale/Scémale, who appears as Schmaleff in the printed version of La Pérouse's voyage, was Vasilii Ivanovich Schmalev, Deputy-Governor of Kamchatka during the visit of Cook's *Resolution* and *Discovery* in 1779. He became Governor during the visit which had aroused his suspicions, as he believed that England planned to take over the Russian settlements and the profitable fur trade – in the light of the Nootka Incident of 1789–90 which resulted in the Spanish losing their claims to exclusive rights on the Northwest Coast, this was not an unreasonable suspicion. Schmalev obtained guns and reinforcements from Irkutsk to strengthen the fort at St Peter and St Paul (Petropavlovsk). The 'commander of the Kamchadals' was answerable to the Governor of Irkutsk who was endeavouring the improve the condition of the indigenous people, so that La Pérouse's description of *commandant particulier* might well be translated as 'protector of the Kamchadals'. See H.H. Bancroft, *History of Alaska 1730–1885*, new edn, Darien, 1970, pp. 310–13.

[3] Magnus von Behm (or Bem) was governor at the time of the visit of the *Resolution* and *Discovery* in 1779. Once he realised the importance of the English expedition he proved himself to be a most generous host. John Gore reported this to the British Ambassador in St Petersburg: 'Behm leads, Ismyloff [Schmalev] follows . . . I have been mosty agreably disappointed in meeting with so much polite civilization in a Country so remote as Kamchatka.' Cook, *Journals*, III, p. 1546. But Behm's generosity was met with little sympathy from his superiors, see *infra*.

CHAPTER XVIII

from Sub-Lieutenant Kaborof[1] who was in charge of the port of St Peter and St Paul, with a sergeant and a detachment of 40 soldiers or Cossacks under him; this officer's helpfulness cannot be described; he and all his soldiers, in a word, everything he had to offer was at our disposal; he did not allow me to despatch an officer to Bolkerets where by a most fortunate coincidence was Mr Coslof,[2] the Governor of Okots, who was carrying out his tour of inspection of the province; he told me that within just a few days that governor was himself due at St Peter and St Paul and that in all likelihood he was already on his way; he added that this voyage was much longer than possibly we thought and that the season of the year did not allow it to be made in sleighs, that the small number of horses to be found in Kamchatka were in Bolkerets and that it was absolutely necessary to travel half on foot half by canoe on the rivers of Awatka and Bolkerets: Mr Kaborof proposed at the same time to send a Cossack with my despatches to Mr Coslof whom he mentioned with an enthusiasm and look of happiness it was difficult not to share; he kept on expressing his satisfaction at the thought that we would meet and deal with an officer whose education, manners and knowledge were second to none compared with officers of the Russian empire or any other country.

Mr Lesseps, our young interpreter, spoke Russian as fluently as French; he translated the sub-lieutenant's words and wrote in my name a letter in Russian to the Governor of Okots to whom I also wrote in French. I pointed out to him that the account of Cook's third voyage had publicised the hospitality of the Kamtschatka

---

[1] There is a brief mention of Kaborov in Bancroft's book (op.cit.), but little information is available on him.
[2] Colonel Grigor Kozlov-Ugrenin, who was in charge of the Okhotsk *ostrog* and therefore district commander, was in the process of reorganising the administration of this isolated region. This involved a great deal of travel through a vast and difficult territory. He was particularly anxious to curb the activities of the independent fur buyers, the *promyshlenniki*, who had often behaved most brutally towards the Aleuts. The establishment of Chelikov's company, and his influence on the Governor of Irkutsk, Ivan Yakobii, gave new impetus to the fur trade. On 15 June 1787, Kozlov-Ugrenin issued two decrees, one ordering traders to respect the rights of the Aleut people who had now become Russian subjects, the other addressed to 'all the chiefs and people of the Aleutian Islands in the Northeast Ocean', promising them the empress's protection. Kozlov's endeavours are praiseworthy, but they were not particularly successful: it is estimated that at the time the first Russians reached the Aleutian Islands in 1741 the native population numbered over 25,000; by 1799 it had dropped to less than half that number, and in the 1830s to a couple of thousands.

administration and that I dared to hope that I would receive in Awatska Bay the same welcome as was enjoyed by the English navigators who like us had travelled for the common good of every sea-going nation. Mr Coslof's reply could not reach us for another five or six days and the good lieutenant told us that he was anticipating his orders and those of his august sovereign by begging us to consider ourselves as though we were in our own homeland and make use of everything the country had to offer. It was evident from his gestures, his eyes and his expression that had it been within his power his mountains and his marshes would have been magically transformed into enchanted places. A rumour spread that Mr Coslof had no letters for us, but that the former governor of Kamchatka, Baron de Stingel[1] (whom Mr Schemales had succeeded as Inspector of the Kamchadals) who lived at Vernessoy,[2] might have some, and on the basis of this quite improbable story he sent him a messenger who had to travel more than a hundred and fifty leagues on foot. Mr Kaborof knew how anxious we were for letters; Mr Lesseps had told him of our sadness when we had learned that no parcel had arrived for us in St Peter and St Paul; he seemed as downcast as we were, and his behaviour seemed to say that he would travel himself to Europe to fetch our letters if he thought we would still be here when he returned; his sergeant and all his soldiers displayed the same eagerness to serve us; Mrs Kaborof was also extremely kind, her house was open to us at any hour of the day, we were offered tea and all the country's refreshments; everyone wanted to give us something and in spite of our inflexible rule to accept no gifts we found it impossible to resist Mrs Kaborof's great insistence, which forced our officers, Mr de Langle and I to accept a few reindeer and fox furs that would have been of much greater use to them than to us who were due to return to the tropics; fortunately we had been provided, back in Europe, with such means of repayment that we insisted to be allowed in our turn to offer what was not available in Kamtschatka, but if we were wealthier than our

---

[1] Colonel Ivan Steinheil belonged to a German family, the Barons Steinheil, who had been in the service of Russia since 1711. The Russianised form of their name was Shteyngel. The colonel's son, Vladimir Ivanovich Shteyngel (1783–1863), who was born in Siberia and brought up in Kamchatka, was implicated in the Decembrist plot of 1825.
[2] The small town of Verncknei or Verkhne-Kamchatsk in central Kamchatka.

CHAPTER XVIII

hosts our manners could not equate their touching kindness which is so much greater than any gift.

I told Mr Kaborof, through Mr Lesseps's intermediary, that I wished to have a small establishment on land to house our astronomers and a quadrant and clock. The largest house in the village was at once placed at our disposal, and as we saw it only a few hours after this request we felt we could accept it without creating any inconvenience because it seemed uninhabited, but we later learned that the lieutenant had removed his secretary, a corporal, the third person in line of importance in the district, to house us there, and Russian discipline is such that these moves are carried out as promptly as any drill and orders are expressed by a mere nod of the head. Hardly had our astronomers set up their observatory when our naturalists, who were no less zealous, wanted to visit the volcano which seemed to be less than two leagues away from us, although one had to travel at least eight leagues to reach the foot of this mountain, which was almost totally covered in snow and whose crater, facing Awatska Bay, gave us endless displays of smoke and, on a single occasion one night, blueish and yellow flames rising slightly above the crater mouth.[1]

Mr Kaborof's zeal was as marked towards our naturalists as it was for our astronomers; eight Cossacks were promptly selected to accompany Messrs Bernizet and the abbés Mongès and Receveur; Mr de Lamanon's health was not strong enough as yet to allow him to undertake such a journey. There may never have been such a strenuous one undertaken for science, and neither the English nor the German scientists who had travelled to Kamtschatka had attempted to carry out such a difficult entreprise. The appearance of the mountain made me think that it was impossible – one could see no greenery, but only a snow-covered rocky surface with a slope of not less than 40 degrees. Our intrepid travellers left with the hope of overcoming all these obstacles; the Cossacks carried their baggage which consisted of a tent, sundry furs and the food they had all

---

[1] Mt Avachinskaya, 2,718 m, had erupted at the time Cook's third expedition was in Kamchatka: 'at daylight discoverd the decks & sides coverd with the finest dust, like Emery, the Air was loaded and darkened with this substance, & towards the Volcano mountain to the N of the Ostrog it was so thick & black that we could distinguish nothing of the body of the hill', 15 June 1779, Cook, *Journals*, III, p. 677. An eruption in 1737 which had destroyed a village was reported by Krashertinnikov who comments on the natives' 'horrible death'. E.A.P. Crownhart-Vaughan, *Explorations of Kamchatka*, p. 99.

taken to last them for four days. The honour of carrying the barometers, the thermometers, the acids, and the various other means of carrying out observations was reserved to the naturalists themselves, who could not entrust these fragiles instruments to men who were to take them no further than the foot of the mountain, a belief that may have been as ancient as Kamtschatka itself making the Kamchadals and their conquerors think that vapours emerge from the mountain which kill all those who have the temerity to climb it. No doubt they hoped that our physicists would stop, as they would themselves, at the foot of the volcano, and a few tots of brandy given to them before departure time had no doubt inspired such kind solicitude for them; they happily set off with this expectation. The first stop was made in the forest 6 leagues from the harbour of St Peter and St Paul; they had so far travelled on easy ground covered with plants and trees, most of these being birches; the pine trees here are stunted and almost dwarfed; one of these small varieties had pine cones whose seed or small nuts are edible and the bark of these birches produces a fairly pleasant liquid which the Kamchadals carefully collect in vases and use a great deal.[1] All types of berries, of every shade of red and black, were also found along the travellers' way; their flavour is in general slightly acid, but sugar makes them very tasteful. At sunset the tent was set up, the fire lit and all arrangements made for spending the night, and this at a speed quite unfamiliar to people accustomed to living under roofs. Great precautions were taken to prevent the fire spreading into the forest; ten thousand blows on the Cossacks' backs would not have expiated such a misfortune, because fire puts all the sables to flight; after such an accident one cannot find any during the winter which is the hunting season, and since it is the country's only wealth, used for every form of trade and for the annual tribute due to the Crown,[2] the enormity of the

[1] The white or silver birch, genus *Betula*, provides a sugary sap drawn from the trunk in spring and fermented into a kind of beer and vinegar. Cook's men were effectively in Kamchatka in sprintime and were able to obtain 'a sufficient quantity of sweet pleasant Sap to mix the Grog with instead of Water which makes a very agreeable drink and an excellent antiscorbutic', Journal of David Samwell in Cook, *Journals*, III, p. 1244. The variety of birch found in this region is the *Betula nana*. The pine tree was probably the *Pinus cembra* whose cones are particularly oily.

[2] This tribute, known as *yasak*, went back to the time of the Mongols. It was a tax of ten per cent, payable in furs, this being the only item of interest to Europeans. Empress Elizabeth Petrovna had decreed in 1748 that the Aleuts should also pay this tax, a form of extortion which they violently resisted.

crime that destroys this benefit is quite evident. And so the Cossacks were careful to cut down all the grass surrounding the fire and to dig, before leaving, a deep hole for the coals which they put out by covering them with soil which was then well watered. The only quadruped seen during this day was a white hare, but no bear was seen, and no reindeer which are quite common in these parts. The next day at dawn they continued their journey. It had snowed a great deal in the night, but what was much worse was a thick fog covering the mountain of the volcano, the foot of which our physicists did not reach before three in the afternoon; in accordance with their practice their guides stopped where the vegetation ended; they set up their tent and lit the fire. This night's rest was essential before they began the next day's endeavours. Messrs Bernizet, Mongès and Receveur started their climb at 6 a.m. and did not stop before three p.m., at the very edge of the crater, but below it. They often needed their hands to hold on between these broken rocks which left highly dangerous precipices. The whole of this mountain consists of lava of various degrees of porosity; at the summit they found gypseous matters and sulphur crystallisations much less attractive than those of the Peak of Tenerife, and in general the schorls they brought back and all the other stones struck us as less beautiful than those of that ancient volcano which had not erupted for a century, whereas this one had sent up material in 1778 during Captain Klerk's stay in Awatska Bay;[1] however, they did bring back some fairly nice samples of chrysolite,[2] but they struck such bad weather and their route was so difficult that one still remains astonished that they were able to add to the weight of the barometers, thermometers and their other instruments. They were never able to see beyond a musket shot, except for just a few minutes when they had a glimpse of Awatska Bay and our frigates which, from this great height, looked smaller than canoes. At the crater's edge, their barometer dropped to 19 inches 11 *lignes* and $\frac{2}{10}$; at the same moment on board, where we were carrying out hourly observations, it read 27 in. 9 ^{lig} 17 hundredths; the thermometer was at 2 degrees ½ above freezing point and showed a

---

[1] The Klerk in question is Captain Charles Clerk of the *Discovery* who took over command of the expedition after James Cook was killed in Hawaii. He died on 22 August 1779.
[2] Chrysolite, a variety of olivine, is a yellow-green magnesium iron silicate.

difference of 12 degrees from the temperature on board ship; therefore, based on the calculations of the physicists who support this method of measuring the height of mountains and adjusting the thermometer readings, our travellers would have still climbed close on 15 hundred *toises*, a prodigious height considering the difficulties they had to cope with; but they were so put out by the fog that they decided to repeat this travelling the next day if the weather was favourable. In a way the difficulties had merely increased their zeal and not weakened it; they came down the mountain with this firm resolve and reached their tents; night had fallen, their guides had possibly prayed for them already and swallowed part of the liquors which they felt dead men would no longer require, but the lieutenant when he learned upon their return of this precipitate action ordered a hundred lashes to be administered to the most guilty party, and this punishment was meted out before we found out about it and were able to beg for mercy on his behalf. The night which followed this journey was a frightful one; the snow fell even more heavily, several feet within a few hours; it was impossible to carry out their plans of the previous day, and they returned to the village of St Peter and St Paul the same evening, after travelling eight leagues, a less tiring journey on the way back because the ground slopes evenly down.

While our lithologists and our astronomers were spending their time to such advantage, we were filling our water casks, storing wood in the hold and cutting and drying hay for the animals we expected to obtain, as we had only one sheep left; but the lieutenant had written to Mr Coslof begging him to collect as many bullocks as possible; he had sadly estimated that there would not be sufficient time to wait for those which the Governor would no doubt have ordered from Vernessoy because the journey would take at least six weeks, and the natives' lack of interest had not allowed them to multiply in the southern part of the peninsula where, with a little care, one could have as many as in Ireland.[1] The best and the

---

[1] The choice of Ireland as a basis for comparison surprises a little, coming from a Frenchman. The fact is that the Cattle Acts of 1663 and 1666 which prohibited the export of live cattle from Ireland were abolished in 1776 and the cattle trade had boomed. Furthermore, French ships had been calling for supplies at Cork on their way to French Canada since the early 1720s. It was estimated that in the last decades of the century upwards of 300,000 head of cattle were slaughtered in Cork each year and more were exported live. As a naval officer, La Pérouse would have been quite familiar with the Cork meat markets and the various breeds of Irish cattle.

CHAPTER XVIII

thickest grass grew in natural meadows to a height of four feet, and one could easily cut an enormous supply of hay for the winter which, to be honest, lasts here seven or eight months, but such work is beyond the Kamchadals: they would need sheds and large, weather-proof stables; they find it easier to live on hunting and especially catching the salmon which, like the manna of the Jews, arrive at the same time each year to fill their nets and ensure their annual subsistence. The Cossacks and the Russians, who are more soldiers than farmers, have followed the same practice. Only the lieutenant and the sergeant had small gardens well stocked with potatoes and turnips; their example and their exhortations had no effect on the others who however were quite happy to eat potatoes, but would not have wanted to do anything more in order to get them than dig them up if Nature had offered them in a wild state in the fields, like the serana, parsnips, garlic and especially the berries which they use to make a quite pleasant drink and some jam for wintertime. Our European seed had kept perfectly; we gave a large quantity to Mr Schemalet, to the lieutenant and to the sergeant, and we hope to learn one day that they have taken well there. In the midst of this work, there was still time for some entertainement and we went several times on hunting parties along the rivers of Awatska and Paratonka and in the eastern harbour but without any marked success, for our ambition was to kill some bears, reindeer and argali, whereas we had to be satisfied with a few ducks or teals that are not worth the exertions and the long walks that were required; but we had better luck with our Kamchadal friends – during our stay, they brought us four bears, one argali and one reindeer with such a quantity of divers and sea-parrots that we gave them to all our crews who were already wearying of fish: a single throw of the seine quite near our frigates would have been enough to supply six ships, but the range was very limited. All we really caught were small cod, herring, plaice and salmon; I gave orders that only a few barrel loads were to be salted as it was pointed out to me that all these fish were so small and so tender that they would not resist the corrosive effect of the salt which it was better to keep for the pigs we would find in the Pacific islands. While we were spending days which seemed so comfortable after the exertions of our recent explorations along the coasts of Locu-yesso and Tartary, Mr Coslof had set off for the port of St Peter and St Paul, but he was travelling slowly because he wanted to see everything and the purpose of his voyage was the establishment of the best possible administration in this province; he knew that one can only draw up an overall plan after examining all the country's products, and those which would benefit, relatively to the climate, would benefit from careful husbandry. He also wanted to know the stones, the minerals and generally all the substances found in the soil of his province; his observations had kept him a few days at the hot springs[1] situated twenty leagues from St Peter and St Paul; he brought back from there different stones and other volcanic matter and a gum which the Abbé Mongès analysed, and he told us quite openly that, having learned from public reports that several able naturalists had sailed with our frigates, he wanted to take advantage of this fortunate circumstance to discover the various substances of the Kamchatka peninsula and thus instruct himself. Mr Coslof's manners, his politeness, were absolutely the same as those of the best people of our largest cities; he spoke excellent French, had some knowledge of every aspect of our researches, whether in the area of geography or of natural history; it was impossible to imagine that an officer whose merit would have been acknowledged in any European nation had been appointed to such a wild and rough region situated at the end of the world; it will easily be realised that close links, intimate even, were soon established between Mr Coslof and us; he came aboard for dinner the day after his arrival, accompanied by Mr Schemalet and the priest of Paratonka. I gave him a 13-gun salute and three 'Long Live the King' shouted by our crew with a gaiety and a strength which seemed to surprise him.[2] After such a long voyage our appearance indicated that we were enjoying better health than when we sailed from Europe; I told him that it was due to a small extent to our care but even more to the extreme abundance of all the necessities of life we had found within his administration. Mr Coslof seemed happy of our fortunate situation, but he expressed his deepest regret at his inability to collect more than seven head of cattle for our departure which was to take place too soon to

---

[1] These hot springs are situated near the village of Nachikin, some 50 miles from Petropavlovsk. There are a number of hot springs in Kamchatka, due to volcanic activity in the region.
[2] The reference to 'Long Live the King' was omitted from the 1797 printed account of the voyage.

CHAPTER XVIII

contemplate bringing some down from the Kamchatka River, a hundred leagues from St Peter and St Paul; he had been waiting six months for the ship due from Okots with flour and other supplies needed for the province's garrison, and he was sadly afraid that that ship had met with some misfortune; our surprise at having received no mail was lessened when we learned that he had had no news of it since his departure from Okots and that he had to undertake the return journey by land around the Sea of Okots, a voyage that is almost as long or at least more difficult than the one from Okots to Petersburg.

The next day the Governor dined on board the *Astrolabe* with all his suite; he was again welcomed with a 13-gun salute and he begged us to desist from this courtesy in future so that we could meet with greater freedom and pleasure.

We were unable to persuade the government to accept any payment for the cattle; it was no use pointing out that, in spite of the close alliance existing between France and Spain, we had met all our costs in Manila. Mr Coslof told us that his sovereign had different principles and his only regret was the small number of cattle he could dispose of in the immediate future.[1] He invited us for the morrow to a ball he wanted to give in our honour, with all the women of St Peter and St Paul, both Kamchadals and Russians. The assembly might not have been large, but at least it was unusual: 13 women, including 10 Kamchadals with round faces, small eyes and flat noses, were seated on benches around the room, dressed in local silks which were wrapped around their heads rather like the mulatto women of our colonies. Mr Duché's drawings will depict these costumes better than I can describe them. We began with Russian dances with very pleasant tunes, very similar to the *Cosaque* that was being danced not long ago in Paris; Kamchadal dances followed – they can only be compared with those of the

---

[1] The English had received the same reply during the visit of the *Resolution* and *Discovery*. This was in line with official policy. In 1775 Empress Catherine II had instructed the Governor of Kamchatka to come to the assistance of any vessels from friendly nations which might call at Petropavlovsk and to accept in payment for any supplies provided nothing more than a receipt. This was more beneficial to visitors than to Russian officials, for Major Behm, who was particularly kind and generous to Cook's expedition, was never reimbursed for the cost of supplying the ships; some years later he found himself in severe financial difficulties and appealed, without success, for assistance from the English government. Emperor Paul I eventually rescued him from penury.

convulsionaries at the famous tomb of St Médard,[1] only arms and shoulders and almost no legs are needed for the guimarts or henels[2] of this part of Asia, they indulge in convulsive movements and contractions that are displeasing to all the spectators; this lack of appeal is heightened by the painful shout coming deep from the chest of these dancers who have no other music for their movements to keep in time with. They become so exhausted during these exercises that they are covered in perspiration and lie prone on the ground without the strength to get up. The powerful effluvia they give out perfume the room with a smell of oil and fish to which European noses are insufficiently accustomed to appreciate. Since in every country dances are always imitative and are in a way merely pantomimes, I enquired what the two women who had just carried out such a violent exercise intended to depict, and I was told they had mimed a bear hunt: the woman rolling on the ground was the animal and the other, turning around her, the hunter, but bears have much to complain about at being so uncouthly imitated. This dance, almost as tiring for the spectators as for the actors, was hardly over when a joyful shout announced the arrival of a courier from Okots; he was carrying a large suitcase filled with our mail. The ball was interrupted, the dancers dismissed with a glass of brandy, a worthy refreshment for these Terpsichores, and Mr Coslof, noticing that we were most impatient to obtain news of everything that interested us in Europe, urged us not to delay this pleasure; he settled us in his quarters, retired with the utmost courtesy in order not to get in the way of the various emotions

---

[1] This refers to a series of events in Paris which began in 1727 with the death of a Jansenist deacon of the parish of St Médard, François de Pâris, the brother of a Paris councillor, a wealthy and pious man. Groups of devout parishioners crowded around his death bed, carrying off a number of relics; at his funeral a woman threw herself on the coffin and shouted that she had been cured of paralysis. Other miracles were reported and his tomb became a place of pilgrimage: the sick lay down on the stone slab, went into convulsions and foamed at the mouth. This hysterical cult spread into the provinces, attracting up to five thousand followers. In July 1731, the Archbishop of Paris attempted to close the cemetery and obtained the Pope's condemnation of the cult, but the pro-Jansenist Paris Parliament refused to act; it was left to the lieutenant of police to enforce the closure in 1732 on the grounds that traffic was being impeded by the crowds. Evidence that several charlatans had taken advantage of the situation, especially in the provinces, subsequently came to light.

[2] The reference is to Madeleine Guimard (1743–1816) and to Anne-Frédérique Hainel (1752–1808), both famous dancers from the Paris Opera. Hainel married the even more celebrated Gaétan Vestris (1728–1808), 'the god of the dance'. It is likely that La Pérouse went to the Opera when he lived in Paris.

CHAPTER XVIII

which might be felt by each of us in the light of news received from family or friends. We all had happy news, but I more especially, being promoted to commodore, being favoured to an extent I dared not hope for. The congratulations I received from everyone soon came to Mr Coslof's notice, who desired to mark this event by the noise of all the town's artillery; I shall always remember with emotion and the greatest interest the marks of friendship and kindness I received from him on this occasion. I did not spend a single moment with the Governor without its being marked by some attention or kindness, and it goes without saying that since his arrival everyone in the place was fishing or hunting for us, and that we could not consume so much; he added gifts of all kinds for Mr de L'angle and me; we had to accept a Kamchadal sleigh for the King's collection, two royal eagles[1] which I brought back to France for the royal zoo and a great many sable; in our turn we offered him what we thought might be useful to him or which might appeal to him, but our only wealth was in items intended to be bartered with natives and we had almost nothing that was worthy of him.

The account of Cook's 3rd voyage seemed to give him great pleasure, and he had with him all the actors the English editor brought on stage, Mr Scemalet, the good priest of Paratonka,[2] the unfortunate Iwaskin;[3] he translated for them all the passages that

---

[1] 'Aigle royal' was a term generally used for the Golden Eagle (*Aquila chrysaëtos*) which is still widely found today in the northern hemisphere, especially in Asia and North America. The reference to their return in France was deleted by Milet-Mureau.

[2] The 'good priest of Paratonka' was Romaan Feodorovich Vereshagin (Vershagen); he was in charge of the small village of Paratunka at the time of the visit of the *Resolution* and *Discovery* in 1779. He was born at Bolscheretsk, the son of a Russian father and a Kamchadal mother. According to Beaglehole, he died not long after their departure (Cook, *Journals*, III, p. clxvi, n.2), but La Pérouse's narrative makes it clear that this is incorrect; the confusion may have arisen as a result of Paratunka becoming totally abandoned around the year 1801. At the time of La Pérouse's visit, Vereshagin would have been just of 50 years of age.

[3] The unfortunate Iwaskin was Pieter Matteioss Ivashkin (Evashkin). The English officers recorded his life story variously in 1779, because he spoke very little English. According to David Samwell, Ivashkin was born in 1723, had joined the imperial guard and been exiled in 1748 (Journal, 16 September 1779), but King has him as being born in 1725, disgraced in 1741 and sent to Siberia in 1748; both agree that he came from a good family and that his father was a general (see Cook, *Journals*, III, pp. clxvi, 706, 1276–7). La Pérouse had the advantage over the earlier informants of having Lesseps as an interpreter and of Ivashkin himself knowing a fair amount of French. If it is correct that he was present at the burial of La Croyère, which took place in October 1741, he must have been in Kamchatka by then;

357

concerned them and they repeated each time that it was perfectly true. Only the sergeant who was in command at the harbour of St Peter and St Paul when Captain Cook arrived had died,[1] all the others enjoyed excellent health and still lived in the country except for Port who had returned to Irkuts[2]. I told him of my surprise at finding the elderly Iwaskin in Kamchatka, as the English accounts stated that he had at last obtained permission to live in Okots; we could not avoid expressing our great interest in this unfortunate gentleman when we learned that his only crime was a few indiscreet comments on Empress Elizabeth after a dinner party where wine had affected his faculties, he was not yet twenty, a guards officer from a distinguished Russian family, with features of a noble cast that time nor misfortune had not changed; he was reduced to the ranks, sent into exile in the depths of Kamchatka and scourged with the knout after having his nostrils split; Empress Catherine whose vigilance extends even to the victims of the previous reigns pardoned this unfortunate gentleman some years ago[3] but more than fifty years spent in the midst of the vast forests of

---

Kruzenshtern who met him in 1805 believed him to be then aged 86, in which case he would have been born in 1719 and aged 68 when he met the French. The reasons for his disgrace are obscure, and he did not dwell on them. If it took place in or before 1741, it could have occurred in the tumultuous days of Ivan IV, at a time when the palace guards were often involved in conspiracies – and Kruzenshtern puts forward the theory that Ivashkin had been engaged in some aborted plot; he may well have been on the losing side of the conspiracy which brought Elizabeth Petrovna to the throne in 1741; or he may simply, as the French believed, have been careless in his comments about the new empress. In time, he became something of a minor government official in eastern Siberia, well treated and probably not unhappy in his final years.

[1] The sergeant in question was named Surgutski; he was most useful to the officers of Cook's expedition; however, this expedition was not led by Cook himself who had been killed in Hawaii. The editor of the printed narrative excised this phrase which represents a slip of the pen by La Pérouse.

[2] 'Port' was Iachan Daniel Pote; he spoke German and may have been of German origin. He 'knew the sound of french words & understood a little of Dutch' and because of this proved most useful to the officers of the *Resolution* and *Discovery*. See Cook, *Journals*, III, pp. 649–55, 1242–53, 1256. Among the earlier 'actors', of course, Major Behm had also left.

[3] Kruzenshtern, who naturally could more easily communicate with the Russian than either the English or the French, denies that Catherine had pardoned him, because he was a violent man, had once killed someone in a fit of anger and had been unjust towards Yakut people for whom he was responsible. Kruzenshtern, *A Voyage round the World in the Years 1803, 1804, 1805 & 1806*, London, 1813, II, pp. 207–8. But whether Ivashkin spoke the truth to his fellow Russian or was fairly reported on by him is still open to question.

Kamchatka, the bitter memory of the shameful punishment he underwent, and possibly a secret feeling of hatred towards an authority that had so harshly punished a crime which circumstances would have excused, these various reasons have made him unresponsive to such tardy justice and he proposes to die in Siberia; we begged him to accept some tobacco, powder, lead, cloth and generally anything we thought might be useful to him. He had been educated in Paris, could still understand French and rediscovered many words to express his gratitude; he loved Mr Coslof like his father and accompanied him on his journey out of affection, and this good governor treated him in a manner that would lead him to forget everything about his misfortunes. He helped us by letting us know of the tomb of Mr de la Croyere whose burial he had witnessed in Kamtschatka in 1741. We attached the following inscription to it, engraved on copper, written by Mr D'agelet who, like him, is a member of the Academy of Sciences: 'Here lies Louis de l'isle de la Croyere,[1] of the Royal Academy of Sciences of Paris, died 1741 on his return from an expedition carried out by order of the Tsar to explore the coasts of America; astronomer and geographer, emulator of two brothers famous in the sciences, he deserved his fatherland's sorrow. In 1786 the Count de la Perouse, commanding the King's frigates *Boussole* and *Astrolabe*, honoured his memory by giving his name to an island close to where this savant had landed.'

We also asked Mr Coslof's permission to engrave on copper the inscription on Captain Klerck's tomb,[2] which was only painted on

---

[1] Delisle (or de l'Isle) de la Croyère who is also mentioned in La Pérouse's Chapter IX had sailed with Chirikov to the northwest coast of America but died on board the *Sv Pavel* before the expedition arrived back at Petropavlovsk on 10 October 1741.

[2] Charles Clerke was born at Weathersfield, Essex, in 1743, served under Byron in the *Dolphin*, had sailed under James Cook on each of his three great voyages, finally serving as captain of the *Discovery*. Affected by tuberculosis, he could not stand the cold and foggy regions through which the third expedition had to sail and he died on board on 22 August 1779, two days before the expedition reached Petropavlovsk. He was buried under a tree in a small valley on the northern side of Avatcha. There were two memorials put up by his fellow officers. One was attached to a tree overhanging the grave and is the one to which La Pérouse refers. The other was in the church at Paratunka, where Clerke wanted to be buried, but could not be, 'it not being customary among the Russians to bury in the Church', as William Bayly, the astronomer, comments, although he might have reflected on Clerke's being an Anglican and thus unsuitable, in the priest's eyes, for burial in an Orthodox church, a point made by David Samwell in his journal. However, a new church was due to be built, and the Russian seems to have suggested that the valley site would

a piece of wood, too fragile a material for the memory of such a worthy navigator.

The Governor was kind enough to add to the permissions he gave the promise of having a more worthy monument erected as soon as possible to the memory of these two famous men who died of their exertions such a long way from home. He told us that Mr de la Croyere had married at Tobolsc and that his descendants enjoyed a great deal of respect; he was quite aware of the navigations of Bering and Captain Ticolof;[1] he stated in this connection that he had left Mr Beling[2] at Okots who had been required by his sovereign to build two ships to continue Russian discoveries in the northern seas; he had issued orders for every means at his disposal to be used to speed this development, but his zeal, his willingness to help, his intense desire to meet his sovereign's wishes could not overcome the obstacles one must find in a country that remains as wild as when it was first discovered, where the harsh climate forces

---

be the location for the new building and that Clerke's remains would finally lie within the new building. See Cook, *Journals*, III, pp. 703n, 1273. This does not quite tally with La Pérouse's information that the new church was to be built at Petropavlovsk, and indeed Paratunka seems to have been abandoned during the next decade or so. Here again, language difficulties may have created the confusion. In 1805 Kruzenshtern tidied the graves and erected a small monument to them. The monument, such as it was, was found by Beechey in 1827, but removed to the governor's gardens 'for better preservation'. F.W. Beechey, *Narrative of a Voyage to the Pacific and Beering's Straits*, London, 1831, II, p. 245. By 1865, however, when another traveller attempted to find traces of it, it had vanished under weeds and brambles. Frederick Whymper, *Travels and Adventures in the Territory of Alaska*, London, 1868, p. 95.

[1] Aleksei Ilich Chirikov (or Tchirikov) was appointed in 1725 as one of Vitus Bering's lieutenants. He returned to Siberia with him in 1737 for a second expedition on which he commanded the *Sv Pavel* (*St Paul*); sailing from Kamchatka in June 1741, he became separated from Bering, went on east and northeast until 15 July when he reached the northwest American coast, the first European to do so. He returned to Petropavlovsk in October, with his crew badly affected by scurvy. He went out again May 1742 to search, unsuccessfully, for Bering. Chirikov died in Moscow in November 1748. *Who's Who*, pp. 56–8.

[2] 'Beling' is Joseph Billings, born at Turnham Green, Essex, in 1758. He sailed with James Cook on his third voyage as an able-seaman. On his return to England he served on various merchant ships, one of which brought him to St Petersburg in 1783 where he offered his services 'as a former companion of Cook'. He was accepted in the Russian navy, rising rapidly from midshipman to lieutenant and then to captain-lieutenant. He was in Okhotsk in July 1786 and led a voyage to the estuary of the Kolyma River in the East Siberian Sea; a second expedition in 1790 took him to the Aleutians which were then of considerable commercial importance to the Russians. He died in 1806. *Who's Who*, p. 26.

CHAPTER XVIII

work to be suspended for eight months of the year, and he felt that it would have been more economical and much quicker to send Mr Beling out from a Baltic port where he could have obtained all he required for several years.

We drew a plan of Awatska Bay or, to be more accurate, we checked the English chart which is very accurate,[1] and Mr Bernizet made a very stylish drawing of it which he gave to the Governor. Mr Duché also offered him a drawing of the ostroc, and the abbés Mongès and Receveur presented him with a small box of acids to analyse the water and the various substances which make up Kamtschatka; Mr Coslof is not without a knowledge of chemistry and mineralogy; he himself liked to dabble in chemistry, but he told us with understable justification that the first duty of a wise and enlightened administration is to obtain bread for these people by instilling a desire for agriculture among the natives; the vegetation indicated that it was extremely fertile, and he did not doubt that instead of wheat, which might not germinate because of the cold, at least rye or barley would produce an abundant harvest; he drew our attention to the pleasing appearance of a few small potato fields, seed for which had come from Irkuts a few years earlier and he planned to adopt a gentle approach, but a sure one, to transform the Russians, the Cossacks and the Kamchadals gradually into farmers. Smallpox in 1769[2] had reduced this nation by three-quarters and it is now down to fewer than four thousand inhabitants in the entire peninsula, and it will soon disappear entirely on account of the continual intermarriage of Russians and Kamchadals; a mixed race, harder working than the Russians who are only any good as soldiers, and much stronger and less plain-featured than the Kamchadals, will arise out of these marriages and replace the former inhabitants; the latter have already forsaken the yourtes[3] in which they went to

---

[1] This English chart of Avacha Bay was drawn by the *Discovery*'s midshipman Edward Riou.

[2] Cook's surgeon David Samwell records that 'In the short Excursions we made during our stay we every where met with the Ruins of large Villages with no Traces left of them but the Foundations of the Houses. The Russians told us that in the Year 1769 above 10,000 of them were destroyed by the small Pox which was brought among them from Ochotzk in Siberia. These poor people are dwindling away fast & it is probable that in a few Generations there will be none of them left.' Cook, *Journals*, III, p. 1252.

[3] The yurt was the tent used by nomadic people in Asia. It consisted of a lattice framework rising to about a man's height, with a sloping frame forming the roof, covered with thick felts and anchored by a ring of stones around the bottom. The

earth like badgers throughout the winter and where they breathed a foul air that was responsible for many illnesses; today the wealthier build themselves ibas or Russian-style wooden houses – they have absolutely the same shape as our poor peasants' cottages and are divided into three small rooms heated by means of a brick stove that maintains a heat of thirty degrees, unbearable for people unused to it. The others spent the winter as well as the summer in balagons[1] which are kinds of wooden dovecotes, with a thatched roof, raised up on stakes 12 to 13 feet high into which both men and women go up by means of very awkward ladders, but these latter buildings will soon disappear – the Kamchadals are great imitators and adopt almost all their conquerors' practices; already the women do their hair and largely wear their clothes in the Russian manner, and the Russian tongue dominates in all the ostrocs, which is quite fortunate because each Kamchadal village had a different jargon, the inhabitants of one hamlet did not understand those of a neighbouring ostroc[2] and the confusion which occurred in the tower of Babel was not more inconvenient; one can say to the Russians' credit that although they have set up a despotic form of government in these harsh climes it is tempered by principles of gentleness and fairness that negate all its drawbacks, and mankind cannot reproach them for any atrocities[3] (as it can with the greed of the English in Bengual or of the Spanish in Mexico and Peru); the tribute they impose is so light that it can only be regarded as a form of recognition of Russian sovereignty and half a day's hunting in one year

---

yurt could be dismantled in less than an hour. When the inhabitants of eastern Siberia abandoned their roaming existence, the yurts became more solidly built. The Russian *isba* was far more comfortable, being constructed of logs, and its growing use reflected the changing lifestyle of the Kamchadals under Russian influence.

[1] The term *balagan* was used to refer to various kinds of huts or basic homes. However, there was also a Kamchadal word, *barabara*, which meant a hut built for the summer and commonly used in Russian Alaska for the temporary or simple homes put up by fishermen and hunters. The surgeon Samwell mentioned 'Kamchadale Houses called Balagans'. Cook, *Journals*, III, p. 1258.

[2] Originally *ostrog* meant a military post. La Pérouse reflects here the growing use of the term in the sense of a settlement or township established in Siberia.

[3] Like most Europeans of his day, La Pérouse had a very sketchy knowledge of the Russian advance into eastern Siberia and Russian expansion into Kamchatka and the Aleutian Islands. The greed and brutality of the fur traders or *promyslenniki* had become a serious embarrassment to the Russian authorities and to such men as Kozlov and warrant a far more severe judgment than is expressed by La Pérouse in this passage.

CHAPTER XVIII

settles their debt towards the Crown; it is surprising to find in these huts, more wretched than the smallest hamlet of our mountain country a quantity of coin all the more impressive in that it exists among a very small number of inhabitants who consume so little in the way of Russian and Chinese goods, that the terms of trade are entirely in their favour and that the balance of what is due to them has to be paid in roubles;[1] as for furs they fetch here a much better price than in Canton, which shows that until now the markets of Kiakia[2] have not been affected by the new outlet in the south, and that the Chinese merchants have been skilful enough to dispose of them in a manner that did not disturb the equilibrium which will have brought them a vast fortune since they paid us only ten piastres for what is worth a hundred and twenty in Pequin. Be that as it may an otter skin is worth 30 roubles at St Peter and St Paul, a sable 3 or 4. The price of the fox furs cannot be determined without mentioning the black foxes which are too rare to be included and sell for more than a hundred roubles. The grey and blue ones range from twenty to two roubles, depending on whether they are closer to black or red. The latter do not differ from those of France by the colour, but really by the softness and thickness of their fur.

The English who, as a result of the sound constitution of the Company, leave to private traders in India all the development they can undertake, despatched a small vessel to Kamtchatka last year; sent by a firm in Bengal and commanded by Captain Peters[3] who had handed over to Colonel Coslof a letter in French which he allowed me to read, they requested in the name of the close links between the two European Crowns permission to trade in Kamtschatka by bringing in all the goods of India and China, cloth as

---

[1] The value of money depends on what one can buy with it. The *Discovery*'s sailors, finding nothing to spend their roubles on in Petropavlovsk, played football with them on the decks.

[2] Kiakhta on the Russo-Chinese boundary provided a valuable opening for trade between the two countries, but La Pérouse could not know at the time that the frontier had recently been closed and would not reopen until 1790.

[3] William Peters had been despatched by the East India Company to sound out possibilities for direct trade with Kamchatka, and sailed from Macao under a Portuguese flag. The English had learned that Russia was beginning to organise the rather haphazardly run fur trade and was contemplating setting up a trading company with monopoly rights rather similar to those enjoyed by the English company. Peters's shipwreck on Copper Island, which is Mednyj, east of Bering Island, and Russian apprehensiveness led to the abandonment of the plan. There were several attempts to open up trade with Kamchatka between 1782 and 1787, including those of William Bolts and James Trevenen.

well as sugar, tea, arrack &, and offered to take local furs in payment; Mr Coslof was too wise not to realise that such a proposition would be ruinous for Russia which sold these items very profitably to the Kamchadals and made even greater profits on the furs the English would export, but he also knew that sometimes permission was granted, with certain restrictions, to the detriment of the mother country and the particular benefit of a colony which later, when it has reached a point when it is no longer dependent on foreign trade, brings wealth to the motherland; be that as it may Mr Coslof was not prepared to make a decision on this matter and had allowed the English to send this request on to the Court at St Petersburg, but he believed that, even if it was successful, the country used too few goods from India and China and found for its furs an outlet on the Kiakia markets that was too advantageous for the Bengual merchants to be able to engage in any profitable speculation. Moreover this very ship that had brought his trading proposal had been wrecked on Copper Island a few days after sailing from Awatska Bay, and only two men had been saved, to whom I spoke and gave some clothing which they very badly needed. So the only ships which until now have safely landed in this part of Asia are those of Captain Cook and our own. I would owe our readers a few more specific details on Kamtchatka if the works of Mr Cox, and those of Steler &[1] left anything to be desired. The editor of Captain Cook's 3rd Voyage drew from these sources and retraced in an interesting manner everything that concerns this region on which much more has already been written than on several internal provinces of our own Europe, and which in respect of its products and its climate can and should be compared to the coast of Labrador in the neighbourhood of the Strait of Belle-Isle[2] but its fauna, like its people, is quite different; the Kamchadals gave me the impression of being the same people as those of the

---

[1] The works in question are those of William Coxe, *An Account of the Russian Discoveries between Asia and America* (London, 1780, 3rd edn 1787), and of Georg Wilhem Steller, a naturalist who sailed with Bering in 1741–2; Steller died in 1746, but his *Georg Wilhem Stellers Ausfürliche Beschreibung von Sonderbaren Meerthieren* was published in Halle in 1753, while in 1774 *Georg Wilhem Steller... Beschreibung von dem Lande Kamtschatka, deren Sitten, Nahmen, Lebensart und verschieden Gewönheiten* appeared in Frankfurt; Steller was also made known by Müller and Krasheninnikov.

[2] The Strait of Belle Isle separates northwestern Newfoundland from the Canadian mainland. La Pérouse became familiar with these waters during the American War of Independence.

CHAPTER XVIII

Baye de Castries on the Tartary coast:[1] their gentleness and their honesty are similar, and physically they also differ very little; thus they can no more be compared to the Eskimos than the sable can to the mink of Canada; Awatska Bay is certainly the finest, the most commodious and the safest one can come upon anywhere in the world: the entrance is narrow and ships would be forced to pass within the range of guns that might be set up there, the bottom is muddy and holds very well; two large harbours on the eastern and western shores could take all the ships of the French and English navies; the rivers of Awatska and Paratonka flow into this bay, their mouths are obstructed by shoals and one can only enter when the tide is high; the village of St Peter and St Paul is situated on a tongue of land which, like a man-made jetty, forms a small harbour behind the village, closed like a private home, where three or four vessels could be laid up and winter: the opening of this kind of basin is less than 25 *toises* wide and nature cannot offer anything that is safer and more convenient; it is on the shore of this basin that Mr Coslof plans to lay out a town which one day will be Kamtschatka's capital and maybe an important centre of trade with China, Japan, the Philippines and America; a superb soft water pond lies to the north of the projected city, and only three hundred *toises* away, and various little streams which it would be simple enough to bring together would procure all the requirements of a large settlement. Mr Coslof knew the value of all these advantages, but before anything is done, he repeated a hundred times, we need bread and workers, and we have very little of either. Nevertheless his orders had been issued to link up various ostrocs with that of St Peter and St Paul where he planned to have a church built in the near future. The Greek religion has been established among the Kamchadals without any persecution or violence and very easily; the priest from Paratonka is the son of a Kamchadal and a Russian; he says his prayers and his catechism in a good-natured way that appeals greatly to the natives; they pay for his attentions with offerings or alms, but do not pay him any tithe. The Greek rite allows priests to marry, from which one can conclude that the priests live more moral lives; I believe them to be very ignorant, and I cannot suppose that they could ever need any greater knowledge than they

---

[1] The natives of Kamchatka are of Siberian origin, predominatly Mongols. Other racial groups include the Koyaks and the Lamuts.

have; the priest's daughter, wife and sister were the best dancers and the most healthy of the Kamchadal women invited to Mr Coslof's ball. This good priest knew that we were very much Catholics, which earned us a good sprinkling of holy water and he also got us to kiss the cross that was carried by his clerk; all these ceremonies took place in the middle of the village because there is still no church; his presbytery was under a tent and his altar in the open air, but he normally lived at Paratonka and had come to St Peter only because of our presence.

He gave us various details on the Kurils whose pastor he also was, and where he went once a year. The Russians have found it more convenient to replace the old names of these islands, which authors do not agree over, with numbers; accordingly they say the 1st, the 2nd &c and up to the 21st which is the limit of the Russian claims; according to the priest's report this last one could be Marikan, but I am not entirely sure because the worthy priest was very verbose and yet we had an interpreter who could understand Russian as easily as French, but Mr Lesseps thought that the priest was bad at explaining things. Be that as it may, the following is what he was firm about and what one can more or less accept as fact: of the twenty-one islands belonging to the Russian crown four only are inhabited, the 1st, the 2nd, the 13th and the 14th.[1] The last two could be reckoned as forming only one because the inhabitants of the 13th spend the whole winter on the 14th and return to the 13th in summer; the others are completely uninhabited and the islanders merely land there in canoes to hunt the otter and the fox; several of these latter islands were no more than islets or large rocks and there is no timber on any of them, the currents are quite fierce between the channels and several are obstructed by rocks at water level. The priest has only ever travelled between Awatska and the Kurils in canoes the Russians call baidars,[2] and he told us that he had on

---

[1] There are in fact 36 islands in the chain. Krasheninnikov counted 22, including 'Matmai' which he described as large and belonging to the Japanese – it was probably Matsumae, northern Hokkaido. See his *History of Kamchatka*, pp. 61–3. The islands mentioned as being inhabited at the time of La Pérouse's visit were Urup, Iturup and Kunashir. The Paratunka priest had never ventured beyond the islands that were clearly under Russian influence, and there were undoubtedly other inhabited islands at the time about which he knew nothing.

[2] *Baidara* was a light craft used for river work. In the Kuril Islands and in Alaska the term was used for a craft made largely of animal skins which could take up to fifteen people. There was also an Eskimo or Aleut word, *baidarka*, which meant a kayak that could take at the most three people.

CHAPTER XVIII

several occasions almost been wrecked and especially nearly starved, having been driven out of sight of land, but he believed that his holy water and his stole had preserved him from danger. Altogether the population of the four inhabited islands adds up at most to fourteen hundred people; they are very hairy, wear long beards and live only on seals, fish and hunting; they have just been exempted from the tribute for a period of ten years because otters have become very scarce on these islands; in addition they are good, hospitable, and docile and have all converted to the Christian religion. The islanders further south who are independent sometimes cross in their canoes the straits separating them from the Russian Kurils to barter a few items from Japan against furs. These islands are part of Mr Coslof's administrative district, but as it is very difficult to land on them and they are of little interest to Russia he did not propose to visit them, and he dearly regretted having left a Russian map of these islands in Bolkerets, in which however he did not appear to put much faith; he showed so much trust in us that we would have liked to give him in turn information about our campaign. His extreme discretion in this matter deserves our utmost gratitude.

However, we did give him a short summary of our voyage, and let him know that we had rounded Cape Horn, visited the Northwest Coast of America, landed in China and the Philippines from where we had gone to Kamchatka; we did not allow ourselves to divulge other details, but I assured him that if the King ordered the narrative of our campaign to be published I would send him one of the first copies; I had already obtained his permission to send my journal to France by Mr Lesseps, our young Russian interpreter who was no longer of value to our expedition since we were about to sail in the southern hemisphere. My trust in Mr Coslof and the Russian governement would have not left me in any state of anxiety had I been forced to hand my despatches to the mails, but I felt that this was a good opportunity to enable Mr Lesseps to become acquainted with the various provinces of the Russian empire,[1] where it is likely he will one day replace his father, our

---

[1] Lesseps's travels in winter across the length of the Kamchatkan peninsula and along the coast to Okhotsk were an epic undertaking. He left with Koslov from Petropavlovsk on 7 October, stopped at Nachikin, where he carried out a chemical analysis of the hot springs, crossed the Bolschaiareka River on a locally constructed raft, and arrived a few days later at Bolscheretsk on the west coast of Kamchatka. In

consul-general at Petersburg. Mr Coslof kindly told me that he was taking him on as his aide-de-camp as far as Okots where he would make it easier for him to reach Petersburg, and that from now on he was part of his family; such gentle and friendly courtesy is easier felt than expressed, we regretted the days we had spent in Awatska Bay while he was at Bolkerets and the winter which was already warning us that it was time to think of leaving – the soil which we had found when we arrived on 7 July[1] to be a brilliant green was as yellowed and as parched-looking by the 25th of the same month as it is around Paris at the end of December; at the same date all the hills were snow-covered two hundred *toises* above sea-level. Mr de Lamanon had found the thermometer at nought before sunrise and he had seen a light crusting of ice along the edge of the pond behind the village. I gave instructions for everything to be readied for our departure and we set sail on the 29th. Mr Coslof came on board to farewell us and, as a calm had forced us to drop anchor in the middle of the bay, he stayed for dinner; I accompanied him back to land with Mr de l'Angle and several officers; he gave us another ball and an excellent supper, and the next day at dawn, the winds having shifted north, I gave the signal to weigh anchor. We had hardly let out our sails when we heard a salute from the entire artillery of St Peter and St Paul; I returned this salute. It had to be done all over again when we reached the narrows where Mr Coslof had despatched a small detachment to arrange another salute from the small battery situated north of the entrance light; we returned this second salute, and we could not leave Mr de Lesseps without a feeling of deep emotion: his great qualities had endeared him to us

---

late January 1788, he left with a caravan of 35 sleighs for Verkhne-Kamchatsk, struggling in the midst of almost constant snowstorms. He left Kozlov at Okhlann and went on to Jamsk where he arrived on 23 April, finally reaching Okhostk on 8 May. His hopes of faster travel in the spring had been dashed by the effect of the thaw which caused tracks to disappear into marshlands and river crossings into roaring torrents. After resting for a few weeks, he left Okhostk on 6 June and reached St Petersburg on 22 September just as the winter was setting in. He handed La Pérouse's despatches to the French Ambassador, the Comte de Ségur, and almost immediately set off across central Europe to Paris where he arrived on 17 October. La Pérouse's hope that the young man would one day replace his father as French consul in Russia was realised in 1802, but at the end of 1812 he was forced to return to France with the remnants of the French army in the great retreat from Moscow. He was never to return.

[1] This is a slip of the pen for 7 September. On 7 July La Pérouse was near Pic Lamanon, able to see, when the fog cleared, nothing but bleak rocks.

CHAPTER XVIII

and we were leaving him on a foreign land about to undertake a voyage that was as strenuous as it would be long. We took away from this country the happiest of memories and the certainty that nowhere else, and in no century, has hospitality reached greater heights.

CHAPTER XIX

Brief details on Kamtschatka. Landmarks for safe entry and departure from Awaska Bay. Very strong westerly winds defeat the plan I had made to explore and survey the Kurils as far as Marikan Island. We sight land birds and ducks in 38 degrees of latitude and 146° 30 of longitude. Sailor fallen overboard from the Astrolabe *whom we were unable to save &c. We cover a distance of three hundred leagues on the parallel of 37^d 30' seeking a land which some claim was found by the Spanish in this latitude in 1620. We come upon several signs of land but without sighting any land although the horizon was quite clear and we spent the night with greatly reduced canvas. The sudden transition from cold to heat affects our health, all the more so because we were totally without fresh food to maintain us in a good state of health, and we find fewer fish in the sea than during our crossing from Concepcion to the Sandwich Islands where we followed much the same route from south to north. We cross the Line for the third time since our departure, and soon find in the southern hemisphere north-westerly winds which cause the sea to become very rough and did not leave us until we were in the 12 degrees near the Navigators Islands which we saw on 6 December after passing over the position assigned to Biron's Dangerous Islands whose longitude must be wrongly determined. We set our course to pass in the strait formed by Opouna Island and Léoné. Numerous canoes come out to us and we are able to barter with them. We proceed on our route towards the island of Maouna fifteen leagues west of the first of the Navigators; description of this island where we anchor on 9 December; we obtain water and visit these people in their villages where we buy a very large quantity of fruit and pigs; practices, customs, skills and practices of these islanders. The swell is so bad in this anchorage that we are forced to weigh anchor on the 10th. We find that our cable had already been half cut away on the bottom. We spend the night of the tenth to eleventh on various tacks, and Mr de L'angle sets off on the 11th at midday with 4 armed longboats to fetch water in a bay he had seen the day before, he is murdered there with 11*

CHAPTER XIX

others, our two longboats being grounded. Detailed account of this event &c.

Russia does not owe its discoveries and settlements on the coasts of eastern Tartary and of the Kamchatkan peninsula to navigators. The Russians, as eager for furs as the Spanish were for gold and silver, have for many years undertaken the most lengthy and arduous of overland voyages to obtain the skins of sables, foxes and sea-otters, and being more soldiers than hunters it seemed more convenient to them to subjugate the natives and impose a tribute on them than to share the exertions of the hunt with them. They did not discover the Kamchatkan peninsula until the end of the last century[1] and their first expedition against the freedom of these unfortunate inhabitants dates from 1696. Russian authority was completely and fully recognised throughout the peninsula only in 1711. All the inhabitants then accepted the conditions of a fairly light tribute which is hardly enough to meet the sovereign's costs, three hundred sables, two hundred grey or red foxes, a few sea otter skins, these represent the Crown's income in this part of Asia where it maintains about four hundred soldiers, all Cossacks or Siberians, with a very small number of officers as commanders in the different districts.

The Russian Crown has changed the form of government of this peninsula on several occasions, the one the English found in force

[1] Fedor Alexeev left with Semen Dehnev from the mouth of the Kolyma River in 1648, passed through Bering Strait and reached the coast of Kamchatka. This discovery remained largely unknown until the eighteenth century, but the name Kamchatka appeared for the first time on a 1673 map. R.H. Fisher, *Bering's Voyages: Whither and Why*, Seattle, 1977, p. 28. Luka Morozko visited the peninsula in 1696, but the first real voyage of exploration was Vladimir Atlasov's in 1697-9. His report on the fur trade and the arrival in Moscow of a Japanese he had met in Kamchatka attracted the attention of Emperor Peter I who, however, lacked the resources needed for investigating such a distant part of the world. Atlasov was provided with a few guns and an escort and set off for Kamchatka in 1707, but he was murdered by his Cossacks in 1711. Contrary to La Pérouse's opinion, Russian penetration and colonisation attempts were marked by a series of wars, especially between 1714 and 1716, and a general uprising by Kamchadals in 1731. Okhotsk was the most secure of the scattered Russian outposts, and it was from there that the *Okhota*, the only seaworthy vessel the authorities disposed of, inaugurated a direct sea route to Bolsheretsk. In 1727 Bering crossed the peninsula from the west coast to the east to build the *St Gabriel* at Nizhne-Kamchatsk in which he set off in the spring of 1728 on his first voyage of exploration. As for Petropavlovsk, it was not founded until 1740.

in 1778 only lasted until 1784. At that time Kamchatka became a province of the Okost administration which is itself a dependency of the sovereign Court or tribunal of Ircuts.

The Bolkerets ostroc, formerly capital of Kamtschatka where Major Beem lived when the English arrived, is today simply under the command of a sergeant named Martinof. Mr Kaborof, a sub-lieutenant, commanded at St Peter and St Paul, Major Eleonof at Nisnei Kamtschatka or in the Lower Kamtschatka ostroc, and finally Verssessay or Higher Kamtschatka was under the orders of Sergeant Mamayef. These various commanders are not answerable to each other but each one directly to the governor of Okots who has appointed an officer as inspector with the rank of major to have particular charge of the Kamchadals and no doubt to protect them from possible vexations by the military government.

This preliminary sketch would give a very inadequate picture of the advantages which Russia obtains from its east Asian colonies if the reader did not know that navigations to the east of Kamtschatka towards the coasts of America followed voyages by land; those of Berings and Ticricow are known throughout Europe, but after these names, made famous by the renown gained by their expeditions and the misfortunes that ensued, there were others which added to Russia's dominions the Aleutian Islands, the groups further to the east known under the name of Onolaska and all the islands to the south of the Olaska peninsula. Finally Captain Cook's last campaign led to expeditions further to the east, but I learned in Kamtschatka that until now the natives have refused to pay the tribute or even to trade with the Russians who unfortunately were careless enough to let them know that they planned to subjugate them, and one knows how proud and jealous of their own freedom the Americans are.[1]

The Russian Crown spends nothing to extend its dominions; merchants order expeditions to be fitted out at Okots where at enormous cost they build small vessels 45 to 50 feet in length with

---

[1] When the French called at Petropavlovsk Russian expansion was in full swing. Russian control had been established over the Aleutian Islands with outposts on Unalaska, Kodiak and Afognak. The Aleuts who lived in small villages on islands that were some distance from each other and not easy for them to reach were unable to resist Russian colonisation. On the other hand, the coastal Indians in Alaska and on the large island of Kodiak lived on a terrain where resistance was much easier to organise and they put up a fierce opposition to the Russian advance.

CHAPTER XIX

a single mast in the middle roughly like our cutters and sloops, with a crew of 40 or 50 men. They are all better hunters than sailors, they sail from Okots in June, usually make their way out between L'opatka Point and the first of the Kurils, set an easterly course and travel to various islands for three or four years until they have bought from the natives or killed themselves a number of otters sufficient to cover the costs of the expedition and bring the owners a profit of at least one hundred per cent on their outlay.

Russia has still not set up any establishment east of Kamtschatka; each vessel builds one in the harbour where it winters and when it leaves it either destroys it or sells it to some other vessel belonging to its own nation. The Okots government is very careful to require the captains of these sloops to have Russian sovereignty recognised by all the islanders they may call on, and they place on board each vessel a kind of customs officer whose function it is to impose and collect a tribute on behalf of the Crown. I was told that a missionary was due to leave Okots shortly to preach the faith among these inhabitants and in this way settle with spiritual goods what the Russians owe for these tributes imposed simply by the power of might.[1]

It is known that furs are more profitably sold at Kiakia on the frontier between China and Russia but the extent of this trade has only been known in Europe since the publication of Mr Cox's book; the imports and exports total close on eighteen million *livres*[2]

---

[1] La Pérouse appears to have learned of the plans of Grigorii Shelikov, a merchant who hoped to obtain a charter for a trade and colonisation company and had offered as an incentive to help the Church set up a mission in the newly annexed territories. Success finally came to him in 1794, but the compensation offered by means of 'spiritual goods' was greatly delayed. Missionaries found life very hard and the help promised by Shelikov long delayed and quite inadequate. Material gains came relatively quickly, but spiritual benefits did not flow until 1820–30 See H.A. Shenitz, 'Father Veniaminov, the Enlightener of Alaska', *American Slavic and East European Review* (February 1959) and 'The Life and Works of the Most Reverend Metropolitan Innocent', *Russian Orthodox Journal* (July–October 1940).

[2] The *livre* should not be confused with the English pound sterling. The rate of exchange fluctuated considerably in the eighteenth century. The disastrous Seven Years War saw the *livre* sink to less than a shilling or five modern pence; the French economy recovered after that, but although France was able to give substantial help to the colonists in the American War of Independence, the drain on her treasury was severe and at the time of La Pérouse's voyage financial crises followed each other in rapid succession and the *livre* was scarcely worth four modern pence. However, La Pérouse is giving an approximate figure based on information given to him by the Russians as well as on Coxe's book.

a year. I was assured that twenty-five ships with crews of a thousand men, both Kamchadals and Russians or Cossacks, were involved in this navigation to the east of Kamtschatka. Being spread out from Cook River[1] as far as Bering Island, a long experience has taught them that otters hardly venture further north than latitude 60 degrees, which leads all the expeditions towards the region of the Alaska peninsula or further east, but never to Bering's Strait constantly obstructed by ice that never melts.

These vessels sometimes put in at Awatska Bay, but they always return to Okots where their owners live as do the traders who go direct to trade with the Chinese on the frontier of the two empires. As ice allows them to enter Awatska Bay at all times, Russian navigators put in there when the season is too far advanced to reach Okots before the end of September, a very wise order from the Russian empress having forbidden navigation in the Sea of Okots after this time of year when begin the storms and gales which have been the cause of very frequent shipwrecks.

Ice never extends in Awatska Bay beyond three or four hundred *toises* from the shore, and it often happens during winter that the land breeze clears the floes which block the rivers of Paratonka and Awatska and navigation then becomes possible.[2] Since the winter is generally milder in Kamchatka than at Petersburg or in several provinces of the Russian empire, the Russians talk of it as we do of Provence, but the snow that surrounded us as early as the twentieth of September, the white frost covering the ground every morning at the same period, and the greenery that was as faded as in the month of January around Paris, all this told us that the winter was of a harshness the people of southern Europe could not support.

Yet, in many respects, we were less sensitive to the cold than the inhabitants of the ostroc of St Peter and St Paul, both Russians and Kamchadals. They wore the thickest furs, and the temperature inside their isbas where they keep their stoves continually lit was 28 to 30 degrees above freezing point; we could not breathe in such hot air, and the lieutenant was careful to open his windows each time we went into his apartment; but all these peoples have become

---

[1] The Cook River is Cook Inlet, a deep sound behind Kodiak Island, between the Kenai and the Alaskan peninsulas. Bering Island itself is east of Kamchatka.

[2] Petropavlovsk is normally closed or obstructed by ice between November and May; Okhotsk cannot be used in winter; only Vladivostok, much further south, can be used throughout the year.

CHAPTER XIX

accustomed to extremes; it is known that both in Europe and in Asia they take steam baths in sweating-rooms from which they emerge covered in perspiration and then go and roll in the snow. The ostroc of St Peters had two of these public baths which I went into before the fire was lit; they consist of a very low room in the middle of which stands an oven, made of dry stones, they light up like those used for baking bread; the vault of this oven is in the middle of the room and surrounded by tiers of benches for those intending to bathe, the heat is greater or lesser according to whether one is on the tier above or below the vault over which water is thrown when it has become reddened by the heat of the fire below it. This water rises up immediately as steam and causes the most abundant perspiration. The Kamchadals have copied this practice from their conquerors as they have many others and within a few years the primitive nature which distinguished them so markedly from the Russians will have been totally eliminated; today their number does not exceed four thousand souls in the whole of the peninsula although it stretches from the 51st degrees to the 63rd across a width of several degrees of longitude, so one can see that there are several square leagues per inhabitant. They cultivate none of the products of the soil and their preference for dogs over reindeer to pull their sleighs prevents them from raising pigs, sheep, young reindeer or calves because they would be devoured before they had acquired enough strength to defend themselves. Fish is the basic food of these carnivorous coursers that nevertheless cover 24 leagues in one day and get fed only when their run is over.

The reader has already noted that their method of travel is not peculiar to the Kamchadals alone and that the people of Chouta and the Tartars of Castries Bay have the same kind of teams. We were very keen to know whether the Russians had any knowledge of these various countries, and we learned from Mr Coslof himself that vessels from Okots have several times sighted the northern head of the island at the mouth of the Amur River, but they had never landed there because it lies beyond the limits of the Russian empire's settlements on this coast which end at Udskoy Ostroc,[1]

---

[1] The Treaty of Nerchinsk of 1689 had fixed the Sino-Russian border at the Amur River and the Stanovoii Mountains. This compelled the Russians to develop a route from Yakutsk to Okotsk through many hundreds of miles of marshlands and mountain country. The effective area of Russian control extended in the south as far as Santarkije Island and Udskaja Gulf (with the inland post of Udskoje or Udskoy)

and as far as the bear islands where sable hunting is carried out extensively. The bay at Awatska is very much like the one at Brest but is very much better on account of the quality of the bottom which is mud. Its entrance is narrower and consequently easier to defend; our lithologists and our botanists found along its shores only substances or plants that are extremely common in Europe and it seems that Steler and Kraschenninikof[1] have left nothing new to be described by those who have come after them; the English too gave a very good plan of this bay in which one must be careful of two shoals to the east and the west of the entrance which leave a large channel between them for ships to enter; one is certain of avoiding them if one leaves two isolated rocks that are on the east coast open on the light signal side but leaving on the other hand, hidden by the western coast, a large rock on the port side which is less than a cable length from land. All the anchorages in this bay are equally good and one can come closer or further from the ostroc depending on one's wish to communicate often with the village.

According to Mr D'agelet's observations, Lieutenant Kaborof's house at the back of the harbour of St Peter and St Paul lies in 53^d 1 minute of latitude and 156^d 30′ of longitude Paris meridian: the tide which was very regular was high at the new and the full moon at 3.30 o'clock, its height in the harbour being 4 feet. Our regulator No. 19 was observed to be losing 10 seconds a day, which differed by 2 seconds from the daily loss observed at Cavitte six months earlier.

The northerly winds which were so much in our favour when

---

and in the north to the shores of the East Siberian Sea. The Bear Islands lie north of the Kolyma estuary and were discovered in 1740 by Dmitri Laptev. The Russians had little reason to attempt to sail in the southern part of the Sea of Okhotsk since both China and Japan were closed to foreigners, and the furs which formed the basis of their Far Eastern trade came from the north and north-east. On Russia's advance towards the east, logistics problems presented by distance and climate, and Chinese opposition, see R. A. Pierce, *Eastward to Empire: Exploration and Conquest on the Russian Open Frontier to 1750*, Montreal, 1973; and J. R. Gibson, *Feeding the Russian Fur Trade: Provisionment of the Okhotsk Seaboard and the Kamchatka Peninsula 1639–1856*, Madison, 1969.

[1] Georg Wilhem Steller, the naturalist of Bering's expedition, was an able geologist. His narrative has now been translated into English by M. A. Engel: *Journal of a Voyage with Bering 1741–1742*, Stanford, 1988. Stepan Petrovich Krashenninikov was the author of *Opisanie Zemli Kamchatski*, published in St Petersburg in 1755 and available in English: E.A.P. Crownhart-Vaughan, *Explorations of Kamchatka 1735–1741*, Portland, 1972.

CHAPTER XIX

we left the bay forsook us two leagues from the shore; they settled in the west with a stubborness and a strength which did not allow me to carry out the plan I had made when I left of surveying the Kurils as far as the island of Marikan. Squalls and storm followed each other so quickly that I was compelled to heave to with a storm-sail and I was driven 80 leagues from the coast. I did not attempt to struggle against these obstacles because this survey was of little importance, and I set my course to cut the 165th degree of longitude in latitude 37^d $30'$ where some geographers have situated a great island, rich and with a large population, discovered they say in 1620 by the Spanish; a search for this land was included in Captain Vries's instructions, and one finds a memoir with a few details on this supposed island in the fourth volume of the *Collection academique Partie Etrangere* page 158.[1]

I felt that, among all the various searches that were suggested to me rather than insisted upon, this one warranted being given preference. I did not reach the parallel of 37^d 30 until midnight on the 14th; during that same day I had seen five or six small birds of the linnet species[2] that came and perched in our rigging and the same

---

[1] The *Collection académique, composée de mémoires, actes ou journaux des plus célèbres académies et sociétés littéraires étrangères* was edited by Jean Berryat and published in 13 volumes at Dijon between 1755 and 1759. The island of Rica de Oro or Rica del Plata, sometimes referred to as two islands, appeared on maps of the northern Pacific for many years. Its origin goes back to Francisco Gali who while in Macao in 1584 was told of a Portuguese vessel bound for Japan which had been driven by a storm towards an island lying far to the east, inhabited by a wealthy and friendly white race. Pedro de Unamuno sailed in search of it in 1587, without success; and Sebastián Vizcaino who sailed in 1611 from Acapulco was no luckier; but the Portuguese Joao de Gama, who left Macao in 1590 for Acapulco, claimed he had seen land to the north-east of Japan – given the erroneous longitudes that resulted from dead reckoning and guesswork in early days, this may have been one of the Kurils, unless it was simply a bank of clouds. The Dutch sent out Mattijs Quast and Abel Tasman as well as Maarten Vries in the hope of finding these mysterious islands, with the usual lack of results. The Spanish decided not to waste any more time or money on such expeditions, Philip V in particular rejecting a 1741 request by the governor of the Philippines, on the grounds that 'nothing justified this search'. However, Milet-Mureau, the editor of the printed narrative of La Pérouse's voyage, added a lengthy marginal note expressing regret at the navigator's lack of faith. In fact, the name survived into this century: Bartholomew's *Advanced Atlas of Physical and Political Geography* of 1917 has it and the *Times Survey Atlas* of 1922 shows a 'Roca de Plata' on map No. 103. See on these islands: W.L. Schurtz, *The Manila Galleon*, New York, 1959, pp. 237–8; O.H.K. Spate, *The Spanish Lake*, Canberra, 1979, pp 106–8, 110, 112; E. Chassigneux, 'Rica de Oro et Rica del Plata', *Toung Pao*, XXX (1933), pp. 37–84.

[2] Not easy to identify, this bird may be the *Carduelis spinus* of Japan.

evening we saw two flights of ducks or cormorants which never go far from the shore;[1] the sky was very clear, and on each frigate look-outs were constantly on duty in the topmasts; a fairly substantial reward had been promised to the first man to sight land and this cause for rivalry was hardly necessary – each sailor was competing for the honour of being the first discoverer of a land that would bear his name, but in spite of the clearest indications of a nearby land we discovered nothing, even though the horizon stretched out as far as was possible; I assumed that this island was in the south and that the recent fierce winds that had blown from that direction had presumably driven north the small birds we had seen resting in our rigging; following this assumption I sailed south until midnight when, being exactly in 37^d 30', I gave orders to steer east with very little canvas, waiting for daylight with the greatest impatience. Daylight came and we saw two more small birds. I continued east, a large turtle passed alongside that same evening; still following the same parallel in an easterly direction, we saw the next day a bird smaller than a French wren[2] perched on the main topsail yard and a third flight of ducks, and so each instant fed our hopes but we never had the joy of seeing them realised. During this time we experienced a misfortune that was all too real; a sailor from the *Astrolabe* fell into the sea while furling the fore topgallant and whether he was wounded by his fall or could not swim he did not reappear, and all the endeavours we made to save him were unsuccessful.[3]

Indications of the presence of some land continued on the 18th and 19th, even though we had made considerable progress towards the east; we daily saw flights of ducks or other shore birds; a soldier even claimed he had seen a few pieces of seaweed go past, but since

---

[1] The uncertainty revealed by the reference to 'ducks or cormorants' suggests a flight of small cormorants, possibly the *Phalacrocorax pelagicus* which is quite common in the northern Pacific and among the Kurils. At the time, La Pérouse was approximately in 37°30'N and 162°40'E. On the other hand, the uril cormorant, *Phalacrocorax urile*, is common among the Aleutian Islands. The point to be noted is that there was no land in the near vicinity, and that the presence of such birds is not a satisfactory indication of land: the ducks or cormorants could have been migrating or, more likely and as La Pérouse subsequently surmises, were victims of the recent storms who had been blown away from their usual haunts, whether the Kurils or the Aleutians.

[2] This is insufficient information to identify this orphan with any finality. It might have been an Unalaska wren, *Troglodytes troglodytes petrophilus*, but La Pérouse later expresses the opinion that these birds came from the south.

[3] The sailor was Gilles Henry from St Brieuc, Brittany. The date was 15 October 1787.

CHAPTER XIX

this was only seen by one person and since it was not likely that only one packet of weed would have broken away from a nearby island, we all rejected this soldier's claim while nevertheless retaining the thought that it was very likely that there was some land close by; but hardly had we reached the 175 degrees of longitude east of Paris when all indications of nearby land disappeared; however I kept to the same route until midday on the 22nd when observations based on No. 19 showed us to be 20 minutes beyond 180 degrees, the limits I had been given for the search for this supposed land. I set course for the south in order to find calmer seas, as since leaving Kamtschatka we had continually sailed in the midst of the wildest swell; a sudden wave had even carried away our small boat which was tied up in the gangway, and had thrown on board more than a hundred barrels of water: these minor problems would hardly have been noticed if events had turned out better and we had finally come upon the island we were wearing ourselves out to discover and which surely exists in the neighbourhood of the route we followed – indications of land were too frequent and too plain for us to doubt it.[1] But I am tempted to believe that we followed too northerly a parallel, and if I had to undertake this navigation again I would sail along the 35th degrees of latitude from the 160th to the 170 degrees. This is the area where we saw most land birds, they seemed to be coming from the south and the violent gales coming from this direction had presumably driven them towards the north; the later plan of my campaign did not allow me to verify this theory by continuing toward the west the same voyage we had made towards the east. The almost constant westerly winds would not have enabled me to cover in two months the distance I had travelled in a week. I turned my attention towards the southern hemisphere in that vast field of discovery where the routes of modern navigators are continually crossing the tracks of Quiros, Mendana, Tasman &c where each one had added some island to those that were already known and where the ancient information leaves the reader quite unsatisfied. It is known that there is in this vast area of the Southern Ocean a zone of some twelve to fifteen

---

[1] The indications of land were more tenuous than he thought. Currents and storms can carry seaweed considerable distances and birds blown away from their home territories can survive at sea for days. The nearest land to the south of the French was Midway, which was not discovered until the next century, in 18°13′N and 177°22′W.

379

degrees from north to south and 140 degrees east to west filled with islands, which in a way are the globe's equivalent of the Milky Way in the heavens. The language and customs of the various islanders who live there are no longer unknown to us, and the observations made by the most recent travellers even allow us to hazard guesses on the origin of these people which can be linked to that of the Malays[1] just as the origin of the various colonies of the Spanish and African coasts can be to the Phoenicians. This was the archipelago through which my instructions required me to sail during the third year of our campaign.

The southern part of New Caledonia, discovered during Captain Cook's second voyage, the south coast of the land of the Orsacides of Surville,[2] the part of the Louisiade which Mr de Bougainville did not have the opportunity to explore but where he had been the first to sail along the south-eastern coast,[3] these various geographical questions had mainly attracted the King's attention, and I was required to determine their limits and their precise latitudes and longitudes. The Society Islands, the Friendly Islands, the Hebrides &c no longer held any interest for Europe, but they offered the chance of obtaining food, and I was allowed to call there according

---

[1] La Pérouse is referring to the origin of the Polynesian people which has long been argued over by navigators and ethnologists. A wide range of theories have been put forward: a lost continent now sunk below the waves with the exception of a scattering of islands, American Indians migrating west from island to island, migrants or the descendants of explorers sent from as far as Egypt, and so on. South-Asiatic or Malay elements can be identified in the Polynesian race, but nothing really points to the migration of a cohesive group, whether gradual or not, as against the emergence of a distinct culture from a new racial mix in a discrete island environment. The most recent summary of current theories, with its own conclusions, will be found in K.R. Howe, *Where the Waves Fall*, Sydney, 1984.

[2] In October 1769 Jean-François-Marie de Surville (1717-70) discovered the northern coast of the Solomon Islands from Choiseul to San Cristobal. He was attacked after putting into Port Praslin in western Santa Ysabel, suffering several casualties among his crew, as a consequence of which he named his discovery 'Land of the Arsacides' after the Moslem Hashashin of the Levant, members of an Ismaili sect founded by Hasan ibn al-Sabbah, known as the Old Man of the Mountains, who established himself in a fortress at Alamut in 1090; members of the sect were sent out to murder Christian crusaders until the capture of Alamut by Mongol troops in 1256. J. Dunmore (ed.), *The Expedition of the St Jean-Baptiste to the Pacific 1769–1770*, Cambridge, 1981, pp. 105–6.

[3] In June 1768 Louis-Antoine de Bougainville (1729–1811) sailed along the southern coast of a line of islands east of New Guinea, which he named the Louisiades archipelago; he then sailed north towards the Solomon Islands. Taillemite (ed.), *Bougainville*, I, pp. 362–4.

CHAPTER XIX

to my needs, and it had been correctly surmised that when I left Kamtschatka I would have very little left in the way of the supplies that are nevertheless essential for the preservation of the health of sailors.

I was not able to progress south rapidly enough to avoid a gale which blew up in that area. On 23 October the sea was very rough and we were compelled to spend the night hove to with a storm-sail. The winds were very variable and the sea very disturbed until we reached the parallel of 30 degrees on 29 October, and generally everyone's health was somewhat affected by too rapid a transition from the cold to the greatest heat, but we all merely suffered slight discomforts which required none of us to keep to our beds.

On 1 November being in 26^d 30' of latitude and 185 of western longitude Paris meridian, we saw numerous birds, including curlews and plovers, species which never go far from land.[1] The weather was overcast and squally, but all parts of the horizon cleared gradually except for the south where heavy clouds remained constant which led to believe that some land might lie in that point of the compass. I sailed in this direction on the 2, 3 and 4th. We kept on seeing birds; the indications of land gradually ceased but it is quite likely that we passed fairly close to some island or shoals which we did not sight, and which chance may perhaps one day offer to some navigator. We then began to enjoy a clear sky and we were at last able to obtain lunar readings, observations we had been unable to make since our departure from Kamtschatka, they differed from those of our timekeeper No. 19 by one degree west. We also caught a few dorado[2] and two sharks that provided delicious meals for us because we were all reduced to salted pork which was beginning to feel the effects of a burning climate. We carried out the same distance observations on the following days and the difference remained the same; we had at last reached the tropics, the sky was becoming finer and our horizon stretched for a considerable distance; we saw no land but every day we sighted

---

[1] La Pérouse had now reached the neigbourhood of Midway Island. Curlews are not uncommon in these waters, particularly *Numenius tahitiensis* Gmelin and *N. madagascariensis* Linn. Among plovers we can choose between the *Pluvialis dominica fulva* and several members of the *Charadridae* family, such as the *Charadrius dubius papuanus*. These birds do not normally go far from land, but at certain periods they migrate long distances, to Siberia or Alaska.

[2] The dorado is the *Coryphaena hippuris*, closely related to the dolphins.

birds one never meets far from the shore; on 4 November in 23ᵈ 38′ of latitude and 184 of longitude (according to a series of distances observed the same day)¹ we caught on board a golden plover² which was still fairly plump and could not have strayed for very long across the sea. On the 5th we cut the track of our crossing from Monterey to Macao, on the 6th that of Captain Clark from the Sandwich Islands to Kamtschatka, the birds had totally disappeared and we were very strained by a heavy easterly swell which, like the western one in the Atlantic Ocean, dominates this vast sea in which we saw neither bonito nor dorado and merely a few flying fish, our supplies of fresh food were totally exhausted and we had counted rather too much on finding fish to soften the austerity of our diet. On the 9th we passed over the southern point of the Villa-lobos Shoal³ according to the position shown for it on the chart given me by Mr de Fleurieu; I set our sails so that I would pass its latitude during daylight, but as we saw neither birds nor seaweed I am led to believe that if this reef exists it must be allocated a more western position, the Spanish having always placed their discoveries in the great ocean too close to the American coasts; the heavy seas calmed down a little at this period and the winds moderated but the sky filled with thick clouds and we had hardly reached the latitude of 10 degrees when we were assailed by

---

¹ The expedition was now sailing west of French Frigate Shoals and Necker Island in 23°45′ and 23°35′ respectively.

² This is the Pacific Golden Plover, *Pluvialis dominica fulva*. It nests in the northern hemisphere and spends the summer months in the South Pacific.

³ In November 1542, Ruy Lopéz de Villalobos (?–1546) set out with six ships from Navidad, Mexico, for the Moluccas. He sailed towards the Revilla Gigedo group, discovered a number of low islands on 25 December, and on 23 January an island he named Los Matelotes – which is very probably Fais in the Carolines. He reached Mindanao on 2 February 1543 and in August of that year he sent Bernardo de la Torre in the *San Juan de Letran* to inform the Viceroy of Mexico of his discoveries. It was Torre who came upon a very low and dangerous island in roughly 16°N which he called Abreojos, meaning 'Open your eyes' – it was probably one of the Marianas. La Pérouse crossed this latitude on 8–9 November in longitude 180°, which is approximately where it appeared on many charts of the day, including Robert de Vaugondy's map drawn for Charles de Brosses's *Histoire des navigations aux terres australes*. Spanish longitudes were however very unreliable and cartographers did not help matters by placing doubtful island discoveries rather at random, or indeed, on occasion, where they looked best on their map. If, as is now generally accepted, Abreojos is Farallon de Medinilla which lies in longitude 146°04′E, La Pérouse's comment that it should be allocated a much more westerly position on the charts is quite correct. See A. Sharp, *the Discovery of the Pacific Islands*, Oxford, 1960, pp. 29–32; *Who's Who*, pp. 247, 259–60.

CHAPTER XIX

heavy rain but only during the day,[1] as the nights were very much better. The heat was stifling and the hygrometer[2] has never recorded a higher level of humidity since our departure from Europe; we were breathing an air that was slack and which, combined with the bad food, weakened our strength and would have almost made us unable to do any strenuous work if circumstances had required it of us; I doubled my attention to the crews' health during this period of crisis, the transition from cold to hot and damp had been too quick. I had coffee given out daily at lunch; I had the below-decks dried out and aired; rainwater was used to wash the sailors' shirts and in this way we turned to good use the rigours of the climate we were being forced to sail through, and whose effect I feared more than that of the high latitudes where we had already travelled. Nevertheless on 6 November[3] we caught for the first time eight bonito which provided a good meal for the entire crew and for the officers who like me had no other food than what we had in our hold. These rains, these storms and the heavy seas came to an end on the 15th when we reached the 5th degrees of latitude, we then enjoyed the clearest of skies and sailed across the calmest of seas; a very wide horizon at sunset reassured us about our run during the night which furthermore was so clear that we would have seen dangers almost as well as in full daylight. The fine weather stayed with us beyond the Line which we crossed on 21 November for the third time since we left Brest. On each occasion we had gone 60 degrees north and south of it and the later plan of our voyage was to bring us back into the northern hemisphere only in the Atlantic Sea when we were on our way back to Europe. Nothing broke the monotony of this long crossing. We had followed a route that was more or less parallel to the one we had taken the previous year going from Easter Island to the Sandwich Islands, where we had been continually surrounded by birds and bonito which had provided us with an abundant supply of healthy food; but on the contrary a vast solitude surrounded us and the air and the

---

[1] Bad weather prevented accurate observations of the height of the sun which were normally taken at midday. The area of ocean through which the ships were sailing at the time is quite empty.
[2] The hygrometer was used to calculate the humidity of the air. The term hygroscope is now used.
[3] This date is a slip of the pen. On 6 November, La Pérouse had recorded that there were no bonito or dorado to be seen.

waters in this part of the globe were uninhabited. However on the 23rd we caught two sharks which provided two good meals for the crews and the same day we killed a very thin curlew which seemed very tired, we thought he might have been driven off from York Island[1] which lay about a hundred leagues from us; he was eaten at my table in a salmi and tasted hardly better than the sharks; as we advanced into the southern hemisphere, boobies, frigate birds, sea-swallows and tropic birds flew around our vessels and we saw them as advance signs of some island which we were most impatient to come upon; we were complaining about Fate which had made us travel over such a long distance since leaving Kamtschatka without making the slightest discovery.

The birds who grew to a great number once we reached the fourth degrees of latitude south continually raised our hopes of finding some land, but although the horizon was very extensive nothing appeared. To be frank we were not making much progress; the winds dropped when we were in the second degrees of latitude south and they were followed by very light airs from the north to the W.N.W. which enabled us to gain a little to the east because I was afraid we might be carried to the leeward of the Friendly Islands; during this period of calms we caught a few sharks which we preferred to the salt meats and we killed some seabirds which we ate in salmis and although they were very thin and they both tasted and smelled of fish to an unbearable extent they seemed to us (in our state of severe scarcity) almost as good as woodcocks: sea-swallows, black or totally white,[2] are peculiar to the South Seas and I have never seen any in the Atlantic Ocean. We shot many more of them than we did boobies and frigate birds, they flew in such numbers around our ships, especially during the night, that we were deafened by their noise and it was difficult to hold a conversation

---

[1] The reference to York Island can create confusion. This was the name Samuel Wallis had given to the island of Moorea in 1767. As it is situated in 17°32S and 149°50'W, La Pérouse who at the time was in 2°47'S and 173°30'W was nowhere near it. This is really a reference to Atafu, the northernmost of the Tokelau group, in 8°33'N and 173°50'W, discovered by John Byron on 24 June 1765 and named by him Duke of York's Island. La Pérouse was closer to the Phoenix group, small islands uninhabited except for birds, that were still waiting to be discovered: La Pérouse missed them and they had to wait another half-century before making their appearance on the map.

[2] The white birds are probably *Gygis alba pacifica*; the black one could be *sterna fuscata*, a noisy seabird; but neither is particularly widespread in this part of the ocean.

CHAPTER XIX

on the upper deck, so that our fairly successful hunts provided us with some revenge for their screeching as well as a bearable meal, but they disappeared when we passed the sixth degree. The northwest to west winds, which had started around the 3rd degree of southern latitude, but very light and cloudless, then rose fiercely and did not cease until the 12th degrees; a strong westerly swell made our navigation very tiring, our ropes rotten by the damp on the Tartary coast kept on breaking and we did not replace them until the last moment in case we ran short of them; squalls, storms and rain stayed with us constantly until 10 degree 50 minutes where we arrived on second December. The winds, still blowing from the west, moderated and were no longer accompanied by rain; we carried out observations of distances which corrected the errors of our timekeepers; since our departure from Kamtschatka they seemed to have lost five minutes in time or placed us one degree fifteen minutes too far east; according to the longitudes we obtained by lunar readings which gave 189 degrees 53 minutes east of Paris we passed exactly over the position given to Biron's Dangerous Islands, as we were in their latitude, and since we did not sight any land, or the smallest indication of any proximity of it, it is obvious that a different longitude should be assigned to these islands, Commodore Biron having sailed merely with the faulty method of dead reckoning;[1] the next day 3 December we were in 11 degrees 33 minutes of latitude and 189 of longitude, according to our distances, exactly on the parallel of Quiros's Island of the Beautiful Nation[2] and one degree to the east; I would dearly have

---

[1] John Byron in the *Dolphin* discovered a group of small islands on 21 June 1765 which he called Islands of Danger; this is Pukapuka, which consists of three small islands, rocks and sandbanks, in latitude 10°53S and longitude 168°09'W. They probably had been seen by Alvaro de Mendaña in 1595 and named by him San Bernardo. La Pérouse had therefore cut the route of both these navigators during the evening of 2 December, but he was to the west of Pukapuka. Low lying and uninhabited, the little group was not easy to see, nor of much use to a navigator.

[2] Gente Hermosa, elegantly translated as *Isle de la Belle Nation*, had not been easy to rediscover since Quiros first came upon it in February 1606. Juan de Torquemada called it thus in his 1615 *Primera (Segunda, Tercera) parte de los veynte y uno libros Rituales y Monarchia* (Seville, 1615, 2nd edn, Madrid, 1723), but Quiros's name for it was La Peregrina – the Pilgrim, which in retrospect seems appropriate in view of the various longitudes assigned to it on different maps. Most historians identify it as Rakahanga, a low island which forms part of the Cooks, in latitude 10°02'S and longitude 161°06'W, further north than La Pérouse's reported position at the time, but the charts he had available to him showed Gente Hermosa closer to the eleventh than the tenth degree of latitude.

liked to run a few degrees west to come upon it but the winds were blowing straight from that direction. This island was too vaguely shown to look for it with such contrary winds, and I thought I should take advantage of these westerlies to reach the parallel of Bougainville's Navigators Islands[1] which is a real French discovery and where we hoped to obtain some refreshments of which our health was beginning to be in great need; we sighted the easternmost on sixth December at three o'clock in the afternoon.[2] We set course to approach it until eleven at night and stayed on short tacks for the rest of the night; since I planned to put in there if I could find an anchorage I passed through the channel that separates the large island and the small one which Mr de Bougainville[3] had left to the south – it is narrow and hardly a league wide but it seemed safe without any dangers. We were in the pass at midday and we observed $14^d\ 7'$ one mile from the coast. At that moment the south point of the island bore from us South 36 West, thus the southern head of this island lies in 14 degrees 8 minutes and its longitude 189 (according to our timekeepers corrected by the difference we found on 5 December between the results of our No. 19 and the lunar distances).

We only saw canoes once we were in the channel; when we were to windward of the island we had seen some houses and a fairly large group of Indians seated in a circle under some coconut trees who seemed to be enjoying quite calmly the spectacle we offered them.[4] They did not put any canoes in the water and did not follow us along the shore; this land, some two hundred *toises* in height, was very steep and covered with trees up to the top, among which we could make out a large number of coconut trees, the houses

---

[1] The Navigators of Bougainville are the Samoan Islands, originally named the Petites Cyclades by the French navigator. He sailed to the north of Rose Island and Tau and south of Tutuila, three islands actually discovered by Jacob Roggeveen in June 1722. Bougainville eventually renamed the group the Navigators Islands 'warranted on account of a dozen canoes with large sails'. Taillemite, *Bougainville*, I, pp. 334-40 and II, p. 250. The name is truly French, but the discovery was Dutch – posterity has settled the argument by keeping the Polynesian name.

[2] La Pérouse's longitude was 168°40'. Tau's longitude is 169°30'. The latitude is 14°15'.

[3] The largest of the three is Tau, the smaller ones are Olosega and Ofu which in fact, being close to each other, often give the impression of forming only one. The strait is just over 6 miles wide.

[4] The southern point of Tau is Siufa'alele Point. There are several villages on the west coast.

CHAPTER XIX

were approximately half-way up where the islanders can breathe cooler air. One could see nearby a few small cleared spaces which presumably were planted with sweet potatoes or yams but overall this island did not seem very productive and anywhere else in the South Seas I would have believed it to be uninhabited; but my error would have been all the greater in that even the two small islands forming the western side of the channel through which we sailed also had inhabitants,[1] we saw five canoes leave them which joined eleven others coming from the eastern island, which after turning round our two ships several times with an appearance of deep suspicion finally decided to risk coming alongside and carry out a little barter with us, but so little that all we obtained was some twenty coconuts and two blue sultana hens.[2] They were like all the South Sea islanders untrustworthy in their trade and when they received payment in advance for their coconuts it was unusual for them not to row away without handing over the agreed compensation. In truth these thefts were of very minor importance and a few bead necklaces with small pieces of red cloth were hardly worth complaining about. We took several soundings in the channel, a hundred-fathom sounding-line did not reach the bottom less than a mile from the shore. We continued on our way so as to round a headland behind which we thought we could find a shelter, but the island was not of the width indicated on Mr de Bougainville's chart, on the contrary it ended in a point and its greatest diameter is at most one league. Thus we found the easterly breeze blowing along this coast which was bristling with reefs and it was obvious that we would look in vain along there for an anchorage; then we set a course outside the channel to haul along the two western islands which together are about the size of the easternmost; a channel of less than a hundred *toises* separates them, and at the end one can see an islet which I would have called a large rock had it not been tree-covered,[3] but before rounding the two southern points of the channel we were becalmed, with a fairly heavy swell which caused me some concern lest we struck the *Astrolabe*, fortunately a

---

[1] One cannot venture a guess at what the population might have been at the time of La Pérouse's visit, but today there are approximately 3,000 inhabitants on the three islands that form this group, known as Manua.

[2] This would seem to be the Samoan water-hen, *Porphirio porphirio samoensis*, whose local name is *Manu'alii*.

[3] The islet is Nuu, very close inshore to Ofu.

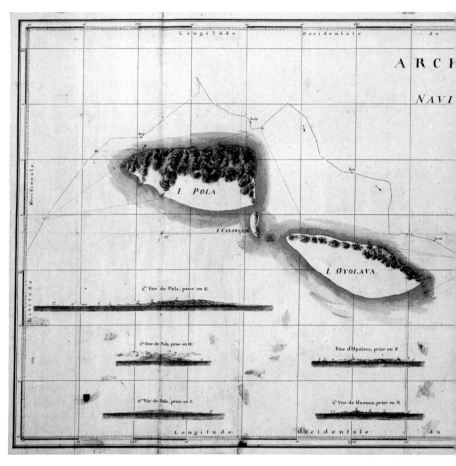

23. Navigators archipelago (Samoas). Unsigned. AN 6 JJ1:45.

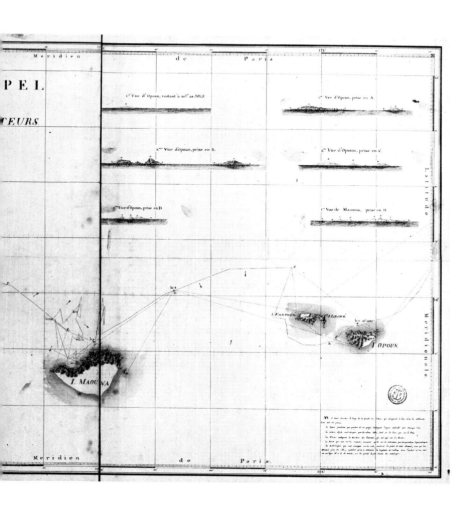

few light airs soon led us out of this unpleasant situation which had made us take little notice of an elderly Indian's harangue, who was holding a branch of kava[1] in his hand and making a fairly lengthy speech. We knew through having read several accounts of voyages that this was a sign of peace, and throwing him a few pieces of cloth we answered him with the word Tayo which means friend in the idiom of several nations of the South Seas,[2] but we were not practised enough to be able to distinguish the words of the vocabularies we had extracted from Cook's Voyages.

When at last we reached the breeze we let out sails in order to get away from the coast and from the band of calms. All the canoes then came up to us; they progress fairly well with a sail but very poorly with paddles; these craft would not be of any use to people who were not such good swimmers as these were, they topple over at any moment and this happening surprises them and worries them less than a hat falling off does with us, they lift the submerged canoe on their shoulders and after emptying it of water get back in, quite sure that they will have to repeat the same operation within a half-hour, one's equilibrium being almost as difficult to maintain in these frail boats as an acrobat's on a tightrope. These islanders generally are tall, and their average height seemed to me to be five foot sept or eight inches.[3] The colour of their skin is approximately like that of Algerians or other people of the Barbary Coast, their hair is generally long and tied back over the top of the head, their features seemed to me not very attractive: I saw only two women whose features revealed no greater delicacy, the youngest who might be 18 years of age had an awful and repulsive ulcer on one leg, several islanders had large ulcerations and it may have been the

---

[1] Kava is actually a drink made from the root of the Polynesian peppertree (*Piper methysticum*). La Pérouse may have been able to identify the branch being waved, although he may be using the term in the general sense of a piece of greenery imbued with a sacred or ceremonial character; however, in Vaujuas's report on the events of 11 December, the branches are clearly identified as *Piper methysticum*.

[2] Vocabularies provided by Cook and other navigators related to Polynesian languages and were of little value in the Samoan Islands. *Taio* in Samoan means a bird and more particularly the wedge-tailed shearwater. There are a number of words for 'friend', namely *aumea*, *uo*, *soa* and *paaga*, none coming close to 'tayo' which must have puzzled the elderly Samoan.

[3] The comment that the Samoans at 5 ft 7 in. or 5 ft 8 in. were tall may surprise today's reader, but man's average height in the eighteenth century was appreciably less than his present-day counterpart. However, La Pérouse later meets islanders who are a couple of inches taller and, like most Samoans, very strongly built.

CHAPTER XIX

beginning of leprosy because I noticed two men whose legs, ulcerated and as big as their bodies, could leave no doubt about the type of sickness they had,[1] the latter approached us fearfully, but without weapons, there is every indication that they are as peaceful as the inhabitants of the Society Islands or the Friendly Islands. We thought we had taken our leave of them and felt little regret over it since they had struck us as very poor, but the wind having greatly moderated during the afternoon the same canoes, which were joined by several others, came two leagues from shore to offer new exchanges, they had gone back to the land when they left us and were coming back a little better supplied than on the first occasion. This second time we obtained five hens, several items of their clothing, six sultana hens, a small pig, but above all the most charming turtle dove we had ever seen, it was white, its head was of the most beautiful purple, its wings green and its front speckled with small red and white spot like the leaves of the anemone;[2] this little animal was tame, ate in your hand and from your mouth, its

---

[1] The grotesquely swollen legs observed on this occasion were more an indication of the presence of elephantiasis than of leprosy. However, it should be noted that the term leprosy in earlier times was applied to a number of ulcerous and deforming diseases and that leprosy was once called *Elephantiasis Graecorum*. As for the islanders' unattractive appearance, it was commented upon by Bougainville when he sailed among these islands: 'They were fairly tall, although smaller and less handsome than those from Cythère [Tahiti]. A woman who came in one of the canoes was awful.' Taillemite, *Bougainville*, I, p. 335. The point to note was that the comparison was being made with people of eastern Polynesia such as Tahitians and Hawaiians. In Tutuila, however, the women proved to be far more attractive.

[2] This is the multicoloured Samoan fruit dove, *Ptilinopus perousii perousii*. The one referred to was probably a young bird, as the head later acquires a distinctive reddish tinge. This dove was named after La Pérouse by Titian Peale, a member of the 1838–42 United States Exploring Expedition and author of volume 8 of the expedition's record, 'Mammalia and Ornithology'. The bulk of this edition was destroyed by fire and a second, much revised edition was prepared by John Cassin who quotes Peale as follows: 'La Pérouse, in the journal of his melancholy voyage, notices beautiful Doves, of various colours, when at the Navigator's Islands, in the same harbor where our first specimens were obtained, probably alluding to this very species to which we apply his name. The native name, Manu-ma, means shame, or modest bird.' J. Cassin, *United States Exploring Expedition during the years 1838, 1839, 1840, 1841, 1842, under the command of Charles Wilkes U.S.N.*, vol. VIII, Mammalogy and Ornithology, Philadelphia, 1858, p. 276. Peale's description appears on p. 195 of his 1848 edition of this book. Two other birds were named after La Pérouse: a megapode found by the 1817–20 Freycinet expedition near the Mariana Islands was named *Megapodius laperouse laperouse* by the naturalist and surgeon Joseph Gaimard, and the *Megapodius laperouse senex* from the Palau Islands was so named by Hartlaub in 1867. [Information kindly obtained by Dr J. A. Bartle, Museum of New Zealand, Wellington.]

character and its plumage made it worthy of being presented to the Queen, but it was unlikely that we would succeed in keeping it until Europe where all we could bring back was its plumage which had lost all its brilliance; as the *Astrolabe* had always sailed ahead of us on this route, all the canoes had begun their trade with Mr de Langle who had bought two dogs from the Indians, which were judged very tasty.

Although these people's canoes were very artistically constructed and they provide evidence of these islanders' skill in working with wood, we never persuaded them to accept our axes or any iron tool and they preferred a few glass beads which could be of no practical use to them to anything we offered by way of cloth or iron; they sold us a wooden vase filled with coconut oil, which had exactly the same shape as one of our earthenware pots and which a European worker would never have believed could be made without a turning-lathe; their ropes were round and woven exactly like several of our watch-chains; their mats were very fine but their cloth inferior in respect of the colour and texture to those from Easter Island or the Sandwich Islands, moreover it would seem that they are quite rare because they were all quite naked and only sold us two pieces, we were sure that further west we would find another island, infinitely larger where we expected that we would at least find some shelter: even if there was no harbour. We put off more detailed observations until we reached that other island which according to Mr de Bougainvillle's chart should only be separated from the last islet that lay athwart of us at nightfall by a strait of eight leagues. After sunset I made only three or four leagues westward and stayed with light canvass on short tacks; at daybreak I was very surprised not to see any land to leeward and only sighted it at six in the morning because the channel is much wider than is shown on the chart I used as a guide and it would be desirable for the maps of a voyage which is second only to Captain Cook's in respect of its astronomical observations, and the extent and importance of its discoveries, it would I repeat be desirable for these particular charts to be drawn with greater care and on a larger scale.

We did not reach the northern point of the island[1] until five

---

[1] La Pérouse had now reached the island of Tutuila. As we shall see later, he became very confused over the islands' names. The north-eastern point of the island is Cape Matalula. Bougainville had sailed south-west, whereas La Pérouse proceeded to sail along the northern coast.

CHAPTER XIX

o'clock in the evening and as I intended to seek an anchorage I signalled to the *Astrolabe* to hug the wind in order to stay on short tacks to windward of the island during the night, and to dispose of all the following day to explore it in every detail. Although we were three leagues from land, three or four canoes came alongside that very evening, they brought us pigs and fruit which they bartered for beads, which gave us a better impression of this island's wealth. On the morning of the 9th I came closer to the land and we hauled along it half a league distant; it was surrounded by a coral reef over which the waves broke angrily but this reef almost touched the shore, and the coastline formed various small coves in front of which we could see breaks through which the canoes could pass and possibly even our boats and longboats; we could see numerous villages at the back of each little bay from which a large number of canoes had come laden with coconuts, pigs and other fruit which we bought with beads; this great abundance increased the desire I had of anchoring. We could see waterfalls coming down from the top of the mountains down to the foot of the villages. So many advantages led me not to be difficult over the anchorage, I coasted along as near as possible and at 4 o'clock having found a bank of rotted shells and a very small quantity of coral in 30 fathoms one mile from the shore we dropped anchor there, but we were tossed about by a very heavy swell driving us towards the land although the wind came from the coast.[1] We immediately lowered our boats and that same day Mr de Langle with several officers and three armed boats from the two frigates went to the nearest village, where they were received by the inhabitants in a most friendly manner; night was falling as they landed, the Indians lit a great fire to light up the assembly, they brought birds, pigs and fruit, and after staying an hour our boats returned to the ships; everyone seemed delighted with the way they had been received and our only regret was to find our vessels anchored in such a bad roadstead, where the frigates rolled as if they were out at sea even though we were sheltered from the northerly to the south-easterly

---

[1] There are two good harbours in Tutuila, but both are on the south coast: Pago Pago now the capital of American Samoa and one of the best harbours in the South Pacific, and Leone which is available for small vessels but seldom used. La Pérouse was unlucky in that he sailed along the north coast which is rough and precipitous with few settlements. The village, Fagasa, one can assume, is not much larger today than in his day.

winds. The calm on its own would be enough to expose us to the greatest danger if our cables were cut, and there was no possibility of sailing against a breeze of any strength coming from the northwest. We knew from the accounts of travellers who had come before us that the trade winds are not regular in these parts, and that it is as easy to sail up eastward as to go down to the west, which makes these people's navigation easier on the lee side; we had even experienced this inconsistency of the winds, and the westerlies had only left us in the twelve degrees. These thoughts gave me a very uneasy night, all the more because a storm was building up in the north, from where the winds blew fairly strongly. Fortunately the land breeze prevailed. The sun rose on a magnificent day. I decided to take advantage of the morning to explore the country, observe the inhabitants in their own homes, obtain water and sail in the afternoon, caution not allowing us to spend a second night in such an anchorage. Like me, Mr de Langle had found this anchorage too dangerous to spend a second night in. We agreed that we would sail in the afternoon, but that the morning which was very fine would be spent obtaining water and buying fruit and pigs. At break of day a hundred canoes were around the frigates with all kinds of provisions which the islanders were prepared to barter only for beads. For them these were priceless diamonds and they scorned our axes, our cloth and all our other trade goods; while part of the crew were employed holding the Indians in check and trading with them, the others were filling up our boats and longboats with empty casks to go and get the water, and our two longboats, armed, commanded by Messrs de Clonard and Colinet and those of the *Astrolabe* by Messrs de Monty and Belle-garde left at five o'clock in the morning for a bay distant about one league and to windward, a fairly convenient location because our boats, laden with water, could sail back with a quartering wind. I was following Messrs de Clonard and Monty close behind in my biscay boat and reached the shore at the same time; unfortunately Mr de L'angle decided to go in his small boat for an excursion to a second cove approximately one league from our watering place, and this pleasure-trip from which he returned delighted, enchanted by the beauty of the village he had visited, was the cause of our misfortunes. The cove towards which our lucky star made us send our longboats was pretty, large and convenient; boats and longboats stayed afloat at low tide half a pistol shot from the shore, the watering place was attractive and

CHAPTER XIX

easy. Messrs de Clonard and Monty established the most satisfactory order, a line of soldiers was placed between the Indians and the shore, we invited them all to sit down under the coconut trees lining the coast less than 8 *toises* from our longboats. They numbered about two hundred, with among them many women and children, each one had with him some hens, pigs, pigeons, parrakeets, fruit and they all wanted to sell them at the same time, which created a little confusion.

The women, some of whom were very pretty, offered with their fruit and poultry their favours to anyone who was prepared to give them beads; soon they crossed the line of soldiers who pushed them back too weakly to stop them; their behaviour was gentle, merry and beguiling; Europeans who have sailed around the world, and especially Fenchmen, have no weapons against such attacks; they went through the ranks, the men came closer, then there was some little disorder, but Indians armed with sticks, whom we took to be chiefs, re-established order; each one returned to his post and trade began anew to the greatest satisfaction of buyers and sellers. However an event had taken place in our longboat which was a real act of hostility which I wanted to repress without any bloodshed although possibly an example was needed to impress these people who seemed to have little regard for us because they were unaware of the effect of our weapons and because their height of 5 feet 10 or 11 inches, their broad limbs whose proportions were colossal made them think that we presented little danger. Be that as it may, I did not feel that I should teach them to have a better opinion of us by punishing this Indian for his insolence; he had climbed onto the back of our longboat, picked up a mallet and struck several blows on the arms and backs of our sailors. I ordered four of the strongest to throw themselves at him and hurl him into the sea, which was immediately carried out. The other islanders seemed to disapprove of their compatriot's behaviour, calm was completely restored and in order to keep them in this happy disposition I had three pigeons bought which were thrown up in the air and shot down in front of the crowd, which seemed to instill some fear in them.[1] I expected

---

[1] This rough handling may have been satisfactory from the French point of view since no blood was shed, nor were any blows struck in retaliation, but if the individual was a chief the affront of hurling him into the sea in this manner may have been more serious than La Pérouse imagined.

more out of this feeling that I did from goodwill – which uncivilised man seldom displays.

While all this was going on quite peacefully, and our water casks were being filled, I thought I could walk a couple of hundred paces away to visit a charming village situated in the middle of a forest of trees that were heavy with fruit and which one could call an orchard; the houses were placed along the circumference of a circle some 150 *toises* in diameter, the centre of which was empty forming a wide public place covered with the finest grass; the trees shading it and the houses preserved a delightful freshness; women, children, old men had accompanied me, they all pressed me to enter their houses, and stretched out the finest and freshest mats on the ground beneath their roofs, made of a selection of little pebbles and raised about two feet above the ground to protect them from the damp; I went into the best hut which presumably belonged to the chief and I was extremely surprised to find a vast latticed room as well and indeed better made than any in the environs of Paris. The best architect could not have given a more elegant curve to the two ends of the ellipse ending this hut, a row of columns five feet from each other ran along the edge, these columns were only tree trunks very elaborately worked between which the Indians had placed some fine mats that could be raised or lowered with ropes like our roller-blinds and arranged with the utmost skill like fish scales, the rest of the house was covered with coconut-tree leaves.

What imagination could conjure up the happiness one would find in such an enchanting site, a climate requiring no form of dress; breadfruit trees, coconuts, bananas, guavas, oranges, &c growing quite naturally offered these fortunate inhabitants a pleasant and healthy nourishment; hens, pigs and dogs living on surplus fruit allowing them to vary their diet. They had such wealth and so few needs that they scorned our iron tools and our cloth, and wanted only beads – with a surfeit of real goods they hankered only after frivolities.

They had sold on our market over two hundred wood-pigeons as tame as puppies, that wanted to eat only from one's hands; they had also bartered the most charming turtledoves and parrakeets as tame as the pigeons,[1] and we commented that these islanders are

---

[1] *Columbidae* are common in the Samoas. The *lupe* or Pacific pigeon, *Ducula pacifica pacifica*, is still common in Tutuila, but the white-chested pigeon, *Columba vitiensis castaneiceps*, is now much more scarce. The parrakeets were *sega vao*, or blue-crowned lory, *Vini australis* Gmelin.

the happiest inhabitants of the earth, they spend their days in idleness surrounded by their wives and have no other care than to adorn themselves, to tame birds and, like the first man, to pick fruit growing above their heads without any effort on their part. We saw no weapons, but their bodies were covered with scars which was evidence that they were often warring or quarrelling and their features indicated a ferocity one did not see in the women's appearance; Nature no doubt had left this mark to warn that, in spite of all the academies that crown the philosophers' paradoxes,[1] man in an almost savage state and living in anarchy is a more malevolent being than the wolves and tigers of the forests; this first visit took place without any incident likely to have serious consequences, I learned however that there had been a few private quarrels which a great prudence had brought to nought. Stones had been thrown at Mr Rollin, our senior surgeon; an islander pretending to admire a sword belonging to Mr Moneron had tried to snatch it from him and getting only the sheath had fled terrified at the sight of the naked blade, dropping the scabbard. All in all I realised that they were very turbulent and not very much under their chiefs' control,[2] but I was expecting to leave in the afternoon and was glad that I had attached no importance to the minor injustices we had suffered. At about midday I returned to the ship in my biscay boat and the longboats followed very close behind. I had some difficulty in getting alongside because the canoes were surrounding our two

---

[1] This is a dig at the Dijon Academy which in 1749 had organised a competition for an essay 'On whether the progress of the sciences and the arts has contributed to the corruption or the improvement of morals'. It was won by Jean-Jacques Rousseau with a *Discours* which made him famous and crystallised for him his feeling that man was originally virtuous, free and happy, but had become corrupted by society, and that in order to restore mankind to some measure of happiness a return to some more natural state was essential. This was followed by a second *Discours* on the origin of inequality. The phrase was deleted by Milet-Mureau in his printed version of the narrative. The ongoing debate on Rousseauism (towards which La Pérouse felt little sympathy) is discussed in Michèle Duchet, *Anthropologie et histoire au siècle des lumières*, Paris, 1971, and Taillemite, *Bougainville*, I, pp. 45–57 has provided an analysis of another navigator's reaction, Bougainville, to the reality of 'natural man' in a Pacific Islands environment.

[2] Some modern commentators have pointed to what they see as the friendliness of the Samoans during this first day, and have assumed that the French must have committed some breach of etiquette or some act of hostility which resulted in the attack on Langle and his men. These comments on the turbulence of the islanders and La Pérouse's attempts to conciliate them and impress on them the dangerous superiority of European weapons give a different impression.

frigates and our market was still full. I had put Mr Boutin in charge of the frigate when I went ashore and had left it to him to make the arrangements he considered appropriate by allowing a few islanders to come on board or totally preventing it according to the circumstances. I found on the quarter-deck seven or eight Indians, the eldest of whom was presented to me as being a chief. Mr Boutin told me that he could have prevented them from climbing up only by giving the order to fire on them, that when they compared their physical strength with ours they laughed at our threats and mocked our sentries, and that knowing my principles of moderation he had not wanted to resort to violent means which however was the only way to keep them in check. He added moreover that since the chief had come aboard the other islanders who had preceded him were much quieter and less insolent. I gave a great number of presents to this chief and showed him the utmost friendliness; then, wishing to give a high opinion of our power, I had several pistol shots fired in front of him which went through wooden planks, we shot pigeons with muskets; it seemed to me that the effect of our weapons did not make much of an impression on him and that he believed them only good for killing birds. Our longboats arrived, laden with water, and I had all arrangements made in order to set sail and take advantage of the light land breeze which promised to give us enough time to get away a little from the coast. At the same time Mr de Langle came back from his excursion. He advised me that he had landed in a small harbour ideal for boats, at the foot of a charming village and by a cascade of the clearest water; as he went past he had given instructions to his ship to make ready to sail. Like me he felt the need to do so, but he insisted most vehemently that we should remain tacking within a league of the coast and obtain a few boatloads of water before leaving this island; however strongly I argued that we had no need for it, he had adopted Captain Cook's views and believed that fresh water was preferable to what we had in our holds and as a few members of his crew had small symptoms of scurvy, he thought that we owed it to them to provide every means of relief.

This island furthermore could not be compared to any other in respect of the abundance of food, the two frigates together had already bought more than five hundred pigs, plus hens, pigeons and great quantities of fruit, and all this had only cost us a few glass beads. I felt that these arguments were valid, but a secret sense of

CHAPTER XIX

foreboding prevented me from accepting them, I told him that I found these islanders too turbulent to send ashore boats and longboats which could not be assisted by our ships' guns, that our moderation had inspired little respect for us on the part of these Indians who were colossi and looked only at our physical strength which was inferior to theirs; nothing could shake Mr de Langle's obstinacy and he added that my own would make me responsible for the progress of the scurvy that was beginning to manifest itself with some severity, he added that the port he had been to was a hundred times more convenient than the one where we had obtained our water, begging me to place him at the head of the first expedition, and that he undertook to be back on board with all the boats and the water within three hours. Mr de L'angle was a man of such excellent judgement and of such ability that these considerations above everything else finally caused my own will to bow to his, and I promised him that we would stay on short tacks during the night and would send the next day our two longboats and two boats ashore armed as he considered appropriate and that everything would be placed under his orders. Events proved that it was time we left – raising the anchor we found one of the strands of our cable cut through by the coral and if we had spent two hours more at this anchorage it would have been cut through.[1]

As we did not get under sail until four in the afternoon it was too late to think of sending our longboats ashore and we put off their departure until the next day; the night was stormy and the winds, changing constantly, made me decide to go about three leagues from the coast. At dawn the dead calm prevented me from sailing up to it, it was only at 9 o'clock that a light breeze rose from the N.E. which enabled me to near the coast from which at eleven o'clock I was only a short league away and I then sent my longboat and my launch commanded by Messrs Boutin and Mouton to the *Astrolabe* to place themselves at Mr de Langle's disposal. All those who showed any sign of scurvy were sent with them together with six armed soldiers led by the master-at-arms, and about twenty barrels shared between the two craft. Messrs de Lamanon and Colinet, who were sick, formed part of the 28 people who left the *Boussole*, and Mr Vaujuas, who was convalescing, went with Mr de L'angle in his launch. Mr Gobien, garde de la marine, was in charge

---

[1] A ship's cable usually consisted of three strands.

and Messrs la Martiniere, l'Avau and Father Receveur were among the thirty-three people from the *Astrolabe*. Altogether the expedition comprised 61 individuals, which number included the elite of our crews. Mr de Langle had had six swivel-guns set up, had taken muskets and swords and I had broadly left him to decide what he needed for his safety. Being quite confident that we had not been involved in anything which might have caused these people to harbour any grudges, the great number of canoes surrounding us out at sea, the feeling of gaiety and trust which reigned in the markets where the abundance of supplies was such that they could perhaps not be compared to any mentioned by navigators, everything tended to increase his feeling of safety and I must admit that my own confidence could not be greater, but sending boats ashore that could not be protected or even seen from the ships, in the middle of such a large crowd, ran counter to my principles unless it was absolutely necessary; the longboats left the *Astrolabe* at half past twelve and in less than three-quarters of an hour they had arrived at the watering place. Imagine the surprise of all the officers, of Mr de Langle himself, when they found, instead of a superb bay, a small cove full of corals into which one could only enter by a tortuous channel less than 25 feet in width where the waves broke as on a bar and when they were inside there was not three feet of water,[1] the longboats ran aground and the boats stayed afloat only because they were hauled by hand to the pass entrance a fair distance from the shore; unhappily Mr de Langle had investigated this bay at high tide, he did not expect that the tide rose in these islands by five or six feet. He could hardly believe his eyes, his first reaction was to turn back and no doubt go to the bay where we had already obtained water and which had every advantage, but the peaceful and gentle appearance of the people waiting for him on the shore with an enormous quantity of fruit and pigs, the large number of women and children he noticed among the islanders who take good care to drive them away when they harbour hostile intentions, and finally his destiny which was driving him inexorably towards his destruction, all these conditions together drove away his first cautious thoughts which an unthinkable fate prevented him from following, and he landed the water casks of the four craft with the

---

[1] A'au Bay is still referred to as Massacre Bay. It lies just over a mile south-west of Fagasa Bay. Practically the whole of this coast is steep and rocky.

CHAPTER XIX

utmost feeling of confidence; on land his soldiers established the most orderly conditions possible, they made a hedge that cleared a free space for our workers; but soon this early calm came to an end, several of the canoes which had sold their supplies to our ships came back to land and all landed in the watering cove so that it gradually filled up; instead of 200 inhabitants, including the women and children Mr de l'angle had found when he arrived at one thirty, there were a thousand or twelve hundred by three o'clock. The number of canoes trading with us in the morning had been so great that we hardly noticed that it had lessened during the afternoon, on the contrary I rejoiced at keeping them occupied with the ships in the hope that our longboats would be less bothered. I was quite wrong and Mr de L'angle's situation was becoming more troublesome by the minute; nevertheless he managed with the help of Messrs Vaujuas, Boutin, Colinet and Gobien to load his water barrels but the bay was almost drained out, and there was no hope of getting his longboats afloat before four o'clock in the afternoon. However he got into them with his officers and his detachment and took up a position forward with his musket and his fusiliers, issuing orders to all of them not to fire until he gave instructions, and he was already feeling all too well that it would soon be necessary, already stones were flying and these Indians who had water only up to their knees were surrounding the longboats less than a *toise* away; the soldiers who had re-embarked no longer being able to contain them; if the fear of starting the hostilities and possibly being accused in Europe of barbarous behaviour had not prevented Mr de L'angle from killing twenty or thirty Indians with muskets or swivel-guns, he would certainly have driven this multitude away but he was confident that he could keep them in check without any blood being shed and he fell victim of his humaneness; soon a hail of stones thrown from a distance of five or six feet with the power of a sling hit almost all those who were in the longboat.[1] Mr de Langle had only time to fire his two shots, he was knocked over and unfortunately fell on the port side of the longboat, where there were more than two hundred Indians who massacred him with clubs and stones and when he was dead tied one of his arms to

---

[1] Every navigator who was at the receiving end of stones hurled by Samoans and Tongans commented on the dexterity and strength shown by their attackers; La Pérouse's own phrase, *jeter des pierres*, 'to throw stones', hardly does justice to the men of A'au Bay.

one of the longboat's thole-pin presumably to profit from his remains. The *Boussole*'s longboat commanded by Mr Boutin had grounded two *toises* from the *Astrolabe*'s and parallel to it, leaving a small passage where there were no Indians and by which all our wounded who were lucky enough not to fall on the offside escaped and reached our boats which luckily had remained afloat under the command of Messrs de Vaujuas and Mouton and were consequently able to save 49 out of the 61 men who made up the expedition. Mr Boutin who was in charge of my longboat had modelled all his movements, all his actions on those of Mr de L'angle; his water casks, his detachment, all his people had been embarked at the same time and arranged in the same way, he himself occupied the same position at the bow and although he was worried about the poor outcome of Mr de Langle's moderation he did not allow himself and his detachment to fire until his commanding officer had done so. It can be expected that at a distance of four or five paces each shot must have killed an Indian, but there was no time to reload.[1] Mr Boutin was knocked over like Mr de Langle, fortunately he fell between the two longboats. In less than five minutes there was not a single Frenchman left in either of the grounded boats, all those who escaped by swimming to the two boats had received over ten wounds, almost all head wounds; those who had the misfortune of being knocked over on the side of the Indians were finished off immediately with clubs and the thirst for loot was such that after capturing the longboats and climbing in, three or four hundred of them, breaking up the thwarts and everything else to look for the riches they thought we had, none of the islanders took much further notice of our boats and gave enough time for Messrs de Vaujuas and Mouton to save everyone and ensure that none remained within the Indians' power apart from those who had been massacred and killed in the water with their

[1] Muskets (the French term was *fusil*, or flintlock musket) took twenty to thirty seconds to reload, depending on the soldier's skill, or to be precise 19·51 seconds under test conditions with a 1 in 6 misfire rate: D. Shineberg, 'Guns and Men in Melanesia', *Journal of Pacific History*, VI (1971), p. 76. See also K.R. Howe, 'Firearms and Indigenous Warfare', *Journal of Pacific History*, IX (1974), pp. 21–38. They had nothing in common with the modern rapid-fire rifle. Once the shot was fired, the soldier or sailor was at the mercy of the islanders who were running at him with clubs or spears which they handled with great dexterity and precision. Cook's death occurred under similar circumstances to Langle's and for much the same reason: once they had fired their first shot, the English were effectively disarmed for half a minute and, worse still, preoccupied with trying to reload.

CHAPTER XIX

Patow Patow[1] after being knocked down by the stones. Our boats which until then had been firing at the islanders and must have killed several had no other thought in mind than throwing the water casks overboard in order to make room for everyone, they had spent almost all their ammunition and retreating was not without difficulty with so many people seriously wounded, stretched out on the thwarts and impeding the movement of the oars, and we owe the safety of the 49 members of the two crews to Mr Vaujuas's sagacity, and the good order he maintained, and to the punctual manner in which Mr Boutin who commanded the *Boussole*'s longboat carried out his instructions. Mr Colinet was found unconscious lying across the canoe's mooring rope, with one arm smashed, a finger broken and two head wounds; Mr L'avau, senior surgeon on the *Astrolabe*, so badly wounded that he had to be trepaned, swam unaided up to the two boats, as did Mr de La Martiniere and Father Receveur who received a severely bruised eye; Messrs de Langle and Lamanon remained on the battlefield where they were massacred with unparalleled barbarity as well as Mr Talin, the *Boussole*'s master-at-arms and three other soldiers, four sailors and two servants, twelve people in all on which the Indians satisfied their rage with such a fury that each was clubbed over a hundred times after they died. Mr Gobien who was in charge of the *Astrolabe*'s longboat under Mr de Langle's orders although seriously wounded did not abandon the longboat until he found himself alone in it, after using up his ammunition he jumped into the water on the side of the small channel created by the two longboats which as I have already stated was not taken over by the Indians and escaped to the boats. The *Astrolabe*'s was so overloaded that it grounded, which gave the islanders the idea of still disrupting their retreat and they went in large numbers towards the reefs at the entrance, which of necessity one had to pass at a distance of ten feet. The little ammunition left was spent on these latter, and the boats finally came out of this lair which on account of its situation and the barbarity of the inhabitants was more frightful than a den of tigers and lions.

---

[1] *Patu* is a largely Maori term used to describe a flat hand-held weapon, usually with a thin cutting edge, made of greenstone, whalebone or a similar material. La Pérouse presumably obtained the term from the accounts of Cook's voyages, in which they are usually spelled *patoo patoo*. Similar weapons existed in the Samoan Islands: this one may be a *fa'alautilaga*, a double-headed hand-axe.

They arrived on board the two frigates at five in the evening and we learned of this awful event. At that moment we had a hundred canoes surrounding us, selling provisions with a feeling of security that proved they were not the accomplices of this perfidious action, but they were the brothers, the children and the compatriots of these barbarous murderers, and I must admit that I needed all my powers of reasoning to stop me from giving way to anger, to the rage that burned inside me, and prevent our crews from killing them. They had already jumped to the guns and the weapons, but I halted these movements which however were quite forgiveable and I had a single gun fired with a load of powder to warn the canoes to leave; one small craft sent from the coast no doubt told them of the betrayal and in less than an hour no canoe remained in sight; one Indian who was on my frigate's quarter-deck when our boat arrived was arrested by my orders and put in irons, and the next day having come closer to the coast I allowed him to jump into the sea, the sense of security with which he had come on board my frigate being clear evidence of his innocence. My first plan was to order a second expedition to avenge our unfortunate companions, to destroy that village entirely, and recapture the remnants of our longboats; I went towards the coast to seek an anchorage, but all I found was the same coral bottom with a swell rolling towards the land and breaking on the reefs like on those of the Chaussée des Saints.[1] The cove where the massacre had taken place was moreover very deep and narrow, it was impossible to approach it within a gunshot and Mr Boutin, although seriously wounded and bedridden, remained clear-headed and pointed out to me that the terrain was such that I would risk losing a number of men to no purpose and that if my boats were unlucky enough to become grounded it was very likely that not one man would survive, the trees which almost touched the edge of the sea providing a shelter for the Indians from our muskets and leaving our French if we landed any ashore exposed to their stones which they threw with such skill and strength that they had the same effect as our bullets and had the advantage over our musket shots of following each other much more quickly. Mr de Vaujuas shared exactly the same

[1] The pass between the two small islands of Les Saintes, of the island of Guadeloupe in the West Indies. La Pérouse was present at the 'Bataille des Saintes' against Admiral Rodney in April 1782 when he escorted the French warship *Zélée* to safety through dangerous rock-strewn waters.

CHAPTER XIX

opinion but I did not want to accept it until I had completely satisfied myself of the impossibility of anchoring our vessels within gunshot of the village; I spent two days tacking in front of the bay; I still could see the remains of our longboats on the sand around which there was an immense crowd of Indians, but – what will sound incredible – five or six canoes came from the coast with pigeons, coconuts and pigs to offer to barter with us; I was time and again forced to contain my anger in order not to send them to the bottom, these Indians knew nothing about our weapons other than the effect of our muskets and they believed that as was the case with their stones which could reach no further than thirty feet, they in their turn could not be hit from that distance and they remained fearlessly fifty *toises* from our vessels, offering us coconuts, bananas and pigs; our gestures urged them not to come any nearer and they spent thus a good hour of the afternoon of 12 December. They soon added jeers to their offers to barter provisions, and as they did not know the effect of our guns, and I could see more canoes leaving the shore and realised that I would soon be forced to alter my principles of moderation I had a gun fired in the midst of these craft. My orders were carried out so precisely that the cannonball caused the water to splash over these canoes which immediately made for the land and we did not see another one during the whole day, the others which had left the coast joining those who were returning from the sea: I felt sadness at having to leave such a tragic place and leave behind the massacred corpses of our unfortunate companions. I had lost, through the most frightful act of treachery, my best friend, my friend of 30 years, a man full of wit, wisdom and knowledge, and certainly one of the best officers in any European navy. His humanity had brought about his death and if he had dared to allow himself to fire on the first Indians who went into the water to surround his longboats he would have saved himself, Mr de Lamanon and ten other victims of the Indian ferocity.[1] Twenty individuals from the two frigates were seriously wounded, so that we had 32 men fewer and our two longboats, the only rowing craft that were captable of taking in addition to their crews a fairly large number of armed men in order to carry out a landing. These issues determined my later conduct, the slightest setback would have forced me to burn one of the two frigates to man the other one and I would have had to abandon my campaign;

---

[1] For a discussion of the possible causes of the Samoan attack, see Introduction.

I did have a longboat ready to be assembled at my next place of call; and if all that was needed to ease my anger was the massacre of a few Indians I had the opportunity to destroy, send to the bottom of the sea and break up a hundred canoes in which there were more than five hundred people, but I was afraid to attack the wrong victims and the call of my conscience saved their lives. Those who will think back to the Captain Cook catastrophe must not forget that his ships were at anchor in Karakakoa Bay, that their guns made them the masters of the seashore and that they could establish control by threatening to destroy the canoes left on the shore as well as the villages lining the coast;[1] we on the contrary were away from the coast out of the range of our guns and forced to get away from the coast when we feared the calms, a strong swell bearing us towards the reefs where we might of course have anchored with steel hawsers, but the village would have been beyond gun range and the swell alone would have been enough to cut the cable at the hawsehole and place the frigates in the greatest danger; I exhausted all possibilities before leaving this fatal island, and it was proved to me that it was impossible to anchor, that the expedition would be foolhardy if the frigates could not assist and success pointless since we were only too certain that not one man remained alive in Indian hands, that they had smashed up and beached the longboats and that we had means enough on board to replace them. Consequently I set sail on the 14th for the third island which I could see in the W¼NW and which Mr de Bougainville had also sighted only from the topmasts because the bad weather had driven him away from it. It is separated from Mahouna Island by a strait of nine leagues;[2] the

---

[1] James Cook was murdered at Kealakekua Bay, Hawaii, on 14 February 1779. The circumstances were different in that relations between the English and the Hawaiians had clearly worsened and Cook tried to recapture a cutter stolen by the islanders. But the hail of stones that fell upon the English party and the general confusion which followed was later paralleled in the attack on Langle and his men. Historians, as J.C. Beaglehole pointed out in 'The Case of the Needless Death', in *The Historian as Detective*, New York, 1969, ed. Robert W. Winks, face the problem of reconstructing situations on the basis of records left by witnesses and participants who are no more and whose accounts differ, and who, as Maurice de Brossard pointed out in *Moana, océan cruel*, Paris, 1966, belonged to totally different cultures (see especially his pp. 13–14 and 321–34).

[2] The island is Upolu, separated from Tutuila by a strait of some 50 miles. It forms part of Western Samoa, whereas Tutuila is part of American Samoa. Bougainville's reference to Upolu occurs in his journal entry of 4 May 1768: 'From the topmasts we saw another land bearing WNW and NW¼W, high land.' Taillemite, *Bougainville*, I, p. 336.

CHAPTER XIX

Indians had given us the names of the ten islands that make up their archipelago. They had roughly sketched them on paper and, although we were convinced that one could hardly rely on the map they had drawn, we could not doubt that there existed a kind of confederation[1] between these ten islands, very well known to each other, and that there was a great deal of communication between them, and the subsequent discoveries we made since that time leave us in no doubt that this island group is larger, as well populated and as abundant in foodstuffs as is the Society archipelago; it is more than likely that one would find some excellent anchoring places, but having no longboats and in view of the state of tension of our crews I decided that I would land only at Botany Bay in New Holland where I proposed to build the longboat I had on board. But for the advancement of geography I felt that I ought to explore the various islands I would come upon, accurately determine their latitude and longitude, communicate with these people through their canoes which, laden with foodstuffs, travel two or three leagues from the coast to trade with vessels, and I left to others the task of writing their history which like that of all barbarous people is of slight interest. A 24-hour stay with an account of our misfortunes is sufficient to describe their atrocious ways, their crafts and the products of one of the most beautiful countries in the world; but before continuing the narrative of our route along the islands of this archipelago I feel that I should offer to public scrutiny the account given by Mr de Vaujuas who took charge of the retreat although he had gone ashore only because he was convalescing and was not on duty; the circumstances gave him back his strength, he made the best arrangements possible and left the bay only after he had satisfied himself that not one Frenchman remained within the Indians' power.

'On Tuesday eleventh of December at eleven in the morning, the Count de la Pérouse sent his longboat and his boat loaded with water casks with a detachment of armed soldiers under the command of Mr de Langle to whom in the morning he had sent Mr

---

[1] Confederation is hardly an appropriate term. Samoan society had no unified political structure. Even the islands were split into tribal and sub-tribal groups between which alliances were made and unmade under the influence of chiefs and village assemblies. Language and culture, together with kinship, established a form of unity between the islands, maintained by frequent communications – hence the name 'Navigators' – even though the outcome was as much war as trade.

Boutin to obtain the information needed for the maintenance of order and safety when the boats went ashore; at the same time our captain also lowered his boats and had them loaded with water casks and weapons; at half past twelve being three-quarters of a league from land on the port tack the 4 boats went to get water in a cove investigated by the Viscount de L'angle to leeward of the one where we had already been and which he had found to be more suitable because it had fewer inhabitants and the watering place was as convenient, but the first one had the advantage over it of a much easier entrance and enough depth for longboats not to run the danger of becoming grounded as happened in the leeward one.

Mr de Langle had even wanted to take command today of the longboat expedition, he invited me although I was still weak and convalescent to accompany him to take a stroll and breathe the land air, he took command of the boat and entrusted the longboat to Mr Gobien. Mr Boutin was in charge of the *Boussole*'s longboat, whose boat was under Mr Mouton's orders. Father Receveur and Mr Colinet, both unwell, and Messrs de Lamanon, la Martiniere and Lavo came with us as well as several people from the two frigates. We comprised, including the crews of the two boats, a detachment of six soldiers led by the Master-at-arms, the officers and passengers, thirty-three men from the *Astrolabe*, and the *Boussole*'s boats had 28 men; in total 61 people.

When we went we saw with regret that a large proportion of the canoes which had been alongside the ships were following us and coming to the same cove. We also saw along the rocks separating it from the neighbouring bays numerous natives going there from the other villages; when we reached the reef which makes up the watering cove and leaves only a narrow and shallow pass for the boats, we realised that it was low tide and that the longboats could not enter without grounding, and indeed they touched as soon as we got within half a musket shot from the shore which they could only reach by our pushing on the bottom with the oars. The Captain who had himself selected this bay had seen it the day before from a more favourable point of view because the tide was not so far out.

When we arrived the natives lining the coast, numbering seven or eight hundred, threw into the sea several branches of the tree from which South Sea islanders obtain their intoxicating liquor,[1] it

---

[1] The tree from which the Samoans made kava (*'ava*) was the *Piper methysticum*.

CHAPTER XIX

seemed to us that among them this was a symbol of peace and friendship but on which we learned to our cost one can scarcely depend at least here. As we landed, Mr de Langle gave instructions that each boat was to be guarded by an armed soldier and a sailor while the boat crews saw to the water, protected by a double line formed by the two detachments, going from the longboats to the watering place. We peacefully rolled out, filled and reloaded the water casks, the natives allowing themselves to be fairly well contained by the armed soldiers, there were among them a certain number of women and very young girls who made advances to us in the most indecent fashion, of which several people took advantage. I saw only one or two children there.

Towards the end the number of natives increased and they became more troublesome. This circumstance caused Mr de Langle to give up his earlier intention of buying some provisions, he gave the order to get back at once into the boats. But before then (and this I think is the primary cause of our misfortune) he had given a few beads to some kinds of chiefs who had helped to keep the islanders at a little distance; we were sure however that this pretence at policing was only play-acting, and if these alleged chiefs have any authority it is only over a very small number of men. These gifts being made to five or six individuals aroused the others' displeasure. From that moment a general murmur arose and we were no longer able to control them; however they let us get back into the boats, but a number of them followed us into the water while the others were picking up stones from the shore.[1]

As the longboats had grounded some distance from the beach we had to get into the water up to our waist, and during this wading several soldiers got their weapons wet, it is in this critical situation with the longboats grounded that the most frightful scene of horror began as the last people were getting back on board the longboat. Mr de Langle ordered the grapnel to be raised and to get afloat; several of the strongest islanders tried to prevent this by holding on to the cablet. Seeing this, with the tumult increasing and stones flying, the Captain fired in the air, which far from frightening the natives was for them the signal for a general attack; they hurled at us a hail of stones which they throw with surprising strength. Then

---

[1] This passage would seem to reinforce the view that the attack on the French had not been premeditated.

those whose muskets were able to be fired killed several of these madmen, but the other Indians paid no attention and continued to club us down. A number of them came up to the longboats while the others, numbering six or seven hundred, assailed us with an unremitting storm of stones.

The moment hostilities broke out I had thrown myself into the sea to get to the *Astrolabe*'s boat where I had just noticed there was almost no one and no officer, this circumstance gave me strength enough for the short distance I had to cover in the water and, in spite of my weakness and a few stones that hit me at this moment, I was able to climb unaided into the boat; I saw with a feeling of despair that there was not a single weapon that was not wet and that all I could do was to get it into the water as soon as possible outside the reef. Meanwhile the battle continued and enormous stones thrown by the natives were still wounding a few of us, and as a wounded man fell in the sea he was finished off with paddles or clubs. Mr de Langle was the first victim of the ferocity of these barbarians to whom he had done nothing but good; at the first volley he was knocked over, covered in blood, across the longboat's thwart onto which he had climbed, and had fallen in the water with the Master-at-arms and the head carpenter who were next to him, the fury with which the Captain was set upon saved the latter two who were able to reach the boat. In a flash everyone who was in the longboats suffered the same fate as our unfortunate chief; with the exception of a few sailors who, fleeing, were able to reach the reef from where they swam to the boats. In less than four minutes the natives were masters of the two longboats, and I had the pain of seeing our unhappy companions massacred without being able to help them in any way. The *Astrolabe*'s boat was still inside the reef and I expected at any moment that it would suffer the same fate as the longboats but the natives' greed saved us. Most of them had rushed into the longboats and the rest were satisfied with hurling stones at us, several however came into the channel and along the reef to wait for us; but in spite of their stones although many of us were seriously wounded, although there was a fairly strong swell and the wind was against us we managed to leave this dangerous place and join outside Mr Mouton who was in charge of the *Boussole*'s boat which he had lightened by throwing out the casks to make room for the men who were swimming and reaching his boat. In the *Astrolabe*'s I had saved Messrs Boutin and Colinet who

CHAPTER XIX

were badly wounded and several other people. The unfortunate men who had stayed in the longboats having all been massacred, those who escaped into the boats were more or less seriously wounded, and so the boats were defenceless and it was out of the question to return to a bay from which we were only too lucky to get out in order to face a thousand furious barbarians, it would have meant risking certain death to no useful purpose.

We therefore set our course to return to the two frigates which at three o'clock at the time of the massacre had tacked out to sea, no one on board could imagine that we ran the slightest risk. The breeze was fresh and the frigates far to windward; an unfortunate circumstance for us and especially for those whose wounds urgently required dressing. At 4 o'clock they veered back to land; as soon as we were away from the reef I had the sail raised to get away from the coast, close to the wind on the starboard tack, and I had everything thrown into the sea that might slow the boat which was full of people. Fortunately the islanders who were busy looting the longboats did not follow us, all we had to defend ourselves with was four or five swords and two or three musket shots to fire which by chance were found in a powder-flask, feeble resources against two or three hundred barbarians armed with stones and clubs and who ride on very light canoes in which they keep the distance that suits them; some of these canoes left the bay shortly after we emerged, but they only sailed along the coast from which one of them went out to warn those that had remained by the frigates, they had the effrontery to make threatening gestures as they went past us, having only three musket shots to fire I was forced to delay my revenge, and to keep them to defend ourselves if we were attacked. What is extraordinary is that in spite of the warning several of them nevertheless stayed alongside the *Boussole*, until the moment when that frigate fired a gun loaded with powder to disperse them.

When we were in the open sea I sailed against the wind towards the frigates which had just tacked towards the land, I had a red handkerchief placed at the top of the mast and as I came near I had the three remaining musket shots fired. Mr Mouton also signalled with two handkerchief which was an appeal for help, but we were noticed only when we were close to the ship, then the *Astrolabe* which was the closest sailed towards us; at half-past 4 I passed them those who were the most seriously wounded. Mr Mouton did the same and we went on immediately on board the *Boussole* where I told the sad news to the commander. His surprise (after the precautions his prudence had inspired him to take and the justified confidence he had in Mr de Langle's caution) was extreme and I can only compare his distress to what I felt myself. This disaster sharply put us in mind of the one suffered on 13 July 1786 at Port des Français and increases the bitterness we feel about this voyage, though still lucky that the greater part of those who went ashore today have escaped, one sixth perished and without doubt none of us would have escaped had the fury of the natives not been stopped or diverted by their eagerness for pillage.

We lost from the *Astrolabe* the captain the Viscount de Langle, Jean Nedelec, Laurent Robin, Yves Hamon, François Ferré, sailors, Jean Giraud, servant, and a Chinese, total eight people, the *Boussole* only lost Mr de Lamanon, naturalist, Talin, master-at-arms, Roth and Joseph Rais, soldiers, total twelve men. Everyone was more or less badly hurt, those who were the most seriously are, on the *Boussole*, Messrs Boutin, Colinet and a soldier; and on the *Astrolabe*, Mr Lavo, the master-at-arms, the coxswain of the longboat and a sailor.[1] But it is confidently hoped that none of these wounds will be fatal. It is impossible to describe the effect this tragic event had on the two frigates, the death of Mr de Langle who had the trust and friendship of his crew caused despair throughout the *Astrolabe*. The islanders from the canoes who were unaware of this happening and who were alongside when I arrived almost became its victims and we had all the trouble in the world to prevent our sailors from sacrificing them to their justifiable resentment. The general sorrow which reigned on board is the finest testimonial to this captain; as for me I have lost a friend much more than a commanding officer, he showed a concern for me which will make me regret him for the rest of my life, I would have been only too happy if I could have proved to him my attachment and my gratitude, but this brave man, more prominently placed than the others, had been the first to fall under the blows of the ferocious beasts assailing us; in the weak state caused by my convalescence I had gone to shore without any

---

[1] Vaujuas mentions that eight people had been killed, but lists only seven. The omission is Louis David, an assistant gunner. The absence of a reference to Fr Receveur suggests that his injuries were not serious enough to explain his death a few weeks later, although Vaujuas does mention that Fr Receveur went ashore because he was not in good health and needed a break from shipboard life.

weapons and under the protection of the others; all the ammunition was used up or wet by the time I got to the boat where all I could do was to issue orders which unhappily were useless.

I would be unfair to those who like me had the good fortune to escape if I omitted to state that all of them behaved with utmost bravery and coolness. Messrs Boutin and Colinet who in spite of serious wounds kept their presence of mind were good enough to help me with advice I found most useful; I was also excellently helped by Mr de Gobien who was the last to leave the longboat and whose example, daring and words played an important part in reassuring those of the sailors who showed signs of anxiety; the petty officers, sailors and soldiers carried out with all the zeal and precision possible the orders they were given. Mr Mouton similarly had only praise for the *Boussole*'s crew, he had stopped as soon as he got away from the reef in order to throw overboard all his casks and make room for those who, having escaped from the carnage, had not been able to get into the *Astrolabe*'s boat and who consequently owe him their lives.

Everyone who went ashore can testify as I do that no careless action and no violent act on our part preceded the natives' attack, our captain had given the strictest orders in this respect and as no one ignored them I am sure they were carried out. Signed Vaujuas.'

On the morning of 14 December I set sail for the island of Oyolava[1] which we had sighted five days earlier before we reached the anchorage that proved so disastrous. It is the one of which the southern part was identified from a great distance by Mr de Bougainville and which he indicated on the map of this archipelago he provided ; it is separated from Mahouna (or Massacre) Island by a strait of approximately 9 leagues, and the island of Thaiti can hardly stand comparison with it in respect of its beauty, size, fertility and enormous population; when we got to within 3 leagues of its north-eastern point we were surrounded by numberless canoes loaded with breadfruit, coconuts, bananas, sugar cane, pigeons and sultana hens; but very few pigs. These Indians no doubt assuming that we preferred fruit to them. They were absolutely similar to those of Mahouna Island by whom we had been so

---

[1] The island is Upolu. The strait is somewhat wider than La Pérouse's estimate of 9 leagues: the distance between Cape Tapaga on south-east Upolu and Taputapu at the western extremity of Tutuila is closer to 40 miles (slightly over 11 leagues).

perfidiously betrayed; their costume, their features, their gigantic size were so much the same that our sailors declared they could identify several of the murderers, and I had great difficulty in preventing them from firing on them, but I was sure that they were blinded by their just anger, and a revenge which I had not felt was permissible against canoes from Mahouna itself, who were alongside my ship when I learned of this awful event, could not be taken four days later in another island 15 leagues from the battlefield. I therefore succeeded in calming this excitement and we carried on our trade but with more tranquility and good faith than at Mahouna Island because we punished the slightest act of injustice with blows or threats and angry words, and the islanders of Mahouna had taken our earlier moderation as a sign of weakness. At 4 o'clock in the afternoon we hove to athwart the largest village of any island of the South Seas,[1] or rather opposite a very wide plain filled with houses from the hilltops down to the edge of the sea; these mountains are situated in roughly the middle of the island where the land slopes down very gently and offers vessels the sight of a superb amphitheatre covered with trees, huts and greenery. One could see the smoke of fires, as in the centre of a large town, and the sea was filled with canoes trying to get near our vessels; several were rowed by men who were led by curiosity, having nothing to sell us, and went round our vessels, seeming to have no other aim than to enjoy the spectacle we provided.

As there were women and children among them it was an almost sure sign that they harboured no evil intentions, but we had strong reasons not to trust in this any longer, and our arrangements were made to repel the slightest hostile act in a way that would have caused these islanders to be fearful of any future navigators. I tend somewhat to believe that we are the first to trade with these people, they had no knowledge of iron, they continually scorned it and they preferred a single bead to an axe or a six-inch nail; in general they looked for nothing in their bartering that might be of any use to them; they are a wealthy people who require only superfluities and luxury items. Among a fairly large number of women I noticed two or three who were very pretty and who one could have

---

[1] Apia is today the capital of Western Samoa; it grew rapidly in size during the early to mid-nineteenth century, but even in the eighteenth century it was an important settlement.

CHAPTER XIX

thought had served as a model for the charming drawing of the Present Bearer of Cook's third voyage,[1] their hair was adorned with flowers and a green ribbon like a head-band plaited with grass and moss, their shape was elegant, their arms rounded and very well proportioned, their eyes, their features, their movements spoke of gentleness whereas those of the men depicted ferocity and surprise. In any one sculptor's study the latter would have been taken for Hercules and the young women for Diana, or her nymphs whose complexion would have been exposed for quite some time to the effects of the open air and the sun.

At nightfall we continued on our way, coasting along the island, and the canoes returned to land where we could see numerous breakers and which offered no shelter to vessels because it was the northern coast which is battered by N.E. winds and a very rough sea; it is more likely that if I had intended to anchor I would have found an excellent shelter in the western part and navigators should almost always seek anchorages to leeward of islands situated between the tropics; I was held in a dead calm throughout the next day. There was a great deal of lightning, thunder and rain; very few canoes came out, which convinced me that they had been told at Oyolava about the events that had befallen us in the island of Mahouna and moral science was not sufficiently advanced among these people for them to know that we would have not wanted any other victims than the murderers themselves; and that they had already on the previous day tested our moderation; since it is possible that the storm and the lightning flashes might have kept the canoes in their harbours, my opinion was only a guess which became much more of a probability on the 17th when we were off the island of Pola[2] which we coasted along much nearer land than we had done with Oyolava and no canoe came up to us; this island, a little smaller than Oyolava but as beautiful, is only separated from it by a channel of about 4 leagues obstructed by two fairly large islands[3] one of

---

[1] The reference is to an engraving by Francesco Bartolozzi of a drawing by John Webber, entitled *A Young Woman of Otaheite, bringing a Present*, which appeared in Cook and King, *A Voyage to the Pacific Ocean*, London, 1784, pl. 27. The engraving gives a highly stylised and romanticised image of an island girl.

[2] Pola, which La Pérouse accurately situates between 13°26′ and 13°42′S and 174°16′ and 174°54′W of Paris, is the island of Savaii. This was a major new discovery, as neither Roggeveen nor Bougainville had sighted it.

[3] Both new discoveries, the islands in Apolima Strait are Manono and Apolima. The latter is of volcanic origin, while Manono consists of basalt and sandy soil.

413

which, very low lying and very wooded, is certainly inhabited. The north coast of Pola is impracticable for ships as are the other islands of this group, but when we rounded its western point[1] we saw a calm sea and clear of breakers which gave promise of some excellent roadsteads; as I believe the public is owed some detailed information on these islands I shall make it the subject of the next chapter.

---

[1] The western point of Savaii is Cape Pua'a.

CHAPTER XX

Departure from the island of Mahouna, new details on the customs and practices of these people, their crafts and the country's products. Basis for a belief that they do not all share the same origin, and that the natives of these islands were before the mixing of the two black nations dark and frizzy-haired like the inhabitants of New Guinea and the Hebrides, their form of government maintains their ferocity; we come upon the islands of Cocos and Traitors further east than their location reported by Captain Wallis. We think that they belong to the ten islands which the inhabitants of Mahouna told us comprised their archipelago, the islanders from both islands promptly come on board and offer us coconuts, breadfruit and bananas in exchange for our beads and our iron tools which they did not scorn like the Navigators; a heavy squall which capsizes many of the canoes compels us to put an end to such a satisfactory trade and go on our way towards the Friendly Islands. The islanders of the Cocos and Traitors are not so tall and robust as those of the Navigators Islands, but their canoes and their dress are absolutely similar. We come upon Vavao Island and various adjacent islands forming part of the Friendly Islands which Cap. Cook had known about only from the islanders' reports but which the pilot Morel had sighted before us: precise determination of these islands which he had situated 5 or 6 degrees further east than their true position. Comparison of our longitudes with those of Captain Cook. At the island of Tongataboo they only differed by 6 minutes. The inhabitants of Tongataboo hasten to come on board and trade with us. Quite remarkable difference between the physique of these people and of those we had already seen. The former's size does not differ in any respect from the Europeans' and their behaviour, although noisy, has nothing wild about it. We come upon Pilstard Island which is only a rock less than a mile round and has been badly described by the navigators who have preceded us. We anchor at Norfolk Island, where we find such a rough sea that our boats cannot find a sheltered place to land. Description of this island. Arrival at Botany Bay in New Holland, *where we meet the English fleet commanded by Cap^n Philip. It had arrived a mere 5 days before us.*

We learned from the islanders themselves that the Navigators archipelago consisted of ten islands, Opouna the easternmost, Léoné, Fanfoué, Mahouna, Oyolava, Calinassé, Pola, Shika, Ossano and Ouera.[1]

We are unaware of the last three's position. They placed them (on a map we drew in front of them) to the south of Oyolava, but if they were where they claimed it would be impossible, from Mr de Bougainville's route, that this navigator did not see them and all the patience and sagacity of Mr Blondelas who devoted himself particularly to obtaining some geographical clarification from the islanders did not allow him to hazard any guess on their position, but the remainder of our navigation taught us that two of the three islands we were looking for might be the Cocos and Traitors islands which Captain Wallis's observations had placed one degree fifteen minutes too far west.

Apouna, the southernmost and easternmost of these islands, lies in 14^d $7'$ of latitude and 189^d $51'$ of longitude Paris meridian, the latitude of Pola, the northernmost and westernmost of this archipelago, is in 13 and [*blank*] of longitude; a glance at the chart will give a better impression of their respective positions, their sizes, their distances relative to each other than any details; one point of each island is shown on the same chart with precise latitude and longitude determinations obtained from several lunar distances which were used to correct the errors of our chronometers.

---

[1] These names reveal a state of utter confusion. Some are district names, others those of villages, possibly even of local chiefs. The easternmost island is not Opouna, but Tau – this 'Opouna', later spelled Apouna, might be the island of Aunu'u east of Tutuila; Leone is a village and bay in southern Tutuila; Fanfoue seems to indicate Olosega; Mahouna is not the island to which La Pérouse refers, but is the name (Manua) of the group to which Olosega and Tau belong; the position La Pérouse gives for Oyolava is that of Upolu, and he may have picked up from his informants the name of Aolo'au Bay on Tutuila; his Pola is the island of Savaii, but there was a district called Polo'a in Tutuila. Directions tend to be roughly accurate but, as often happens with local informants, distances are much smaller than the Europeans, accustomed to travelling further afield, realised. All this shows how dangerous it is to interpret information provided by natives in a language the visitors did not understand. La Pérouse had encountered similar problems in the Gulf of Tartary.

PART OF MAOUNA

24. Chart of part of Maouna (Tutuila), showing the track of the expedition in December 1787. By Blondela. Actual size 46 cm × 34 cm. AN 6 JJ1:47.

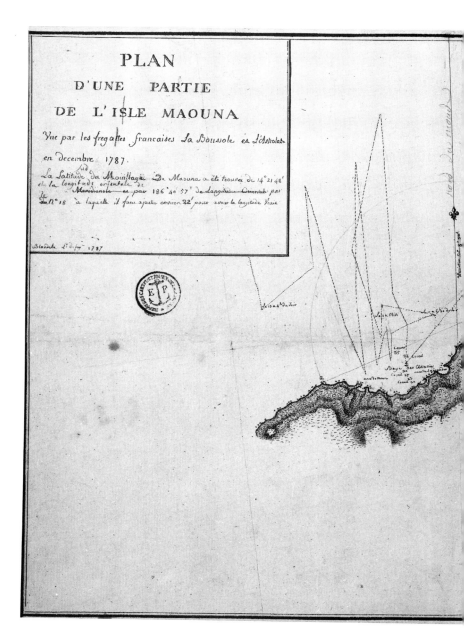

PLAN
D'UNE PARTIE
DE L'ISLE MAOUNA

Vue par les fregattes françaises La Boussole et L'Astrolabe en decembre 1787.

La Latitude Sud du Mouillage de Maouna a été trouvée de 14° 21' 48" et la longitude orientale Meridionale est par 186° 40' 57" de longitude Occidentale par le N° 18 à laquelle il faut ajouter environ 22' pour avoir la longitude vraie.

Blondela Lt. de fr. 1787

Echelle de 5 Lieues Marines

l'aiguade

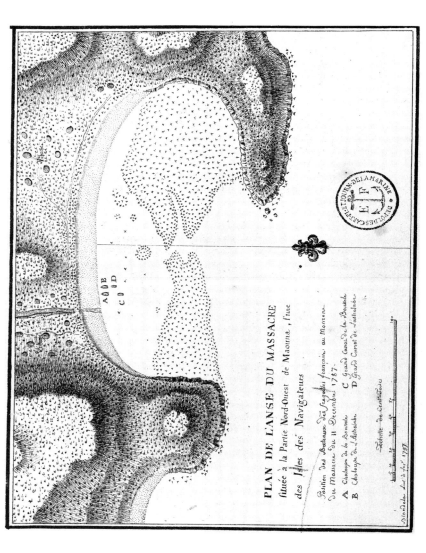

25. Massacre Cove, Maouna (Tutuila), showing the position of the longboats (near the shore) and the supporting boats on 11 December 1787. By Blondela. AN 6 JJ1:46.

CHAPTER XX

Several geographers attribute this discovery to Rogewin who in 1721 named them Beauman Islands, but neither the historical information on these people nor the geographical location the historian of Rogewin's voyage assigns to these islands coincides with this opinion.

The historical account of Rogewin's voyage reported by Président de Brosses was written in the French language in 1739 by a German born in Mecklembourg, a sergeant-major of the troops who sailed with Rogewin's fleet[1] 'we discovered, says the sergeant-major, three islands together in the 12th degree of latitude; they seemed of a very pleasant appearance, we found them covered with fine fruit trees and all kinds of grasses, vegetables and plants. The islanders who came to meet our vessels offered us all kinds of fish, coconuts, bananas, and other excellent fruit, these islands must be well populated, since when we arrived the shore was lined with several thousand men and women. Most of the former carried bows and arrows. All those who inhabit these islands are white and differ from Europeans in no other respect than that they are sunburned; they seemed to be good people, lively, and merry in their conversation, gentle and humane towards each other and there was no sign of savagery in their manners, nor did they have their bodies painted like those of the islands we had discovered earlier, they were dressed from the waist to their heels with a fringed cloth of artistically woven silk, their heads were covered with a similar very fine and very large hat to protect them from the fierceness of the sun; some of these islands had a circumference of ten, fourteen and up to twenty miles. We called them Beauman Islands after the

---

[1] Jacob Roggeveen (1659–1729) sailed from Texel on 1 August 1721 with the *Arend*, the *Thienhoven* and the *Afrikaansche Galey*, entered the Pacific through the Strait of Le Maire and discovered Easter Island on 5 April 1722; in June he discovered the Manua group, Tutuila and Upolu. Tutuila was called Thienhoven and Upolu Groeningen. When he reached the Dutch East Indies, his remaining two ships were seized on the grounds that the voyage infringed the Dutch East Indies Company's monopoly. Cornelis Bouman was captain of the *Thienhoven*, and the Manua group was named after him. The historian in question was Carl Friedrich Behrens, a native of Mecklenburg who took part in the expedition and wrote *Reise durch Süd-Länder und um die Welt*, Frankfurt and Leipzig, 1737, which contains a number of inaccuracies and exaggerations; it was translated into French and widely read under the title *Histoire de l'expédition de trois vaisseaux envoyés par la Compagnie des Indes occidentales des Provinces-Unies aux terres australes en MDCCXXI*, The Hague, 1739. La Pérouse underlined in this paragraph the parts which he considered to be totally at variance with what he had found when he sailed off Tau, Olosega and the rest of the Samoas.

captain of the ship *Tienhoven* who had been the first to sight them; it must be said, adds the author, that it is the most humanised and the most honest nation we found in the South Seas; <u>the entire coastline of these islands provide good anchorages, one can anchor in between 13 and 20 fathoms of water.</u>'

The remainder of this chapter will prove that these details have almost no connection with those we have to give on the people of the Navigators Islands, and since the geographical position is no more applicable to them (there is a German map on which Rogewin's route is traced and which situates it in 15 degrees which is not correct) I believe Beauman's islands not to be the same as those to which Mr de Bougainville gave the name of Navigators and which it seems to me to be necessary to retain for fear of introducing into geography a confusion of names that would be very harmful to the progress of that science. These islands situated between the 13th and the 14th degrees of latitude and the 18th and [*blank*] of longitude Paris meridian, comprise one of the finest archipelagos anywhere in the South Seas, as interesting for its crafts, its products and its population as those of the Society or Friendly islands which the English navigators have made so well known to us;[1] although we remained only a moment among these people, our misfortunes enabled us to analyse their character better than if frightened by our weapons fear had made them hide, and the most cruel experience taught us that it is futile to try to obtain through good turns the goodwill of these ferocious souls which can only be held back by fear.

The people of these islands are the tallest and most robustly built we have met. Their quite normal size is 5 feet 9, 10 or 11 inches, but they are less surprising by their height than by the colossal proportions of their bodies; our curiosity which led us to take quite frequent measurements enabled them to compare their physical strength to ours, which was not in our favour, and we may owe our misfortunes to the idea of superiority which they gained from these various trials; their expression often seemed to me to indicate

---

[1] The English navigators are, for the Society Islands, Samuel Wallis and James Cook, but La Pérouse could have mentioned his compatriot Louis de Bougainville's immensely popular account. As far as the Friendly Islands (the Tongan group) are concerned, the English navigators are also Wallis and Cook, but the first European visitors were Dutch: Schouten and Le Maire in 1616 and Abel Tasman in 1643. The name Friendly Archipelago was bestowed on them by Cook.

CHAPTER XX

a feeling of scorn towards us, which I thought I could dispel by ordering our weapons to be used in front of them, my purpose would have been met only if I had had them aimed at victims, anyhow they looked upon the noise as a diversion and a joke.

A very small number of these islanders are below the size I have mentioned, and I had some measured at 5 feet 4 inches, but they are this country's dwarves, and the latter's strong and sinewy arms, their wide chest, their legs and their thighs are very differently proportioned to ours; it can be stated that they are to Europeans what Danish horses are to those of our various provinces.

The men are painted or tattooed in such a way that one could almost believe they are clothed although they are always absolutely naked, having only a belt of seaweed around the waist and down to the knees which makes them look like these rivers that the fable has personified and our painters adorned with reeds. Their hair is very long, they often tie it up over their heads and thereby increase the ferocity of their features; they always display either surprise or anger, the slightest dispute that arises among them is followed by blows, with sticks, clubs or paddles, which can often cost them their lives, and they are all either wounded or covered with scars that must be the result of these private fights. The women also are very tall and before their springtime has ended they have lost the shapes and that gentle expression, which Nature has never withheld from these uncivilised people but which it seems to leave with them for only an instant and reluctantly. Among a very large number of women I found only three who were very pretty; the rough impudent expression of the others, their indecent gestures, the off-putting way they offered their favours, everything made them in our eyes seem worthy of being the wives or the mothers of the ferocious beings surrounding us. As the story of our voyage can add a few pages to that of mankind I will not omit pictures that might shock in any other kind of book and I shall mention that the very small number of young and pretty island girls I referred to soon attracted the attention of a few Frenchmen who in spite of my orders endeavoured to establish links of intimacy with them; since our Frenchmen's eyes revealed their desires they were soon discovered; some old women negotiated the transaction, an altar was set up in the most prominent hut, all the blinds were lowered, inquisitive spectators were driven off; the victim was placed within the arms of an old man who exhorted her to moderate her sorrow, for she was weeping; the matrons sang and howled during the ceremony,[1] and the sacrifice was consummated in the presence of the women and the old man who was acting as altar and priest. All the village's women and children were around and outside the house, lightly raising the blinds and seeking the slightest gaps between the mats to enjoy this spectacle. Whatever navigators who preceded us might say, I am convinced that at least in the Navigators Islands girls are mistresses of their own favours before marriage, their complaisance casts no dishonour on them, and it is more than likely that when they marry they are under no obligation to account of their past behaviour. But I have no doubt that they are required to show more restraint when they are married.

Some of their crafts are also very advanced, I have already mentioned the elegant shape of their houses, and they rightfully rejected all our iron tools, since they managed so well all their wood working with axes made of a very fine and very compact basalt shaped into adzes; and they sold us for a few glass beads wooden dishes affixed to three feet holding them up like a tripod and which seemed to be painted with the finest varnish, a good European worker would have needed several days to carry out such a task which on account of their lack of tools must take them several months of work;[2] they set almost no value on it because their time has none, trees and roots of [blank] and haro[3] growing wild ensure their subsistence, that of their pigs, their dogs, their hens, and they spend their days in idleness or engaged in tasks that have no other purpose than their clothing and their luxury; they know and they make some paper-cloth similar to that of the Society and Friendly Islands; they sold us several lengths of a single reddish-brown colour. It seems that they do not prize it very much and have little use for it, the women prefer mats that are extremely well plaited and I saw only two or three men whom I took to be chiefs who had instead of a grass skirt a length of material wrapped around them like a skirt, this cloth is woven with a true thread drawn no doubt

---

[1] This may be the wedding chant or *tigi*.

[2] La Pérouse is presumably referring to the dishes used in kava ceremonies, some of which are very ornate.

[3] Milet-Mureau, presumably perplexed by the word 'haro', omitted it from the printed account of the voyage. It is an error for taro (in Samoan *talo*), the *Colocasia antiquorum* which is widespread throughout Polynesia.

CHAPTER XX

from some ligneous plant, like a nettle or flax,[1] it is made without a shuttle and the threads are woven through absolutely as with the mats, this cloth has both the suppleness and the strength of our own, is very suitable for their canoe sails and cannot be compared in respect of its advantages to the paper cloth of the other islands which they also manufacture but seem to disdain.

At first we had found no similarity of language with our vocabularies from the Society and Friendly Islands but more careful study convinced us that they speak a dialect of the same tongue;[2] which adds further confirmation of the view held by the English on these people's origin; that is because a young servant from Manila who was born in Cagayan province north of Manila could understand and explain to us the greater part of the islanders' words; it is known that Coyayon, Talgale[3] and in general all the languages of the Philippines are derived from Malay, which language more widely spread than Greek or Latin is used by the innumerable peoples who live in the South Sea in the islands of both hemispheres; I consider it proved that these various nations are merely Malay colonies which in very remote periods conquered these islands, and the so-called antiquity of the Chinese and Egyptians, the reigns of Sesotris &c may well be recent events compared with these which I would not be rash enough to give a date for, but I am convinced that the indigenous people of the Philippines, of Formosa, New Guinea, New Britain, the Hebrides, the Friendly Isles &c in the southern hemisphere: and of the Carolines, the Marianas, the Sandwich Islands in the northern hemisphere were these frizzy-haired men who still live deep in the interior of the Luzon islands and of Formosa, whom it was impossible to subjugate in New Guinea, New Britain and the Hebrides and who, defeated in the islands further east, which were too small for them to find a refuge

---

[1] The leaves of the *ti* (*Cordyline terminalis*) were normally used for this. La Pérouse may be referring to the *'ie tutu pupu'u*, a high-quality loin-cloth reserved for chiefs. See on this Te Rangi Hiroa, *Samoan Material Culture*, Honolulu, 1930, p. 261.

[2] In fact, Samoan can be considered as one of the purer forms of the Polynesian language. A number of specialists consider the Samoan Islands to have been the cradle of Polynesia, from where migrating parties left for the Marquesas and other archipelagos within the Polynesian triangle. However, allowance must be made for the natural evolution of a language throughout the centuries in a closed environment, even if one sees it as the mother tongue from which different dialects developed.

[3] 'Talgale' is Tagalog, the major native language of the Philippines. 'Coyayon' is a mis-spelling for Cagayan in northern Luzon where several dialects still subsist.

in the centre of these said islands, intermarried with the conquering people and gave rise to that very dark race of men whose colour retains ten shades more than the skin of those families presumably distinguished in their countries who made it a point of honour not to marry beneath them. We were particularly impressed by these two very distinct races in the Navigators Islands and I do not attribute any other origin to this.[1] These people have acquired in these islands a vigour, a strength, a size and proportions which they did not inherit from their fathers and no doubt owe to the abundance of food and the ease with which it is obtained, to the mildness of the climate and to various physical causes that generally influence all living beings after a great number of generations. The crafts they may have brought with them have been lost on account of the lack of the appropriate raw materials, and so they no longer have any knowledge of iron or shuttles; their marine soon reduced to canoes has not enabled them to make long voyages, they became isolated, and forgot an origin about which it would be foolish to speculate if the similarity of language did not provide in this labyrinth the Oriane's thread[2] that enables all the twists and turns to be followed. The feudal form of government has also been kept here and it still

---

[1] This lengthy analysis indicates that La Pérouse and no doubt his officers with him thought at length about the origins of the peoples of the South Seas. He notices the presence of two different races, the Polynesians and the Melanesians, but also the wide range of features and skin shades which made it difficult for scientists and ethnologists to establish clear dividing lines. The French show an awareness of the influence of the physical and cultural environment in which people live and, long before the theories blossomed into Darwinism, speculate on evolution. He is clearly of the view that migrations, in successive waves, occurred in distant times, disagreeing with those who held the view that it was a cohesive movement of relatively recent date. Current theories are that, after a slow and progressive advance along the Indonesian chain and from East Asia, skirting the more southerly lands earlier occupied by the Melanesians' Austronesian ancestors, a new race emerged, centred on the Tongan-Samoan archipelagos, which some four or thousand years ago developed a culture adapted to its island environment and in time spread to other distant island groups. The question cannot easily be resolved and differing views are still strongly defended. A number are discussed in recent works, such as P. Bellwood, *Man's Conquest of the Pacific*, Auckland, 1978; R. & M.E. Shulter, *Oceanic Prehistory*, Menlo Park, 1975; and K.R. Howe, *Where the Waves Fall*, Sydney, 1984, especially his pp. 5–24 and his useful detailed bibliography.

[2] 'Oriane' is Ariadne, daughter of King Minos of Crete, who enabled Theseus to escape from the Minotaur's labyrinth by providing him with a ball of string he unwound as he entered and wound up again to retrace his steps. The reference to canoes making long voyages impossible is not quite correct. Travel to distant islands was possible, but apart from migrations due to overpopulation the incentive for long voyages, e.g. for trade, did not exist.

CHAPTER XX

exists in all its force in the motherland.[1] This form of government which minor tyrants might regret, which tarnished Europe for some centuries and whose remnants still complicate our laws, and are the tokens of our former barbarousness, this government is the one most likely to preserve the ferocity of daily life, because the slightest interests lead to warfare between one village and the next and these kinds of wars are fought without magnanimity, without courage, sudden attacks and treachery being used in turn, and in these unfortunate countries instead of noble warriors one finds only murderers. The Malays are still today the most perfidious nation in Asia and their children have not degenerated because the same motivations have retained and produced the same results. It will be argued perhaps that it must have been very difficult for the Malays to travel from west to east to reach these various islands, but in the neighbourhood of the equator within a belt of seven or eight degrees north and south the westerly winds are at least as frequent as those from the east and they are so variable among these different islands that it is hardly more difficult to sail towards the east as to go down to the west; moreover these conquests did not occur at the same period, they spread from one group to the next and everywhere introduced this form of government which still exists in the Malay peninsula, in Java, Sumatra, Borneo and in all the countries that are under the sway of that barbarous nation.

Among the fifteen or eighteen hundred islanders we had the opportunity of observing at least thirty declared themselves to be chiefs, they exercised a form of policing and doled out great blows with their sticks, but the order they seemed intent on maintaining was upset in less than a minute, never were sovereigns less obeyed, and never were insurbordination and anarchy more tumultuous.

Mr de Bougainville was quite justified in naming them the Navigators; all their travel is made by canoe and they never walk from one village to another. These villages are all situated in coves at the water's edge and tracks are all they have to penetrate into the interior of the island which is covered up to the hilltops with fruit-laden trees where the most charming birds are perched, such as wood-pigeons, turtles-doves, green, pink or of various other

---

[1] The reference to feudalism surviving in the motherland, i.e. France, was naturally deleted from the printed edition which was published in 1797 after the French Revolution had swept away most of the old privileges of the aristocracy.

colours; charming parakeets, a type of blackbird, we even saw some partridges[1] and they sold us more than three hundred sultan hens with the finest plumage; they relieve the boredom caused by their idleness by taming these birds, their houses were filled with these doves I mentioned earlier which they bartered by the hundred and which seemed to eat only when fed by hand.

Their canoes are fitted with an outrigger[2] and are very small although they can take up to 14 people but most only take five or six; they did not seem to us to deserve the praise the travellers lavished on their speed; I do not believe it exceeds seven knots under sail, and with the paddles they were unable to keep up with us when we were doing four miles an hour. These Indians are such good swimmers that they use their canoes only to rest in, since they get swamped with the slightest movement they are forced time and again to jump into the sea, hold them up on their shoulders and empty them; they sometimes join two of these canoes together by means of a wooden crosspiece[3] in which they make a mast-hole, in this way they capsize less frequently and they can keep their supplies for long voyages; their sails made of mats or plaited cloth are spritsails and do not warrant any special description.

They only fish with a line or with a cast-net; they sold us nets and fish-hooks that were were very artistically fashioned out of mother-of-pearl or of a white shell in the shape of a flying fish concealing a hook made of a turtle shell that was strong enough to hold a tunny fish, a bonito or a dorado.[4] They bartered the largest fish for a few glass beads and we could see they were not worried about any food shortage.

These islands are of volcanic origin and all the stones along the shore were only pieces of lava and smoothed basalt or of the coral which surrounds the whole island and on which the sea breaks with

[1] The blackbird is probably the *Tutumalili* (*Turdus paliocephalus samoensis*) which is a black bird with a yellow beak. There were no partridges in the Samoas: La Pérouse presumably saw a *ve'a* (*Rallus philippensis goodsoni*) a small bird with parallel brown and whiteish stripes, or some *puna'e* (*Pareudiastes pacificus*), formely widespread in Tutuila and Savaii, but now extinct.
[2] Samoan canoes were of different sizes and types. They ranged from the simple hollowed-out tree-trunk, the *paopao*, to the double-outrigger *va'ale*, and the double decked *taumu'ala*.
[3] The double canoe, *'alia*, was used for inter-island voyaging. The sail was of a quadrangular shape, suspended from a large crosspiece and lashed to the mast.
[4] The bonito hooks were known as *pa'atu*; their shape and composition varied according to the specific fishing requirements and often from district to district.

CHAPTER XX

a fury that sends the waves up to a height of more than 50 feet; this coral leaves in the middle of almost all the coves a narrow pass for canoes or even boats and longboats and thus creates very adequate small harbours for the islanders who never leave their canoes in the water; when they arrive they store them near their houses as we do with our carriages: they place them in the shade under the trees and they are so light that two men can carry them on their shoulders. The liveliest imagination could hardly conjure up more agreeable sites than those where their villages are located: under trees laden with fruit that retain a delicious coolness, on the banks of a stream trickling down from the hills, along which they have made for half a league into the interior a small path lined with houses; all their architectural style is designed to preserve them from the heat and I have already stated that they have not overlooked the need for elegance; the houses are surrounded by roll-up blinds which they raise up on the side of the breeze and lower on the sunny side; they are large enough for several families; they sleep on very fine and very clean mats laid on the soil which has been raised up at least two feet above ground level, and quite protected from the damp; we saw no *moraï* and we cannot report anything on their religious ceremonies.

Pigs, dogs,[1] hens, birds, fish abound in these islands which are covered with coconut trees, breadfruit trees,[2] guava, grapefruit, bananas and another tree which produces a large almond one eats cooked and which we found tasted like chestnuts. Sugar canes grow wild along the riverbanks[3] with plants whose roots are very similar to those of yams or camagoncs;[4] the sugar canes are watery

---

[1] The Samoan dog, the *'uli*, was small and had pointed ears.

[2] La Pérouse uses the term *rimas* which so puzzled Milet-Mureau that he left it out of the printed version. Rima was a vernacular term for the breadfruit tree in use in the Philippines. It appears in P. Sonnerat, *Voyage à la Nouvelle-Guinée*, Paris, 1776, which La Pérouse no doubt had on board and which includes a section on the Philippines. The term had a very limited use.

[3] Sugar cane, or *tolo*, was used as food, but more for roofing. Among other products mentioned, the coconut is the *cocos nucifera*, which was very widespread in the Samoas and later formed the basis for a substantial copra industry; the banana is the *fa'i*, now one of Samoa's major exports; the grapefruit is the *citrus maximus* – La Pérouse uses the term *pomples moussiers* which Milet-Mureau could not identify and left out of the printed edition: the French name *pamplemousse* is now in general use and comes from the Dutch term *pompelmoes*.

[4] Yams or *ufi* (*Dioscorea*) were widespread in the islands. La Pérouse may also have seen the masoa (*Tacca pinnatifida*), sometimes called the Polynesian arrowroot.

and less sweet than those of our colonies, no doubt because they grow in the shade on too heavy a land that has never been worked over. Dangerous though it was to venture into the interior of the island, Mr de la Martiniere and Colignon obeyed more the impulses of their zeal than the rules of caution, they collected several plants and Mr de la Martinière who was attacked with stones swam towards our boats keeping his bag on his back. The Indians wanted one glass bead for every plant he picked up and threatened to knock him down with their stones if he refused to pay; until the 11th we saw no other weapons than clubs or *patow patow*, but Mr Boutin assured me he had seen that afternoon several bundles of arrows without any bow;[1] I am rather inclined to think that these arrows were spears used for fishing, and they would be a hundred times less dangerous than the stones weighing two or three pounds they throw with extreme skill and the power of a sling. These islands are very extensive and extremely fertile, and I believe they have a very considerable population; the eastern ones, Opouna, Léoné, Fanfoué, are small, the last two especially have a circumference of scarcely five miles, but Mahouna where we suffered such a great misfortune, Oyolava in particular and Pola must be reckoned among the largest and the finest in the South Sea. There is nothing in the various navigators' accounts to compare with the beauty and the great size of the village to leeward of which we hove to on the northern coast of Oyolava; it was a little late, however the sea was filled with canoes which the desire of seeing us or of trading with us brought out from their harbour; several brought nothing and were simply coming to enjoy the spectacle we provided; some were very small, containing only one man, the latter were highly decorated with at the end of every peg a shell of the porcelain variety;[2] since they were going around the ships without attempting any barter we called them the gigs – they had their drawbacks, at any moment the slightest choc from other canoes caused them to capsize; we had no intercourse with the large and superb island of

---

[1] The observation is correct. Samoans did have arrows (and bows) which they used mainly for fishing in rivers or lagoons. The bow or *aufana* was made with the wood of the *fisoa* (*Columbrina asiatica*); the arrows were made by fixing two or three slivers of hardwood to a reed or bamboo stalk, and their length was variable.

[2] These decorations were often signs of rank or wealth. The white shell of the *pule* (*Ovulum*) was quite scarce and consequently most valuable and sought after as an indication of status.

CHAPTER XX

Pola which we hauled along very close in, which made me suspect that the news of our disaster had reached them; rounding the western extremity of this latter island we saw a calm sea that seemed to offer a promise of good anchoring places at least when the winds were north to south by east; the tension among our crews was still too great for me to decide to anchor there; it was impossible to send them ashore after what had happened to us, without arming each man with a musket and each boat with a swivel-gun and everything the islanders did would have seemed an unjust action which they would have repressed with musket shots, which could have been fatal to both sides; moreover in these mediocre anchoring places a ship always runs the risk of destruction when there is no boat capable of carrying an anchor which one can use for warping the ship, and I had taken the firm decision to anchor only in Botany Bay where I planned to built a new longboat for which we had the material on board, but I had decided to sail through this labyrinth of islands along routes which might make me discover islands for the advancement of geography, and I planned to trade with these people while hove to near their islands. On the evening of 17 December we rounded the western coast of Pola Island and lost sight of land; there were only three islands missing which the islanders had called Shika, Ossamo, Ouera, and which they had located to the south of Oyolava; I endeavoured to sail south-south-east. The E.S.E. winds at first were unfavourable, they were very weak and we were making only eight or ten leagues a day; they finally veered north and then north-west, which enabled me to progress toward the east and to sight on the 20th a round island precisely to the south of Oyolava but at a distance of almost 40 leagues. Mr de Bougainville who had sailed between the two islands did not see it because he was a few leagues too far north; the lack of wind did not allow me to approach it the same day, but on the morrow I came to within two leagues and I saw two other small islands south of the first one, which I identified easily as the islands of Cocos and Traitors of Schouten or of Boscawoent and Kepel of Walis, the island of Boscawoent or Cocos is only a very steep sugarloaf covered right up to the top in trees, of a diameter not exceeding one league, separated from Traitors Island which is flat and low lying and has only a fairly high hill in the centre, by a strait of three miles approximately, which is even blocked by an tiny islet which we sighted off the north-east

427

point of Traitors Island, but we had the opportunity of checking and of being certain that the latter is divided in two by a small channel 150 *toises* wide which neither Schouten nor Walis had a chance to see because one needs to be absolutely in line with this strait's opening and we would not even have suspected its existence if we had not coasted along the island very close in this area. We had no further doubt that these three islands which have to be counted as merely two were those of Shika and Ossamo of the Navigators group. As there was a strong N.W. gale, it was late and the weather looked bad,[1] we were not really surprised not to see any canoes coming up to us and I decided to spend the night on short tacks in order to examine these islands more closely the next day, trade with these people and obtain some refreshments; the weather was squally and the winds varied only between N.W. to N.N.W. I had seen a few breakers on the N.W. point of Traitors Island, which caused me to tack out a little to sea. At daylight I sailed closer to this latter island, which being much flatter and larger than Cocos Island seemed to have more inhabitants[2] and at 8 a.m. I hove to two miles W.S.W. of a large sandy bay on the western side of the large Traitors Island where I did not doubt there was an anchorage but in fact impracticable with the north-westerlies blowing at that moment but well sheltered from the east;[3] twenty to twenty-two canoes came out at once from the coast and came up

---

[1] Bougainville had encountered bad weather in these waters, 'Cloudy stormy weather with rain at intervals... Never was there a darker night.' Journal entry of 6–7 May 1768, Taillemite, *Bougainville*, I, p. 338. Cocos Island was discovered by the expedition of Willem Schouten and Jacob Le Maire; on 10 May 1616 a high island was seen, with a longer and lower one to the south of it. They stopped at the former, but were attacked by the inhabitants of the other island which they accordingly named Verraders (Traitors). Cocos is the island of Tafahi, a volcanic island some 2,000 ft in height; Verraders is Niuatoputapu, about five miles from Tafahi. They are outliers of the Tongan group, lying some 150 miles north from Vavau. Samuel Wallis reached them on 16 August 1767 and named them Keppel (Niuatoputapu) and Boscawen (Tafahi). Their position is 15°57′S and 173°46′W. It is probable that a third expedition sighted them, on 12 August 1772, that of the *Mascarin* and the *Marquis de Castries*, led by Le Jar du Clesmeur following the death of the commander, Marion du Fresne, in New Zealand. See A.M. Rochon (ed.) *Nouveau Voyage à la Mer du Sud commencé sous les ordres de M. Marion*, Paris, 1783, p. 171.

[2] This assumption is correct: only 60 or so people live on Tafahi, but Niuatoputapu has a population of close on 1,500.

[3] La Pérouse was wise not to waste time on seeking a better anchorage: the island does not offer much shelter for larger vessels. The anchorage on the west of the island is not good, but it has long been a convenient calling point for Tongan vessels travelling to and from the Samoas.

CHAPTER XX

to the frigates to barter with us; several had come from the strait which separates the two islands from Traitors, they had only coconuts, the finest I had yet seen, a very small quantity of bananas, a few yams and a dozen *chadeqs* or grapefruit,[1] only one had a small pig and three or four hens. It was noticeable that these Indians had already seen or heard about Europeans, they came up without any fear, traded fairly honestly, never refused as those from the Navigators had done to hand over their fruit before receiving payment; they accepted pieces of iron and nails as enthusiastically as beads; moreover they spoke the same language, had the same clothes, the same tattoos and differed only in that they all had the two joints of their left-hand little finger cut off,[2] and I had seen only two individuals in the Navigators Islands who had been operated on in this way. Furthermore one could have no doubt that they were the same people, same language, same canoes, same ferocious expression, they were simply not so tall and less gigantic in their build, no doubt because these islands are less fertile, and the soil there is less suitable for the growth of the human race, which becomes much more dependent on the influence of the soil and its productions and on the climate as it moves further from the civilised condition than the nations of Europe whose institutions have no doubt greatly reduced their physical strength, but this loss is a thousand times compensated for by their arts and moral character. Every island we saw reminded us of one aspect of the islanders' treachery. Rogewin's men had been attacked and stoned at Refreshment Island,[3] east of the Navigators, Schouten's at Traitors Island[4] which was

---

[1] No doubt again puzzled by an unfamiliar word, Milet-Mureau omitted it from the printed version of the voyage. The term is 'shaddock' and refers to the pummelo (*Citrus grandis*) which is considered the parent of the modern grapefruit (*Citrus paradisi* and other varieties). Shaddock is understood to be the name of the English captain who introduced the plant to the West Indies and thereby the fruit to England.

[2] The amputation of one or two finger joints was a sacrifice addressed to the gods to plead for a sick relative. Called *tutunima* ('hand cut'), this practice was widespread in the Tongas.

[3] On 2 June 1722 Roggeveen called at Makatea which he named 'Verquicking' – Refreshment Island. The Dutch obtained a supply of anti-scorbutic plants; hoping to get some more they returned the next day, but ventured into the higher bush, possibly close to some village, and were attacked. Two of Roggeveen's men were wounded. A Sharp (ed.), *The Journal of Jacob Roggeveen*, Oxford, 1970, pp. 137–40.

[4] Willem Schouten was attacked in May 1616. A chief had come on board with his followers and gifts were exchanged, but after he left the islanders attacked the *Eendracht* and the Dutch responded with grapeshot.

before us and south of Mahouna where we had ourselves been so shamefully murdered in such cowardly fashion; these thoughts altered our behaviour towards the Indians; we struck out with sticks at the slightest theft and unfair treatment, we showed our muskets to indicate that flight would not save them from our resentment, we refused to allow them on board and made them clearly understand that we would punish with the death penalty those who would be bold enough to climb up in spite of our orders. These attitudes were a hundred times better than our former moderation, and we only regretted having arrived among these people with principles of moderation and patience which the philosophers may well preach but which navigators must not adopt unless they are to forsake their own safety; common sense and reason tell us that one has every right over a man whose well planned intention would be to kill you if fear did not hold him back,

At midday on the 23rd while we were trading for coconuts with the Indians a strong W.N.W. squall assailed us which dispersed the canoes, several capsized and after being righted they all hastened back to land; the weather was threatening, nevertheless we sailed until four in the afternoon around Traitors Island to see all its features and draw up a precise chart. Mr D'agelet had carried out some very good observations of the latitude at midday, and of the longitude during the morning which enabled him to correct the position assigned to these islands by Captain Walis who places them 1 degree 15 minutes too far west, and at four o'clock I gave the signal to sail S.S.E. towards the Friendly archipelago, whose islands I proposed to survey, which according to Captain Cook's narrative must be north of Inahomooka[1] and which he did not have the opportunity of exploring.

The first night after our departure from Traitors Island was frightful, the winds veered west, wild stormy gale with heavy rain;

---

[1] The island is Nomuka, discovered by Tasman in 1643 and shown as Anamocka on the expedition's charts but actually named Rotterdam by him. Cook used the name Anamocka; he visited the main Tongan group during his second and third voyages, sending parties ashore and setting up observatories. The translation follows the logic of the argument rather than the holograph text: La Pérouse wrote 'doivent être au nord Dinahomooka' ('must be in the north Dinahomooka') which seems to be not merely a strange mis-spelling of Anamocka but makes nonsense of his argument. Nomuka is not one of the northern islands of the Tongan archipelago but in the centre; the Vava'u group in the north was indeed an area which Cook did not explore on his visit to the widespread Tongan archipelago.

CHAPTER XX

as we had not had one league of visibility at sunset, I stayed athwart until dawn head on to S.S.W. The westerlies continued to blow, they were fairly strong and especially accompanied with much rain; all those who showed the slightest symptoms of scurvy suffered greatly from this humidity, and it is somewhat remarkable that none of the crew was attacked by this sickness, but the officers and especially our servants felt its onset; I attribute this to the shortage of fresh food which was not an unusual situation for our sailors to be in but the servants who had never been to sea were not accustomed to this deprivation. David, the officers' cook, died on the 10th of scorbutic dropsy.[1] This is the first man since we sailed from Brest who has died of natural causes on the *Boussole* and if we had done nothing more than an ordinary circumnavigation we would have returned to Europe without losing a single man; in fact the last months are the most difficult as the body weakens and the food begins to go bad, but, if the length of voyages of discovery has limits one cannot exceed, it is very important to assess what they are and I believe that when we reach Europe the experiment will be completed. Out of all the preventives I believe molasses and Prussian beer[2] were the most successful. Our crews did not cease drinking it in hot climates and every day they received a bottle of it with half a pint of wine and a small tot of brandy well diluted with water, which make them feel that the other provisions were bearable; the abundance of coconuts and pigs at Mahouna was only a

---

[1] Although advances were being made in understanding and treating scurvy, medical knowledge of this condition was still sketchy. Furthermore, dropsy was a broad term covering a number of conditions. Poor diet could lead to dropsy – the term 'starvation oedema' has been used. A vitamin deficiency, causing scurvy, could lead to an oedematous condition and lead to cardiac or respiratory failure. In the case of David, the officers' cook whom one might normally have expected not to suffer as badly as others from inadequate food, one should note La Pérouse's comment that officers and their servants showed signs of scurvy whereas the crews seemed to be in better condition; he saw in it a consequence of some difference in diet: one could also argue that the element of stress, which was often a complicating factor, may have affected the officers to a greater extent. Finally, 'David was affected by alcoholic cirrhosis', Tayeau, F., and Kernéis, J.P., 'Les Médecins de Lapérouse au cours du grand voyage', in *Colloque Lapérouse Albi 1985*, p. 332.

[2] Prussian beer, corrected by Milet-Mureau to *sprucebeer* in the printed edition, was a term properly used by La Pérouse. Spruce was an archaic word for Prussia or 'Spruceland', and spruce beer basically means beer from Prussia. James Cook made it from a recipe he had obtained in Newfoundland; it is an extract from the leaves and branches of the spruce fir to which are added molasses and the sap of pine trees; Cook and his officers further improved it by adding a liberal dose of rum. La Pérouse wisely followed Cook's recipe, but favouring brandy over rum.

temporary resource; we could neither salt them because they were too small, nor keep them alive because there was no feed for them; I decided to make two distribution daily to the crew, and swellings on their legs and all the light symptoms of scurvy disappeared; this new diet was equivalent to a long call at a port, obvious proof that sailors are less in need of the land than of healthy nourishment.[1]

The N.N.W. winds followed us beyond the Friendly archipelago, but with rain and often as strong as the westerlies one finds in winter on the coast of Brittany, and although we knew quite well that we were in the rainy season and consequently the season of storms and hurricanes we had not expected such bad weather; on 27 December we saw the island of Vavau[2] whose northern point bore at midday exactly west and our latitude was 18° 34'. This island which Captain Cook never visited but which he knew about from accounts given by the inhabitants of the Friendly Isles, is one of the largest in this archipelago, roughly equal in size to Tongataboo and much better because it is higher and does not suffer any shortage of drinking water; it lies at the centre of a multitude of small islands which must bear the names Captain Cook listed, but which are difficult for us to allocate. We could not in all fairness lay any claim to this discovery which adds to the Friendly archipelago a number of islands approximately equal to those explored by Captain Cook. I had obtained in China an extract of a journal by the Spanish pilot Morel[3] who was sent from Manila in 1781 to take

[1] La Pérouse had been a follower of the widely-held view that scurvy could be cured by allowing sailors to stroll ashore and breathe in the 'land air' of which they were deprived on long voyages. He had done this as recently as early December when he let several sick go ashore at Tutuila; but like others he had begun to harbour serious doubts about this old medical tradition and gradually sided with those who saw scurvy as a disease brought on by a deficient diet.

[2] Vava'u is only one island out of the group of 34 which bears this name. Its existence was known from a list supplied by Cook, and La Pérouse had the Spaniard Maurelle's account of his discovery of the Mayorga Islands which are in fact the Vava'u group. The islands had probably been visited in August 1772 by the Marion du Fresne expedition, then led by Le Jar du Clesmeur.

[3] Francisco Antonio Maurelle (1754–1820) was sent from Mexico to Manila in 1780 in the *Princesa* and left for the return voyage on 21 November of the same year, intending to sail by the central and south Pacific instead of the usual northern route. While making for New Ireland he came upon the Hermit group in January and on 26 February 1781 discovered Fonualei in the Tongan group and went on to Vava'u where he landed. He discovered Toku in April and two of the Ellice Islands (Tuvalu) in May, but he was unable to battle his way further east and veered north to Guam, completing his voyage on 21 September 1781 by the northern route. His later career in the Spanish navy was quite distinguished and he rose to the rank of admiral. *Who's Who*, pp. 176–7.

CHAPTER XX

despatches to America which he planned to reach by way of the southern hemisphere, following roughly Mr de Surville's route[1] and trying to reach the high latitudes where he correctly expected to find westerly winds.

That navigator did not know about the new methods of determining longitude and had never read any of the accounts of modern travellers; he sailed according to Belin's old French charts[2] and made up with the utmost precision in his dead reckoning and his bearings for the imperfection of his methods, his instruments and his charts. Like Mr de Surville he coasted New Ireland, sighted several small islands Messrs Bougainville, Carteret and Surville had already come across, discovered three or four new ones and believing himself to be near the Solomon Islands came upon firstly an island to the north of Vavao which he named Margoura[3] because it offered him none of the refreshments he was beginning to be badly in need of; he did not manage to see a second island to the east of the first one, which we saw perfectly and is only visible from three or four leagues because it is very flat, and finally arrived at Vavao and dropped anchor in a fairly convenient port where he obtained water and a fairly considerable quantity of food; the details of his navigation were so real that it was impossible not to recognise the Friendly Islands or even to be mistaken over the portrait of Poullaho[4] who being the main chief of all these islands lives in several as the need arises, but seems to have his principal residence at Vavao: I

---

[1] Jean-François-Marie de Surville entered the Pacific in the *St Jean-Baptiste* in August 1769 north of the Philippines and sailed south-east to the Solomons, later veering south and east to New Zealand. Maurelle, sailing from Sisiran on the east coast of Luzon, partly followed Surville's route but somewhat more to the west and La Pérouse is mistaken when he states later that 'like Mr de Surville he coasted along New Ireland'.

[2] Jacques-Nicolas Bellin (1703–72), geographer and hydrographer, published a number of charts which he collected into one volume, *Hydrographie françoise*, Paris, 1753. In 1764 he published a *Petit Atlas maritime* in five volumes. He also drew the maps used to illustrate A.F. Prévost's great *Histoire générale des voyages* (Paris, 1753, reprinted in 25 volumes in 1758). His maps, often based on unreliable longitudes, contained a number of errors: Bougainville, threading his way around the islands east of New Guinea, exclaimed, 'Oh Bellin, how much hardship you are causing us!'. Journal, entry of 22–23 June 1768, Taillemite, *Bougainville*, I, p. 360.

[3] The island he named Amargura ('Bitterness') is Fonualei, a volcanic island 40 miles north-east of Vava'u.

[4] Maurelle' Poulaho was is Fatafehi Paulaho, the supreme chief or Tui Tonga of whom Cook wrote, 'If weight of body could give weight in rank or power Poulaho was certainly the most eminent man'. Cook, *Journals*, III, p. 880.

will not allow myself to give further details on this voyage which may one day be printed[1] and to which I have referred only out of a sense of fair play towards Sr Morel. He had named the Vavao group Islands of Majorça after the Viceroy of New Spain, and the Apae group Calvez Islands after the Minister of the Indies' brother,[2] but, persuaded that it is a thousand times better to keep the country's own names, I felt I ought to use them in the chart drawn up by Mr Bernizet from latitudes and longitudes determined by Mr D'agelet and more correct than those of the Spanish navigator who showed these islands some six degrees too far west, from which a new archipelago would have been created which would have been copied by geographers from one century and one age to the next and yet would have had no basis in reality.

We followed various tacks during the 27th in order to approach the island of Vavao from which the west-north-west winds were driving away somewhat and during the night, having tacked northward so as to extend my area of vision twelve or fifteen leagues beyond that island, I saw Morel's Margoura bearing west from me and coming nearer I sighted a second very flat island,[3] covered in trees, which this navigator had not seen; by contrast Margoura Island is fairly high and it is likely that both are inhabited; after taking all our bearings I made for Vavao Island which was visible only from the topmasts. It was rather like a capital completing the Friendly archipelago,[4] the other small islands scattered to the north and west can scarcely bear comparison with the latter. By about midday I was at the entrance of the harbour where the navigator Morel anchored; it is constituted by fairly high small islands leaving between them very narrow but quite deep passes, and provides vessels with a perfect shelter from the winds; I would have found this port, vastly superior to that of Tongataboo, very suitable for a

[1] Maurelle's expedition eventually became known when Milet-Mureau printed it at the beginning of his edition of La Pérouse's voyage.
[2] The islands of the Vava'u group were named after Don Martin de Mayorga, but the Haapai group (Apae) nearby was named after Don José de Galvez, Marqués de la Sonora (1729–86), Secretary for the Colonies under Charles III and uncle of Bernardo, Conde de Galvez, one-time Viceroy of Mexico.
[3] This island is Toku. Encircled by a reef, it is very lowlying, rising a bare 80 ft above sea-level. Its discovery can be credited to La Pérouse.
[4] This sentence, omitted by Milet-Mureau in the printed edition of the voyage, is best understood in an architectural context, as the top feature of a structure or column.

CHAPTER XX

few days, but one anchors two cablelengths from land and in that situation a longboat is often necessary to take an anchor offshore and get away from the coast; I was time and again tempted to give up the decision I had made on leaving Mahouna to put in nowhere until I reached Botany Bay; but reason and prudence made me keep to it. I wanted at least to establish contact with the islands; I hove to fairly close to the shore. No canoe came up to the frigates, the weather was so bad and the sky so threatening that it hardly surprised me, and since the horizon was becoming darker by the minute I myself sailed for Lathé Island[1] which I could see and which is high enough to be visible on a clear day from twenty leagues off; this name of L'athée is included in the list supplied by Captain Cook, and the navigator Morel allocated it in his journal to this same island following information supplied by the islanders of Vavau who told him it was inhabited and one could anchor there: from which it can be understood how important it is for geography to retain local names, for if like navigators of old or like Morel himself we had had an error of seven or eight degrees in longitude, we could have assumed that we were a considerable distance from the Friendly archipelago, and the similarity of language, of customs and of dress would not altogether have eliminated our uncertainty, because it is known that these people are like each other although very far away from each other, whereas an identity of name with the most sketchy description of the island's shape and extent together provide unshakeable evidence; the night was frightful and so black that we could not see two musket shots away. It would have been foolhardy to sail in this darkness among so many islands and I decided to stay on short tacks until daylight at which time the wind was even worse than during the night: the barometer had dropped 3 points and although a storm could be much stronger it could hardly seem more threatening; I set a course for Lathé Islands; I came up to within two miles, certain however that no canoe would risk putting out to sea: near this island I was assailed by a squall which forced me to make for the islands of Kao and Toffoa which were no doubt fairly close, although the fog did not allow us to see them. These two islands were the first two shown on Captain

---

[1] Late, latitude 18°50′S, longitude 174°40′W had been discovered by Tasman in February 1643; he named it Hooch Eijlandeken, or 'small high island'. It is a volcanic island, with a peak rising to 1,700 ft (520 metres). A. Sharp, *The Voyages of Abel Janszoon Tasman*, Oxford, 1968, pp. 150, 172.

Cook's chart, who had passed through the channel two miles wide that separates them[1] and had determined their latitude and longitude with great precision, it was very important for us to compare with them our chronometer's longitudes. In truth I planned to get close enough to Tongataboo to fully complete this comparison. Mr Dagelet considered with good cause Tongataboo's observatory as equivalent to that of Greenwich since its determination was the outcome of more than ten thousand distances taken in a space of four or five months by the indefatigable Captain Cook. At five o'clock in the evening during a break in the weather we sighted Kao Island whose shape is that of a very high cone visible from 30 leagues away; Toffoa Island although it is also very high did not make an appearance, and remained hidden in the fog. I spent the night like the previous one on short tacks with only the main topsail and the foresail, because the wind was so strong we could not carry any more canvas; at daybreak the weather was a little better and fairly clear, we saw both the islands of Kao and Toffoa at sunrise.[2] I came up to within half a league of Toffoa and verified that it was inhabited at least along the three-quarters of its circumference the edge of which I saw close enough to make out the stones along the shore. This island is very mountainous, very steep, covered with trees right up to the top and seems about four leagues round. I think the islanders of Tongataboo and the other Friendly Islands often land there during the fine season to cut down trees, and presumably build their canoes, because they lack timber on their flat islands, where they have kept no trees other than those which like the coconut and breadfruit[3] trees bear fruit that form part of their necessities. Hauling along the island we saw several slides where trees cut down on the sloping hillsides fall down to the

[1] La Pérouse is referring to Cook's second voyage. On 1 June 1774, Cook came upon two islands he named Amataffoa and Oghao: 'the Southernmost and the one on which the Vulcano is or is supposed to be is called by the natives Amattafoa and the other which is round high and Peaked Oghao. We pass'd between the two, the Channell being two Miles wide, safe and without soundings, both are inhabited but neither of them appeared firtile.' Cook, *Journals*, II, pp. 447–8.

[2] Kao is an extinct volcano which rises to 3,380 ft (1,030 m), the highest point in the Tongan group, and less than 5 square miles in area. Tofua, 1,700 ft (518 m) high, is more than three times the area of Kao; its volcano, Lofia, is still active. Both islands are now uninhabited. Tongataboo is the island of Tongatapu, site of the capital, Nuku'alofa.

[3] Milet-Mureau once again encounters the term *rima* and omits it, so that there is no reference to breadfruit in the printed narrative of the voyage.

seashore, but there were no huts or clearings in the forest, nothing indicating a residence: then continuing on our route towards the two small islands of Hoongatonga and Hongahapee,[1] we aligned the island of Kao with the middle of Toffoa Island whereupon the former looked merely like a peak belonging to the latter and recorded it then as bearing North 27 degrees East; Kao Island is three times higher than the other and looks like the vent of a volcano; its base seemed to me to be less than two miles in diameter. We also observed on the N.E. point of Toffoa Island on the side of the channel separating it from Kao a stretch of land that was entirely burned as black as coal, and bare of trees and any vegetation which presumably had been ravaged by lava flows. After midday we saw the two small islands of Hoongatonga and Hoongahapee. They are included in Captain Cook's chart but not on it is a highly dangerous reef running roughly N¼NW and S¼SE, extent two leagues; its northern point situated five leagues north of Hoongahapaee and its southern point three leagues north of Hoongatonga, forming with the two islands a strait of some 3 leagues; we sailed past a large league to the west and we saw its breakers rising like mountains, but it seemed possible to me that in a calmer sea it would be less prominent and then would be more dangerous.

The two little islands of Hoongatonga and Hoongatapaee are merely large uninhabited rocks high enough to be visible at a distance of fifteen leagues. They continually changed shape and any outline one might draw could usefully be done only from a well determined point. They seemed to me to be of equal size, and less than half a league round separated from each other by a channel of one league and situated E.N.E. and W.S.W. They lie 10 leagues north of Tongataboo, but since this latter is a low island one needs to be half that distance away to see it; we saw it from the topmasts on 31 December at 6 a.m.; we could only see the tops of the trees which seemed to be growing out of the sea. As we came nearer the land rose but merely by two or three *toises*. We soon recognised Vandiemen Point[2] and the line of breakers that lies on the offside of

---

[1] The two islands of Hunga Haapai and Hunga Tonga, rising respectively 400 ft (102 m) and 490 ft (150 m) above sea level, lie to the north-west of Tongatapu. They are uninhabited, possess deposits of guano, but offer no anchorage.

[2] In January 1643 Abel Tasman reached Tongatapu which he named Amsterdam and anchored off the north-west coast, calling his anchorage Van Diemens Reed (Van Diemen's roadstead) after the Governor-General of the Dutch East Indies, Antony van Diemen. A. Sharp, *The Voyages of Abel Janszoon Tasman*, Oxford, 1968, p. 153.

that point, it bore from us at midday east approximately two leagues: as the winds were northerly I steered for the southern coast of the island which is very safe and can be approached within three musket shots. The waves were breaking furiously along the entire coast but these breakers were close inshore and beyond we could see the most delightful orchards. All the island appeared to be entirely cultivated, trees lined the fields which were of the finest green because it was the rainy season for in spite of the magic of this sight we could not conceal from ourselves the thought that it was more than likely that during part of the year the most horrible drought must reign over such a flat island, there was not a single hillock to be seen,[1] and the sea itself cannot be flatter.

The islanders' houses were not laid out in villages but scattered through the fields, as in our best cultivated plains. Soon seven or eight canoes put to sea and came near our frigates. These people, more farmers than sailors, handled them timidly, they did not dare come up to our vessels although they were hove to and the sea was very calm. They jumped into the water eight or ten *toises* from our frigates and swam up bringing in each hand coconuts which they bartered with good faith for pieces of iron, nails or hatchets; their canoes were no different from those of the Navigators but none had any sail and it is likely that they would not have been able to work them;[2] the utmost confidence soon established itself between us, they climbed on board, we spoke to them of Poulaho and Finou.[3] We looked like old acquaintances talking about old friends;

[1] By sailing along the south coast, La Pérouse missed the port of Nuku'alofa which lies on the northern coast, but it was already well known from Cook's accounts and he had no wish to be delayed for several days in this island. His comments on the island being flat are correct: Tongatapu is of coral formation, quite flat and has no running streams.

[2] Bad weather on the previous day and too brief a stay at Tongatapu did not give La Pérouse an opportunity to see the various types of canoes used by the Tongans, including single and double canoes, those with outriggers and those fitted with a sail which Cook described at length: 'The sail is of a triangular figure with the upper side, which is a little curv'd, fix'd to a yard nearly three times as long as the mast. This hoists up to the top of the mast, making the sail incline obliquely upwards & backwards.... They carry from forty to fifty people, and some sixty or seventy. The size of one of the largest as measur'd was twenty five paces, the extreme breadth (of one boat) about three feet, and the depth including the rais'd plank five and a half.' Cook, *Journals*, III, pp. 937–8.

[3] For La Pérouse's Poulaho, see note above. Finou is the chief Finau Ulukalala Feletoa who met Cook in 1777. It is not impossible that La Pérouse met one of his sons: they were quite numerous. These Tongans chiefly families were soon to

CHAPTER XX

a young man from one of the canoes gave me to understand he was Finou's son, and this lie or this truth earned him several gifts, he shouted with joy when he received them and urged us with gestures to go and anchor near the shore, explaining that we would find there an abundance of food and that the canoes were too small to bring them out in the open sea; I was only too well aware of this, there were neither hens nor pigs on any of their boats, their entire cargo consisted of a few breadfruit,[1] bananas and coconuts; and since the smallest waves made these frail craft capsize the islanders' gestures said that they would be drowned before reaching the ships: their behaviour was noisy but implied no ferocity, and neither their features nor the proportions of their limbs nor the presumed strength of their muscles could inspire any awe even if they had been unaware of the effect of our weapons; their constitution without being inferior to ours did not seem to have any advantage over that of our sailors and although their language, their tattoos and everything in general indicated that they shared a common origin with the Navigators, it was evident that this was a degenerate race and that the strains of agriculture, the aridity of the Friendly Isles, in brief all the physical properties of the soil and the climate were less favourable for the growth of the human species and I am strongly of the opinion that on its own the island of Oyolava is superior in population, fertility and real strength to the whole of the 150 islands of the Friendly archipelago which for the most part are only uninhabited and uninhabitable rocks and where the islanders are forced to water with the sweat of their brow the fields which supply their sustenance, but these circumstances have caused their state of civilisation to advance, tempered their ferocity and possibly led among them to the development of skills which place them on the same level as other peoples and protect them from invasion by their neighbours. Yet we only saw the same weapons, which were *Patoux patoux*, we bought several which did not weigh a third of

---

become embroiled in bitter civil strife that caused much bloodshed throughout the Tongan Islands at the end of the eighteenth century. La Pérouse's comments on the apparent mildness and peaceful nature of the Tongans could not have been more wrong, for not only was war brewing, but James Cook was very nearly ambushed and killed by Finau and his followers who had been planning to capture both his ships; see Cook, *Journals*, II, pp. ciii–civ.

[1] Once again the word *rima* stumps Milet-Mureau.

those we bought at Mahouna and which the inhabitants of the Friendly Isles would not have had the strength to use.

The custom of cutting off the two joints of the little finger is still very widespread here as in the islands of Cocos and Traitors and this sign of sorrow over the loss of a relative or a friend is almost unknown in the Navigators Islands. I know that Captain Cook thought that the islands of Kepel and Boscawoen, or Cocos and Traitors, formed part of he Friendly group; he based his opinion on Poulaho's report who was aware of the trade Captain Walis had carried out in these two islands and who had in his treasury before Captain Cook's arrival several pieces of iron obtained through barter between the frigate the *Dolphin*[1] and the people of Traitors Island; I thought on the contrary that these two islands were included among the ten whose names had been given to us by the Navigators because I found them precisely in that point of the compass they indicated and further east than the determination given by Captain Walis, and I thought that they could complete with Quiros's Island of the Beautiful People[2] the finest and largest archipelago of the South Sea, but I must admit that the stature and external appearance of the Cocos islanders are much more like those of the Friendly Isles people than of the Navigators from which they are at much the same distance, and I find it a small sacrifice to side with Captain Cook in all those matters, who had spent so much time in the various islands of the South Sea.

All our political exchanges with the people of Tongataboo[3] amounted to one simple visit and seldom has one been made so far; we gained hardly more out of it than the same refreshments one offers one's neighbour by way of snack in the country. But Mr D'agelet had an opportunity to check his chronometers. The great number of observations carried out at Tongataboo by Captain

---

[1] Cook had found one puzzling piece of iron in 1773: 'The only piece of Iron we saw among them was a small tool like a bradawl and which had been made of a small nail.' Cook, *Journals*, II, p. 266. His officers came across others and the question remains, did they come from Abel Tasman or from Samuel Wallis's visit of 1767? Georg Forster thought that the nail was so rusty that it belonged more to the seventeenth century than to 1767, and Beaglehole comments that 'a Tasman origin does not seem at all impossible.'

[2] This is Quiros's Gente Hermosa or Peregrina.

[3] Cook stayed at Tongatapu from 10 June to 10 July 1777 and established an observatory on the western point of Mua Inlet where the longitude was determined by a series of 131 lunar observations.

CHAPTER XX

Cook left him in no doubt about the exact location of the *Resolution*'s observatory and he felt he should in some way use it as a kind of basic meridian from which he assigned the relative positions of the entire Friendly archipelago and even of the other islands we visited in the southern hemisphere; the result of his observations obtained through a very large number of lunar readings showed a difference of less than seven minutes from those of Captain Cook. Thus by accepting that famous navigator's longitudes Mr Dagelet also followed his own and he became increasingly convinced that comparisons based on determined points could increase confidence in the chronometers but were not necessary for their verification, a sequence of lunar distances taken under favourable circumstances leaving nothing to be desired, from which it may be concluded that, if we had had no knowledge of Captain Cook's navigation through the Navigators archipelago, the Vavao Islands would have occupied within five or six minutes the same geographical place on our charts.

At nightfall on 1st January, having lost all hope of obtaining by beating about out at sea in this way enough food to at least balance what we consumed, I decided to bear S.W.S. and make for Botany Bay, following a route taken as yet by no other navigator and it was not part of my plans to sight Pilstard Island[1] which had been discovered by Abel Tasman and whose position Captain Cook had determined, but the winds which veered from the north to S.W.S. forced me to sail south and on the morning of the 2nd I saw this island which is very steep, at least for a quarter of a league along its greatest length, having only a few trees on the N.E. coast and being of value only as a home for seabirds.

This small island bore west from us at ten thirty in the morning and its latitude observed at midday was found by Mr D'agelet to be $22^d\ 22'$, 4 minutes further north than had been assigned by Captain Cook, who working out this latitude only from distant bearings might have made some mistake. The longitude of Pilstard is

---

[1] Pijlstaert was discovered by Tasman on 19 January 1643. It is the island of Ata in latitude 22°20′S and longitude 176°12′W, some 85 miles (135 km) south-south-west of Tongatapu. Pijlstaert was the Dutch name for the tropic bird. Ata, now uninhabited, consists of two volcanic peaks, the highest rising to 1,135 ft (355 m), and was given the name of Vrouwe Borsten (Woman's Breasts) in the Vingboons map of the Tongas; see F.C. Wieder (ed.), *Monumenta Cartographica*, The Hague, 1932, IV, pl. 100 and pp. 138–9, 142–3.

[blank].[1] The calms gave us only too many opportunities of verifying and correcting our observations. We stayed three days in view of this rock. The sun in its zenith maintained the calms, a hundred times more troublesome for sailors than contrary winds. We were awaiting with the utmost impatience the south-easterly breezes which we had expected to find in these waters and which were to take us to New Holland. The winds had blown constantly from the westerly quarter since 17 December and whether the weather was fine or they were in the shape of squalls they varied only from north-west to south-west, so the trade winds are far from regular in these waters; they nevertheless blew easterly on 6 January and varied up to E.N.E. The sky became very overcast and the sea very rough. They continued in this way with a great deal of rain and a restricted horizon until the eighth; we then had a settled gale but very strong from the N.E. to the S.E. The weather was very dry and the sea extremely rough; as we had passed the latitude of all the islands the winds had returned to their usual direction which had been totally interrupted from the Line to the 26th degree south. The temperature had also greatly changed and the thermometer had dropped 6 degrees either because we had passed the sun or, what is more likely, because these strong easterly gales with a whiteish sky reduced its effect because it was only four degree from our zenith, and its rays were almost vertical; on the 13th we sighted Norfolk Island and the two islands[2] that lie off its southern point; the sea was so rough and had been for so long that I had little hope of finding any shelter on the N.E. coast of this island, although the winds were southerly at that moment; however as I approached I found calmer waters and decided to anchor a mile from land in 24 fathoms, hard sandy bottom mixed with just a little coral. I had no other purpose than to send our naturalists and botanists to examine

[1] Dagelet gets it marginally more correct: Cook had calculated Ata's latitude as being 22°26′, an error of 6 minutes, Dagelet was only 2 minutes out. On the other hand, La Pérouse gives no longitude; Cook got it right within 14 minutes.
[2] The two small islands are Nepean and Philip. Nepean Island, just off the south coast, was named by Lieutenant Philip Gidley King in March 1788 in honour of Evan Nepean (1751–1822), Under-Secretary at the Home Office. The other island, larger but further offshore, was also named by King, in honour of the governor of New South Wales Arthur Phillip who had appointed King Superintendent and Governor of Norfolk Island. It will be noticed that King spelled the name with only one l, which by an interesting coincidence was how his own first name was spelled! Norfolk Island had been named by James Cook as an homage to 'that noble family', the Dukes of Norfolk.

CHAPTER XX

the soil and products of this island, because since our departure from Kamtchaka they had had very few occasions of recording new observations in their journals. We could see the sea breaking wildly around the island, but I thought our boats would find some shelter behind some large rocks lining the coast; however since we had learned at our expense that one must never depart from the utmost caution especially when the occasion is motivated by little more than curiosity, I placed Mr de Clonard, post-captain and the expedition's second-in-command, in charge of the four small boats sent out from the two frigates and ordered him not to risk a landing under any pretext whatever if our biscay boats were running the slightest danger of capsizing. The prudence and punctiliousness of this officer left me no reason to be concerned; he was no longer at the age when one feels one should not retreat in front of breakers; and Mr de Clonard to whom I intended handing over command of the *Astrolabe* when we reached Botany Bay deserved my fullest confidence in every respect.

Our frigates were anchored opposite the point where we believed Captain Cook landed between two headlands at the northern end of the N.E. coast; our canoes set off for this kind of bay but found there the waves breaking over some big rocks with a fury that made it impossible to approach. They coasted along the shore half a musket shot away toward the S.E. and travelled in this way half a league without finding a single point where a landing was possible; they could see all around the island a wall of lava which had run down from the top of the mountain and cooling as it came down had left in numerous places a kind of roof jutting out several feet over the side of the island which was almost perpendicular;[1] and if it had been possible to land they would have been able to penetrate into the interior of the island only by following for fifteen or twenty *toises* the course of very fast torrents that had formed ravines. Above this the island was full of pinetrees and the finest greenery among which our gardener presumably would have found some vegetables; this expectation increased even more everyone's desire to visit this land where Captain Cook had landed

---

[1] Norfolk is indeed an island of volcanic origin; its average elevation is 400 ft (115 m) with two peaks rising in the north-west of the island. There is no harbour: landings can only be effected on the southern side where the town of Kingston is now situated, or, when the sea is too rough in that roadstead, on the northern coast off The Cascades.

with the greatest ease,[1] but he had found himself in more favourable circumstances and enjoying several days of good weather, whereas on the contrary we had been sailing through such heavy seas that our portholes and windows had not been opened for a week. I followed the manoeuvres of our boats from the deck with my glass and seeing that they had not found a suitable landing-place by nightfall I signalled to them to return and soon after I gave the signal to weigh anchor. It was obvious that I would have had to wait maybe for a very long time for a moment suitable for a landing and a visit to this island was not worth this sacrifice, but as we were about to set sail a signal from the *Astrolabe* caused me the greatest anxiety by telling me there was a fire on board. I immediately sent my boat to hasten to her assistance but it was hardly half-way there when a second signal told me the fire was out, and shortly after that Mr de Monty told me by hailer that a case of acids or other chemical liquids belonging to Father Receveur and stored beneath the upper deck had caught fire spontaneously and spread such a thick smoke below decks that it was very difficult to discover the source. They had managed to throw this case overboard and the accident had no other consequences. It is likely that some flask had broken inside the case, and the mixture of liquids would then have caused this fire by spreading first to bottles of alcohol that might also have been broken or were badly stoppered; whatever it might have been I felt glad that right at the start of the campaign I had ordered that a similar case of chemical liquids I had on my frigate belonging to the Abbé Mongès be placed in the open air on the forecastle where no danger was likely to result from a fire.

Although it is very steep, Norfolk Island rises only 70 or 80 *toises* above sea-level, it is covered with trees of the pine species, the same in all probability as those of New Caledonia or New Zealand;[2] Captain Cook says that he found there a considerable number of

---

[1] James Cook landed on the west of Dunscombe Bay on 11 October 1774. The north-west point is now called Point Howe and a monument to Cook has been erected on the eastern point of the bay. La Pérouse, irritated by the rough sea and the strong gale, remembered ruefully that the Englishman had enjoyed a moderate SE to ESE breeze which had presented no problem.

[2] The appearance of Norfolk Island has changed greatly since the eighteenth century, but all the early navigators agree that it was covered with forests dominated by the tall pine *Araucaria excelsa* known today as the Norfolk pine.

palm-cabbages,[1] and our wish to get some was not the least motive of our call; it is probable that the palm-trees that produce these cabbages are very small and lower than the pine trees, because we saw no trees of this type; as the island is uninhabited it is full of seabirds and particularly tropic birds all with their long red feather.[2] We could also see numerous gannets and sea-swallows but no frigate birds; a sandbank over which there is 20 to 30 fathoms of water extends three or four leagues to the north and east of this island, and maybe all round, but we did not take soundings in the west. While we were at anchor, we caught on this bank a few red fish of the type called captains in the Isle de France, or sards, which provided us with an excellent meal, and they were if one can put it that way the net proceeds of our call;[3] at eight in the evening we were under sail; I set course to W.N.W. bearing away gradually to S.W.¼W under reduced sails and sounding constantly while over this bank, where it was possible one might encounter some shoal, but on the contrary the ground was very level and the depth increased very gradually as we sailed away from the island; at 11 p.m. a 60-fathom sounding-line could not reach the bottom, we were then ten miles west-north-west of the northernmost point of Norfolk. The winds had settled E.S.E. in squalls, fairly foggy, but the weather was very clear between squalls. At daybreak I crowded on sails for Botany Bay which was only three hundred leagues away, but on the 14th after sunset I gave the signal to heave to and take soundings with a line of two hundred fathoms; the Norfolk Island plateau had made me think the depth might remain the same as far as New Holland, but this conjecture was false and we went on our way with one less error in our minds because I had been very firmly of that opinion.[4] The East S.E. to N.E. winds remained constant until we were in sight of New Holland. We made good progress during the day and very little at night because we

---

[1] This is not the so-called cabbage tree, but the *nikau* or *Rhopalostylis Bauri*, varieties of which are found in New Zealand and in the Kermadec Islands. The *nikau* has an edible root. Much confusion occurs in early navigators' accounts, dating back to the use of the word 'cabbage' for palms.

[2] This was the *Phaeton rubricauda*.

[3] The red trumpeter, *Lethrinus Chrysostomus*, is quite common in waters around Norfolk Island and provides an excellent meal.

[4] Norfolk Island is not linked to Australia by an undersea ridge, but to New Zealand. The long Norfolk Ridge runs from New Caledonia to northern New Zealand by way of Norfolk Island.

had not been preceded by any navigator along the route we were following.

On the 17th in latitude 31^d 28' and longitude 159^d 15' we were surrounded by a multitude of sea-swallows which made us suspect that we were passing close to some island or rock, and several bets were made that we would discover some new land before we arrived at Botany Bay from which however we were only a hundred and eighty leagues away. These birds followed us until we were within 80 leagues of New Holland and it is fairly likely that we left behind us some islet or rock they use as a place of refuge because they are far less numerous close to inhabited land.[1] We took soundings with a line of two hundred fathoms from Norfolk Island until we were in sight of Botany Bay and found bottom in 90 fathoms only when we were 8 leagues from the coast which we sighted on 23 January.[2] It is not high and hardly visible from a distance greater than twelve leagues; at this point the winds became very variable and like Captain Cook we were affected by currents which took us each day fifteen minutes south of our reckoning so that we spent the 24th tacking in view of Botany Bay without being able to round Solander Point[3] which bore north from us one league, from which the winds blew strongly and our ships were not good enough sailers to beat both the strength of the wind and the currents at the same time; but on that day we beheld a spectacle which was quite novel for us since the day we left Manila, that of an English fleet at anchor inside Botany Bay, of which we could see the flags and pennants.

Europeans are all compatriots at such a great distance and we

---

[1] The French passed close to Lord Howe Island which Lieutenant Ball in the ship *Supply* was to discover exactly one month later. The island, of volcanic origin, lies in latitude 31°33′ and longitude 159°05′E (156°45′E of Paris). Approximately 20 miles to the north-east are found Ball's Pyramid and the Admiralty Islets inhabited by thousands of seabirds.

[2] The Australian coast sighted on 23 January 1788 by La Pérouse is the area around Broken Bay. His latitude at midday, 33°43′, places him athwart Narrabeen, south of Broken Bay, near Narrabeen Lagoon. All this area, which forms part of the Hawkesbury Estuary, is low lying, and is now part of Greater Sydney.

[3] Cook named the northern point Cape Banks and the south-eastern one Point Solander, after the naturalists Sir Joseph Banks and Daniel Carl Solander. The currents had carried La Pérouse to the south of Botany Bay, and he was now endeavouring to enter it by rounding Point Solander. The French anchored on the northern shore, where now stands the Lapérouse Museum and Father Receveur's tomb. Opposite, on the southern shore, stand various memorials to Captain Cook: Captain Cook Landing Place Park, Inscription Point and Captain Cook Drive.

CHAPTER XX

were most impatient to reach the anchorage; but the weather was so foggy the next day that we were unable to see the land, and we only reached the anchorage on the 26th at nine o'clock in the morning. I dropped anchor one mile from the northern coast athwart the second bay in seven fathoms of good grey sand. As I was entering the pass an English lieutenant and a michikmane[1] were sent to my ship by Captain Honter[2] commanding the King of England's frigate *Sirius* and they offered on his behalf all the assistance he could give, adding however that circumstances allowed him to give us neither food nor munitions nor sails. And since they were on the point of weighing anchor to go further north their kind remarks amounted merely to good wishes for the ultimate success of our voyage.[3]

I sent an officer to carry my thanks to Captain Honter whose anchor was already apeak and whose topsails were already hoisted; my message was that our needs were limited to wood and water, which we would have no difficulty in obtaining in this bay, and that I knew ships given the task of establishing a colony such a great distance from Europe could be of no assistance to navigators. We learned from the lieutenant that the English fleet was commanded by Commodore Philip[4] who had sailed the day before from Botany Bay in the corvette *Spey* with four transport vessels to go north and seek a more suitable place for his settlement. The English lieutenant

---

[1] This is an interesting French rendering of 'midshipman'.

[2] John Hunter (1737–1821), after spending some thirty years in the Royal Navy, was appointed captain of the *Sirius* for the voyage of the First Fleet under the overall command of Arthur Phillip. The Fleet, with the *Sirius* and the *Supply*, six transports with the first 759 convicts, and three storeships, sailed from England on 13 May 1787 and reached Botany Bay between the 18th and the 20th of January 1788. The *Sirius* was later wrecked at Norfolk Island. After he returned to England, Hunter published his *Historical Journal of the Transactions at Port Jackson & Norfolk Island* (London, 1793). He was promoted vice-admiral in 1810 and died in London in 1821. The mention of the king of England was deleted for the 1797 printed narrative of La Pérouse's voyage.

[3] This well-known passage hides the penury in which the English officers found themselves after their long and hard voyage, having landed in a country which had very little to offer that was of immediate use for their proposed convict settlement. Their decision to move from Botany Bay to Port Jackson had little to do with their inability to assist the French at this stage.

[4] Arthur Phillip (1783–1814), commander of the First Fleet and first Governor of New South Wales, soon realised the inadequacy of Botany Bay and established his settlement at Port Jackson out of which grew the town of Sydney. The '*Spey*' is the ship *Supply*, captained by Henry Lidgbird Ball (1757–1818), the discoverer of Lord Howe Island, after whom Ball's Pyramid is named.

seemed to be creating a great deal of mystery around Commodore Philip's future plan and we did not permit ourselves to ask any questions on this subject, but we could have no doubt that the planned establishment was very near Botany Bay since several boats and longboats were already putting up their sails to go there, the ships not considering it worthwhile hoisting them aboard for such a short journey and soon sailors from the English boat, less discreet than their officer, told ours that they were only going to Port Jackson 16 miles north of Bancs Point where Commodore Philip had personally investigated a very good harbour running ten miles towards the south-west. Vessels could anchor within a pistol shot of the shore in waters as calm as those of a basin. Later we had only too many opportunities of obtaining news of the English establishment whose deserters caused us a great deal of trouble and inconvenience[1] as we will explain in the next chapter.

---

[1] English reports make a number of references to attempted escapes by convicts. The most interesting case is that of Peter (Pierre) Paris: a recent study by Anny P.L. Stuer, *The French in Australia*, Canberra, 1982, takes up a suggestion by E. de Blosseville in his *Histoire des colonisations pénales dans l'Angleterre et l'Australie*, Paris, 1831, p. 134, that Paris was really a Frenchman and that consequently La Pérouse's men took pity on him, hid him on board and sailed away with him; this would mean he died a few months later in the Vanikoro disaster, Stuer (1982), p. 41. However, Paris had been arrested in Exeter and was therefore not a Frenchman who happened to be in London; nor is the name Paris necessarily an indication of French nationality; furthermore his escape is reported as taking place at the same time as that of a woman, Ann Smith, and it is unlikely that the French would have accepted a woman, obviously not a Frenchwoman, on board. A number of convicts succeeded in escaping in those early months, some were recaptured, others fled into the interior where they presumably died or were killed by aborigines, and sadly any precise record of Peter Paris's ultimate fate vanished with La Pérouse's two ships, but speculation on what happened to him is warranted and it may well be that his name should be added to the list of those who lost their lives at Vanikoro.

MAPS OF THE
LA PÉROUSE EXPEDITION

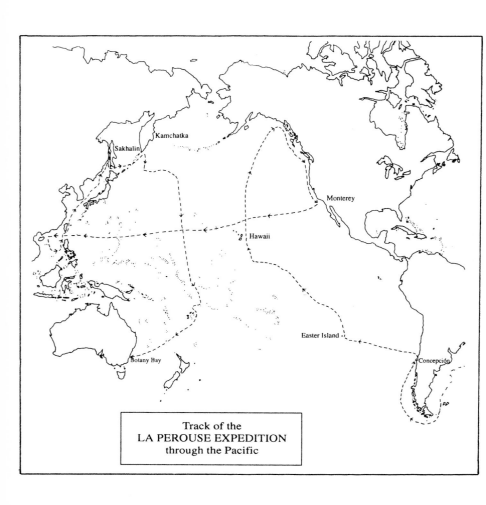

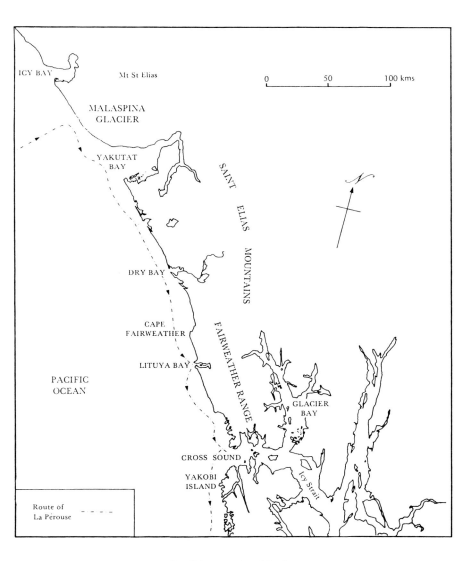

La Pérouse in Alaska

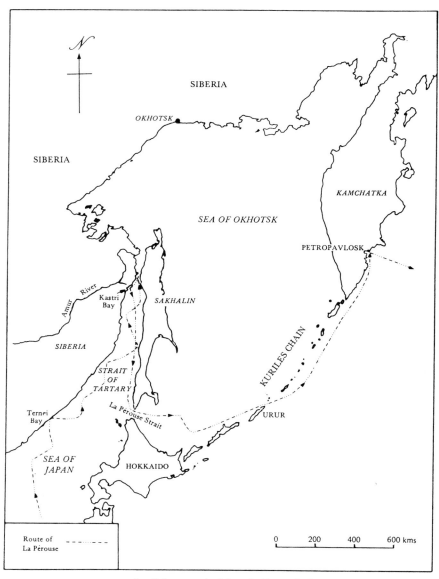

La Pérouse in North-East Asia

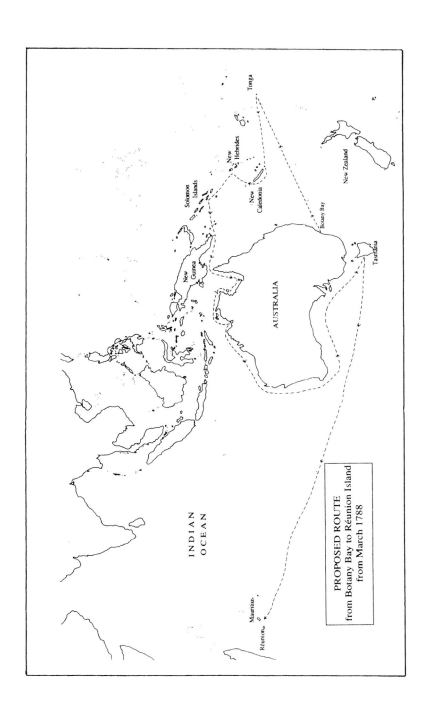

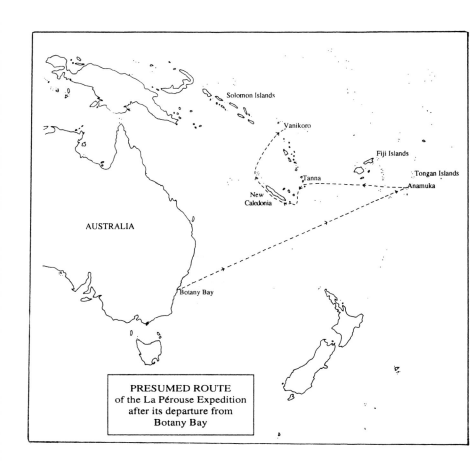

APPENDICES

APPENDIX I

Selected Correspondence

The Archives Nationales in Paris hold a number of letters concerning the expedition, the problems that resulted from the loss of the two ships, and the publication of the account of the voyage in 1797 (references Marine 3JJ, 386 to 389, B4 318 and 319, C7 165). They cover the period 1785 to 1807. The present selection is taken exclusively from letters written during the voyage, from the departure from Brest to the stay in Botany Bay, New South Wales.

A number of letters were reprinted by Milet-Mureau in his 1797 edition, some in full, others with cuts or alterations. In quite a number of cases, the Archives Nationales hold several copies of the same letter, many with slight variations in the text, often in the spelling, but usually bearing La Pérouse's signature. It is evident that several people helped to copy out the letters – La Pérouse complained that he had no one on board appointed to carry out the functions of a secretary. Whenever it has been possible to find the original, identifiably written by La Pérouse or an officer or naturalist, this version has been used; this is shown by the indication 'holograph'. The term 'unpublished' is used to show that it was not included in the Milet-Mureau edition. Most of the letters bear an annotation added at some stage after their arrival in France; this is shown between brackets to help identify letters which were often undated.

Quite often the writers repeat something they have mentioned in an earlier letter. This is understandable since they could not know whether their previous communications had been received or were still making their way to France. As a whole, the letters reveal the daily concerns of the expedition's leaders and add a valuable human dimension to the general narrative. We see how circumstances led La Pérouse to make changes to the overall plan of the voyage, his eagerness to discover new routes and new lands, and to equal the achievements of James Cook. Anglo-French rivalry is clearly reflected in a number of the letters, whether in exploration or trade, in the Pacific or in the Far East. The difficulties he had to face during the voyage are referred to on a number of occasions: his vessels are extremely bad sailers, so that he loses precious time, in spite of all the efforts and goodwill displayed his officers; relations between scientists and officers were often strained (although it is interesting to compare Lamanon's version of the Tenerife problem of August 1785 with La Pérouse's); these difficulties were often due to the length of time the ships were at sea, when the naturalists had little to do, and the exiguity of their quarters on board, so different from their studies at home; the letters in fact do not suggest that these problems were serious enough to justify La Pérouse making anything approaching an official complaint in his correspondence – the only lasting irritant was the older Prévost who seems to have gone on strike for most of the voyage. There are passing comments in the letters on local administrators, mostly Spanish, whom La Pérouse met when he called at the various settlements, but one finds more detailed information in his reports and his journal.

The letters give an interesting insight into La Pérouse's personality and his friendship with Fleurieu. The affection he felt for the latter is shown in the way he ended his letters ('I embrace you and love you with all my heart'). His own modest nature ('Writing is a craft like any other, and I did not learn it', letter of 23 September 1787), his early optimism about his mission and his weariness and probable indifferent health in the later stages, his occasional bursts of irritation, and his views on closet philosophers and ethnologists, are clearly revealed in this correspondence: he would probably have allowed himself more latitude in expressing his feelings had the letters not been of a semi-official nature and intended to be copied by some of his helpers on board.

The spellings of proper nouns is fairly erratic. Better known place names are usually correctly spelt, and the English equivalent, if there is one, has been used in the translation. In other cases, they have been left as they appear in the letters and, where some clarification seems necessary, the correct spelling is provided between square brackets.

APPENDIX I

La Pérouse to the Minister
Holograph, unpublished.[1]

(from Madeira, 16 August 1785)

My Lord,

I dropped anchor at Madeira on the 14th of this month intending to fill thirty wine barrels belonging to each ship, left empty in our holds at Brest, but the information I obtained about the cost of Madeira wines suitable for the crews has made me decide to sail without delay for Tenerif where I shall obtain the same quantity of wine at half the price; it will still be fairly dear and cost at least fifty *écus*[2] a barrel, but I will be confident that it will keep for the entire voyage, an advantage Bordeaux wine does not offer.

The crews of both ships are enjoying the best of health; they can see that nothing is being neglected to provide them at considerable cost with beverages that would have been much less expensive in France, but would have been of a much lower quality. I have not concealed from them the knowledge that my orders require me to take good care of their health and to take all steps necessary to ease the hardships of this campaign. I am quite sure that their zeal and their good will will repay the trouble Mr de Langle and I are taking on their behalf.

I shall not, My Lord, open the despatches you handed me until I sail from Tenerif. I do not know what they contain, but I beg you to authorise Mr de Langle and me to promise a warrant of permanent appointment to each one of our main chiefs if we find ourselves in circumstances where their work deserves such a reward. It is possible that, after some stranding, we may owe our salvation to the master carpenter or the master caulker, at least as much as to the boatswain, and it would only be in such crises that we would make use of the permission which I am taking the liberty of requesting today.[3]

My crossing has been most satisfactory. I shall have the honour of reporting from Tenerif on the operations of our timekeepers and on the other observations we will have the opportunity of making. I shall add to these the first chapter of my voyage such as I hope to have the honour of submitting it to you upon my return. It will be

---

[1] AN.M. 3 JJ 386, 111:2, No. 13.
[2] The *écu* was a silver coin equivalent to 6 *livres*. It fell into disuse after the fall of the monarchy in the 1790s, but the term was revived with the adoption of the European monetary system as an acronym for European Currency Unit.
[3] This request was granted in a letter from the Minister dated 15 October 1786.

completely written up when I arrive and will merely require, in order to be published, for you and the King to give their approval.

At Madeira I have purchased from a French merchant vessel fourteen tierces of brandy for our crews.[1] I have requested the Consul or his representative to settle the account. This is our only expenditure in Madeira, which will not exceed twelve hundred *livres* and will lessen the quantity of wine we will require at Tenerife.

With deepest respect, I have the honour to be,
 My Lord,
  your most humble and obedient servant,
   Lapérouse.
At Madeira, 16 August 1785, day of my departure for Tenerif.

De Langle to the Minister
Holograph, unpublished.[2]

(Madeira, 16 August 1785)

My Lord,
I have the honour of reporting to you on the navigation of the King's frigate the *Astrolabe*, which has been most happy. The strength, good health and good will of my crew, combined with the zeal of the officers and other persons who make up the staff of the two vessels and the good relations that exist between them, give me great hopes for the future. Mr de la Pérouse shares my view that it would be most beneficial if the future of our senior men was under our control and we could determine the date of their formal appointment. We have both allowed ourselves the hope that you will be kind enough to confirm what we may do in this respect. I have the honour to guarantee that I shall exact a high price for these favours, and to renew my assurances of the zeal and eagerness which animate me and all the officers, scientists, artists, soldiers and sailors on board the *Astrolabe*.

With deepest rspect, I remain,
 My Lord,
  Your most humble and obedient servant,
   Chev^{er} de Langle
On board the *Astrolabe*, Fonchal roadstead, this 16 August 1785.

---

[1] A tierce was a cask containing approximately one-third of a pipe or 42 gallons of wine.
[2] AN.M. 3JJ 386, 111:2, No. 14.

APPENDIX I

La Pérouse to Fleurieu
Holograph. Unpublished.[1]

(To Monsieur de Fleurieu, director of the ports, at Court)
I shall write to you more fully, my dear Fleurieu, from Tenerif where I am going this evening. I have spent only 36 hours at Madeira; it is impossible to buy any wine there for the crews, the English have caused the price to rise to such heights that the poorest type cost 30 pounds sterling the pipe here; we will get it for less than half that price at Tenerif, and I am forced to take some because at Brest they left 30 barrels empty in each vessel. Mr Hector was convinced that wine from Cahors[2] could not keep for four years under various climates and that it was absolutely necessary to have a year's supply of Spanish wine, which seems quite reasonable to me.

My crossing was one of the best. Not a single man sick. We left drawing twenty inches of water more at the stem than at the stern, the ship handled fairly badly, but everything has turned out for the best. The worst seas will be behind us in six weeks and by then we shall be roughly on an even keel.

I shall send the 1st chapter of my voyage to the Minister from Tenerif, so that he can see the style and the way the work is constructed which will be entirely completed when I return and will need, to be published, merely to be approved by him and the king. You know what you have promised for the introduction.

Farewell, my dear friend. I embrace you and love you with all my heart.
  Lapérouse.
Madeira, 16 August 1785.

Lamanon to the Minister
Holograph. Unpublished.[3]

      (Mr the Chevalier de Lamanon, undated,
      must be from Teneriffe, 26 August 1785)
My Lord,
As Mayor and First Consul of the town of Sallon in Provence

---

[1] AN.M. 3JJ 386, 111:2, No. 15.

[2] Cahors in south-western France was well-known as a trading centre and renowned for its wine market. Its wine, dark and heavy, was widely exported but regarded as *vin ordinaire*; it was only in the twentieth century that it acquired a reputation of its own, being recognised as an 'Appellation d'Origine Contrôlée' in 1971.

[3] AN.M. 3JJ 386, 111:2, 16.

[Salon-de-Provence], I beg you to render my compatriots a service which lies within your discretion and consists in sending to my colleagues the small bag of seeds which I am taking the liberty of enclosing.[1]

The peasants of lower Provence and those of Sallon in particular depend for their sustenance on haricot beans; some twelve years ago a fog caused this species to degenerate to such an extent that it has been difficult to obtain it ever since: the plant grows quite well, but as soon as fructification begins rust attacks the leaves and spreads to the stem, causing the plant to die. Everything has been tried to cure this problem, except the renewal of the species. My research in Teneriffe in the field of natural history led me to discover a type of white bean that is as prolific as that of Provence and tastes even better. It does very well at a certain height above sea level, where the climate is very close to ours; I believe in consequence that planting these seeds in Provence will lead to a revivification of the tainted species; and we will thereby provide the people with a food it appreciates and which is all the more necessary in that wheat is scarce in our district and potatoes do not do succeed. Moreover potatoes are too bland in taste for the people of Provence who are accustomed to a stronger type of food and to drinking quantities of a wine that does not inebriate them. In hot countries appetite needs stimulation; the salad of white beans our workers eat after their soup and to which they add a strong seasoning represents their only food and they used to gather it in their fields before the fog I mentioned made its appearance.

We are now back, my Lord, from the peak of Teneriffe, a great and superb mountain. We were there on the eve of the feast of St Louis and we drank the King's health there; none of his subjects has ever celebrated his feast is such a high place. Barometric measurements taken with a great deal of care indicate that the peak rises 1950 *toises* above sea level. In 1783 in Switzerland people had not gone higher than 1560 *toises*.[2] In 1784 I reached 1805 *t*. ½ and that is

---

[1] Lamanon became Mayor-Consul of Salon on 26 December 1784, following in the footsteps of his elder brother Auguste who had been appointed in April 1782. Four months after his election, Lamanon was approached by Condorcet to join the La Pérouse expedition and he left for Paris in May 1785. He had had therefore little opportunity to do anything for the people of Salon. See *Le Chevalier de Lamanon et le drame de l'expédition Lapérouse 1785–1788*, Salon, 1986, pp. 8, 11.

[2] Mountaineering as a sport or even for scientific research was hardly known in Lamanon's time. His letter was written a year before the conquest of Mt Blanc in August 1786.

APPENDIX I

the highest anyone has read a barometer in Europe. The peak of Teneriffe is much higher, but it can be reached with less trouble; this is because of the climate; the cold is bearable here two thousand *toises* up, whereas in France and in Switzerland one comes upon permanent ice at 1500 *toises* above the sea. On the other hand as the peak of Teneriffe as well as the entire island is the result of volcanic eruptions the ground is of more recent formation than on our mountains; it has been less excavated by torrents and has fewer precipices than the French and Swiss Alps.

There were 13 of us when we climbed the peak; the sulphurous smoke made 10 of them leave after an hour. A large number of experiments were carried out in physics by me and in chemistry by the Abbé Mongèz. It would be impossible to give details of them in this letter. I will only say that careful observations showed that the magnetic needle dips by 60 degrees at the top and base of the peak, but the needle oscillated at 42 on the mountain and 23 on the seashore. Ammonia retained all its strength there as if it was on a plain whereas it loses all its smell in the highest Alps. When I reached the peak my pulse beat at 119 per minute, and 110 after an hour's rest. I found a large quantity of curious stones, and the island's lithology, an area of research that is completely untouched, will be of interest to naturalists. Mr D'Hermand gave me all the assistance he could in my investigations, and I and all my comrades owe him a great deal.

The day after tomorrow our floating homes will follow a new route and we shall leave Teneriffe. Everyone is well and we look upon ourselves as members of one family of which Mr de la Pérouse is the father. I respectfully remain,
My Lord,
   Your most humble and obedient servant,
      Chev^{er} de Lamanon.

De Langle to the Minister
Holograph, unpublished.[1]
                    (Mr de Langle, undated, must be from Teneriffe,
                                      28 August 1785)
My Lord,
   The deplorable state to which seasickness has reduced Mr Monges during the crossing from Brest to Madeira caused me

[1] AN.M 3JJ 386, 111:2, No. 18.

some anxiety about the remainder of the voyage. As he had regained some strength at Madeira I hoped he could get used to the sea, but the latest crossing has so indisposed him that all my officers and I felt that it was not possible for him to continue on this campaign and I was the first to advise him, as I would have done with my best friend, to return to France He was unable to counter the objections which Mr de la Pérouse and I raised and he decided to leave in the vessel that is taking the Consul to Cadiz. My regret and those of my staff are all the more real in that Mr Monges combines ability with the greatest amiability and we can carry out astronomical tasks on land only in so far as Mr Dagelet can guide us. It is my hope that our zeal will make up for the wide knowledge of the man we are losing. We have been fully provisioned here, My Lord. We are all leaving in a good state of health and full of the zeal and eagerness that are necessary for success. I am sure that I shall not forget for a single moment all that I owe you, My Lord, and the deep interest you are showing in our campaign; nor the affection I have for Mr de la Pérouse.

Respectfully I remain,
My Lord,
Your most humble and obedient servant,
Chev^{er} de Langle.

La Pérouse to Fleurieu
Holograph. Unpublished.[1]

(La Pérouse to Fleurieu, undated but written on 28 August 1785)

Enclosed, my dear Fleurieu, my journal as far as Madeira. I would have liked to append the table reporting on our timekeepers drawn from our observations at Tenerif, but our calculations will not be finished until this evening and I am sailing right away. Be assured that everything will go well. We set out our astronomical tent ashore, with Mr Dagelet, Mr Monges, Descures, chevalier Darbaud, and we are today most satisfied with our timekeepers which have shown a difference of a mere few seconds since we left Brest. But we think the table supplied by Bertoud concerning the barometer is not quite correct. We are also very dissatisfied with

---

[1] AN.M 3JJ 386, 111:2, No. 19.

the work on the Borda circles, particularly the sights, which are detestable.

Mr Monges's health does not allow him to continue on the campaign. This is a great misfortune for us, because he is a talented and amiable man. I would have preferred it if this sickness had affected the chev. de Lamanon who is a troublemaker and displays in particular a greed which has little in common with the distinterestedness mentioned by the Baron de Choiseul. He wanted to travel to the peak, and make others travel with him, at the King's expense, which would have cost over a hundred *louis*[1] because they would not have been as careful with costs as when they knew they would have to meet them; twelve or fifteen mules were dismissed when this sad news was known, and Mr de Lamanon had the impertinence to tell me that he would get his voyage to the peak printed for his own benefit to repay him for his expenses. I replied with a laugh that a lot of authors have been ruined by their books.

He is full of zeal, but as ignorant as a Capuchin friar[2] when it comes to anything outside the realm of systematic physics. He claims to know better than Mr Buffon how the world was formed, and I am convinced they both know as little as each other, but there is not a fifteen-year-old girl in Paris who does not know more about the globe than this doctor who has been searching for the tropic with his spy-glass ever since the apprentice pilots told him it could be seen from a hundred leagues.

My dear friend, we are drawing a bill of exchange for approximately 20 thousand on Mr Fournier because we bought here a hundred and twenty pipes of wine, making 62 barrels Bordeaux-size, to make up for the same quantity of Cahors wine which was to be supplied to us and which we did not take, being persuaded that wine from Teneriff would keep for the whole campaign, which would not be the case with the wine from Cahors.

Moneron had planned to measure the peak's height by taking levels. I thought this operation would be very long and difficult because I saw that surveyors seldom had recourse to it; however

---

[1] The *louis* was equivalent to 24 *livres* or 4 *écus*.

[2] An unflattering remark which reflects La Pérouse's anticlericalism, a trait of eighteenth-century *philosophes*. The Capucins were an order of mendicant Franciscan friars whose vows of poverty were seldom observed but gave them an excuse for being importunate. The order was suppressed during the Revolution and reformed and reinstated in the nineteenth century.

since this is a major part of his work I did not raise any objections, but our scientists were less restrained. He complains a great deal about them and prevailed upon my feelings of friendship for him to insist that I forward to you a copy of the letter he has written on this to his brother. I am carrying out his wishes, but advise you that this small touch of anger has subsided, that they are on the best of terms and that I use a firmness and at the same time a mildness that turns these minor happenings into matters of no consequence. Farewell, my dear Fleurieu, I embrace you and love you with all my heart.

  Lapérouse.

Get the draft of the journal copied out before handing it over to the Maréchal.

La Pérouse to the Minister
Holograph. Unpublished.[1]

          (La Pérouse, Tenerife, 28 August 1785)
My Lord,

  I am taking advantage of the fact that a Spanish frigate is leaving this evening for Cadiz to let you know about my arrival at Tenerif, where I filled, for the *Boussole* and the *Astrolabe*, 60 barrels of wine which had remained empty in our holds. This wine costs approximately 640 *l.* a barrel, which is about the same price as the wine from Cahors which was intended for us; consequently, barring the disadvantage of having to pay to foreigners the sums that we would have had to settle at Bordeaux, the difference is almost nil, but we have the certainty of knowing this wine will keep until the end of the campaign, and it is even so strong that it can be watered up to a third, which greatly increases our supplies. We shall sail on Tuesday August thirtieth. I am entrusting all the papers, containing the most detailed accounts I have the honour to send you, to our Consul who is sailing for France in a few days. I shall append the first chapter of my voyage so that, My Lord, you may see the way in which I propose to write it.

  I am sending this copy to Mr de Fleurieu because it has a hundred crossing outs, and I do not dare to send it to you direct in such a state, but we are leaving tomorrow and there is insufficient time to have it copied.

[1] AN.M. 3JJ 386, No. 21.

APPENDIX I

I had the honour of advising you from Madeira, My Lord, that wine there was so excessively dear that I intended buying my supplies in Tenerif, but I added that I had purchased from a French merchant ship 14 tierces of brandy that would cost approximately 12 hundred *ls*. and that this would be my only expenditure in Madeira. We did buy these fourteen tierces, but as this was a very small transaction we settled it with the funds we had on board, in this way there will be nothing to pay to the Chargé d'Affaires at the French Consulate in Madeira.

With the deepest respect, I have the honour to be,
   My Lord,
      Your most humble and obedient servant,
         Lapérouse.
Teneriff, 28 August 1785.

Lamanon to Condorcet
Holograph. Extract. Partly reprinted in the Milet-Mureau edition.[1]
(from Lamanon)
To the Marquis de Condorcet, of the French Academy, Permanent Secretary of the Academy of Science, Hôtel de la Monoye, Paris.[2]

5 November 1785

This letter, Marquis, is being written at a distance of two thousand leagues from you. After a voyage lasting two months we are reaching the island of St Catherine where [we] shall remain for only as long as it takes to load wood and water. Since we left Tenerife we have seen no other land than the islands of Martin Vas which are uninhabited and the island of Trinity where a Portuguese settlement replaced the English settlement a year ago. It has a garrison of some 200 men, no women. Food supplies are taken to them every six months, and nothing is grown on this island which is nothing but a rock of volcanic basalt. I went up to within hailing distance but the sea is full of reefs and the captain had ordered us not to land.

You will no doubt have received the letter I wrote from Teneriffe,

---

[1] AN.M. 3JJ 386, 111:2, No. 23; Millet-Mureau, *Voyage*, IV, pp. 252–6.

[2] Marie-Jean-Antoine-Nicolas de Caritat, Marquis de Condorcet (1743–94), mathematician and biographer, later republican politician and educational reformer, was arrested during the Terror and found dead shortly after. While in hiding during the Terror, he wrote his major work, *Esquisse d'un tableau historique des progrès de l'esprit humain*.

when the present one reaches you. Forced to write to you before anchoring at St Catherine for otherwise I would not have the time, I cannot give you much in the way of news. Our floating homes are not good sailers, lengthening the voyage which, they say, will last 3 ½ years. There will not have existed navigators before us who remain so long at sea because we spend very little time at our places of call. It is true that we are in a hurry to round Cape Horn during the good season. This lengthy period at sea is not really what one needs for lithological observations, but I make use of it for other purposes. I am in good health, I usually work for 12 hours a day without tiring in spite of the ship's roll; instead of staying in bed until 9 or 10 as was my cosy custom I see the sun rise every day and I do not regret it.

I enclose a memoir on the results I obtained by observing the barometer hourly from one degree north to one degree south. It appears that the combined action of the sun and the moon produces an ebb and a flow in the atmosphere which causes the barometer to vary by one twelfth – it should be only $\frac{1}{5}$ of a twelth according to Mr de la Place (see *Cos. élem.*).[1] It is true that in another book (I think it is in the preface to Exansarideu's work on the temperatures of 1779) it is said that according to Mr de la Place the barometer should at the Equator following the motion of the moon vary by ½ twelth, so there is some doubt. Mr de la Place will be able to verify whether the observation is in line with the theory. Moreover there must be some uncertainty over the basis of calculation if I am to go by the opinions of the leading mathematicians on ebb and flow. Some say that if the sea consisted of mercury the ebb and flow would be the same; others that it would be different. It is up to you top-class mathematicians to re-examine this issue and decide what we should think.

I am carrying out magnetic observations with a great deal of care. It would take too much space to give you an account of this. I observed the inclination of the compass for a full 24 hours to discover the moment at which we would be passing the magnetic equator and found true zero at 7 a.m. on 8 October in latitude 10.46. I observe iron bars I have had placed on the ship, other bars that are not fixed, the movement of the needle horizontally and per- pendicularly, on a weight held by a magnet according to latitudes.

---

[1] Pierre-Simon de Laplace (1749–1827), astronomer and mathematician.

To conclude, I hope that no one has gathered so much information on these matters for a long time. It [...] that results printed in our general account and you will be kind enough to accept the hommage of this small work which grows daily in size. Since we hand over to Mr de la Pérouse a descriptive summary of our journals at each place of call, the narrative of the voyage is made up as we go and it will be sent to the printers on the day we arrive. Until now Mr Mongèz and I have been recording our observations separately and have handed them to Mr de la Pérouse who writes them in with each of our names between inverted commas; I felt that this approach would not be the most attractive to the readers and suggested to the Abbé that we should write an account each in turn: one will give his observations to the other and each will keep his own style, there will be no duplication and we will keep more variety within a unified frame. He has agreed and I am starting with St Catherine.

We have had no sickness on board the *Astrolabe* either, with the exception of Mr de Blondelas who is seriously affected in the lungs. We are all pleased with each other and with Mr de la Pérouse. I in particular am grateful to him and he willingly makes my work easier. Mr Mongèz has taken over the study of birds, microscopic animals and cryptogamy; for my part I have ichtyology, butterflies, coleoptera, and sea, land and river shells; we have not yet drawn a demarcation line in respect of mineralogy....

Lamanon to the Minister
Holograph. Unpublished.[1]

(from Lamanon)

I have the honour to send you, Sir, a small memoir for the Marquis de Condorcet. If you find it in the least unsuitable to reach the academy, you may retain it.

Our crossing has been a happy one, our call at St Catherine very pleasant, our commander will no doubt give you details. All I can say is that there is no finer country on earth in respect of climate. I have found some rare quadrupeds, some strange fish and above all some unknown land shells. The inhabitants are kind, but the slavery of the blacks, the Inquisition's superstition and military despotism are great evils.

[1] AN.M. 3JJ 386, 111:2, No. 25.

We are leaving, I have only the time to assure you of my respectful devotion and the honour I have in being,
My Lord,
Your very humble and obedient servant,
Ch^{er} de Lamanon.
St Catherine roadstead, 16 November 1785.

La Pérouse to Fleurieu
Holograph. Unpublished.¹

(St Catherine, 16 November 1785)
You will find herewith, my dear Fleurieu, the fairly shapeless draft of my letter to the Minister. You will read in it all the information I could have given you. I deeply regret that I have so little time to myself. We are all in the best of health. Do not forget, my dear friend, to write to us in Kamtchatka, Manila, China, the Isle de France; you cannot imagine how your letters will please us and are necessary to us. I get you to enclose all the gazettes you can. Farewell, I love you with all my heart and am attached to you for life.
Lapérouse.
All our scientists behave very well and my courtesies, compensating for my refusal at Teneriff to pay for their excursion to the peak, have restored everything. I am very happy with this. There is some rivalry between them today, each one favouring his own special subject, but I expected this.
Mr Dagelet is a charming man. I am only afraid that he is somewhat lazy and is not keeping up his diary in the way you would expect. Apart from that no one could be more knowledgeable or more pleasant.
At St Catherine, 16 November 1785.

La Pérouse to the Minister
Holograph. Unpublished.¹

(from St Catherine)
My Lord,
I had the honour to advise you of my departure from Tenerif on

¹ AN.M. 3JJ 386, 111:2, No. 31.
² AN.M. 3JJ 386, 111:2, No. 27. Undated, but presumably written on 16 November 1785. A copy of this letter, with slight alterations, may be found in the same file under reference 26, entitled 'Copy of the letter from Mr de la Peyrouse to the Maréchal de Castries'.

APPENDIX I

30 August. We crossed the Line on 29 September. The trade winds left us in 14 degrees north and we had constant west to SW breezes that forced me to sail along the coast of Africa and cross the Line much further east than I wanted, but had it not been for this problem I would not have sighted Trinity because we found the SE winds on the Line, which would have brought us up against the coast of Brazil if we had been further west. Cast your eyes, My Lord, on the map of Captain Cook's second voyage: we followed of necessity the same route and found exactly the same breezes. I must add that we were sailing in approximately the same season. On 16 October I sighted the islands of Martin Vas, and the next day Trinity Island.[1] I planned to get water and wood there in order to continue on my way south. I did not doubt that the English had evacuated the island and that it was quite deserted; but when I rounded the SE point of Trinity I saw the Portuguese flag over a kind of fort surrounded by a few small houses. I sent the *Boussole* and the *Astrolabe*'s boats ashore commanded by Messrs de Vaujuas and Boutin. Only the *Astrolabe*'s landed. Mr Boutin has orders to take soundings and look for an anchorage for our vessels. The sea was so rough and landing so difficult that without the Portuguese's prompt assistance Mr de Vaujuas would have perished. The commander of this outpost met him on the beach and did not allow him to go up to the fort; he had with him some two hundred men, very few of them in uniform and almost all the others wearing shirts, so we do not know whether they were inhabitants or soldiers, but we saw no sign of cultivation.

Mr de Vaujuas learned from the Portuguese officer in charge of the detachment that the Governor of Rio Jeneire had taken over this island a year earlier, that it provided no foodstuffs, there was hardly enough water for the garrison, and even then it had to be brought from quite a distance in the mountains, and there were only a few trees on hilltops a league from the shore. He added that every six months a small vessel was sent from Brazil to bring the food required by the garrison which was to be relieved annually. Moreover he did not know, or pretended he did not, that the English had previously occupied this island. Realising from Mr

---

[1] Trinidade in 20° 30′ 32′ S. and 29° 50′ W. is roughly 4 miles long and 2 miles wide. It now belongs to Brazil, but remains uninhabited. In the nineteenth century it was believed to have some strategic value.

Vaujuas' report that I would not be able to obtain at Trinity the water and wood I needed, that the island did not have any and that in addition the commander would have opposed our landing, I decided to call at the island of St Catherine in 27d 30 of latitude south. This stop did not take me out of my way, and it was very preferable to Rio Janeire where empty formalities would have made me waste a lot of time, wheras at St Catherine all I had to do was to cut the wood, fill my barrels and leave. But since I was taking this decision I wanted to make sure of the existence of the island of Asconçaon[1] shown on Mr D'Après's map 100 leagues west of Trinity and 15 minutes further south. I sailed accordingly. I sailed a hundred and fifteen leagues west and could see ten leagues in front of me, which makes a hundred and twenty-five leagues. I saw no land and in spite of Mr D'Après's authority I am convinced that island does not exist, certainly not along the parallel of 27d 40m. After this fruitless search I sailed for St Catherine where I only arrived on 6 November. No foreigner since Admiral Ansson has landed here except for the Spanish who captured it a few years ago. It is nevertheless impossible to find a better place. Every kind of refreshment is available at the lowest price and in the greatest abundance. The Governor Don Francisco Boro, an infantry brigadier, received us as if our ships had belonged to his own country, and the inhabitants are so hospitable that they gave up their beds to some of my frigate's sailors who had capsized in a small boat which the inhabitants salvaged and brought back to my ship. We found no problem, apart from three days of bad weather during which it was impossible to send ashore for the water and wood we required, which kept me ten days at St Catherine, and I am sailing tomorrow 17 November.

I learned from the Governor of St Catherine and from two Portuguese merchant captains who are here whale fishing that the island of Asconçaon does not exist. The error is caused by the fact that the Portuguese used to refer to Trinity as Asconçaon and that either name can appear on old charts. Nevertheless the Governor of Rio Jeneiro organised a search last year for this island because he had been impressed by the confident tone used by Mr D'Après and

---

[1] Ascençaõ is not to be confused with the island of Ascension in the South Atlantic Ocean. It was an alternative name for Trinidad. Both names relate to religious feasts that are celebrated at various times, depending on the date of Easter.

the extract from the journal of Mr du Poncel de la Haye, but the Portuguese found nothing and have deleted it from all their charts.

My instructions, My Lord, estimated that I would reach the land of Sondowick [Sandwich][1] on 10 November and I am sailing from the island of St Catherine on the 17th of that month. The difficulties in Brest roadstead and the length of the crossings have all created an insurmountable obstacle. I am consequently forced to limit my campaign in the Atlantic to a search for the island of La Roche and then to make without delay for Cape Horn, the season no longer allowing me to visit the island of Georgia and the lands of Sondowick. I do feel however how important this search was. Sondowick land in particular has never been rounded in the south and nothing proves that it is not a headland of the austral continent. If I had dared to change completely the plan of my campaign and begin by the east, I could have sailed around Georgia this year, sailed south of the Sondowick Islands, sighted Cape Circumcision, and put in at the Isle de France whence I could have sailed for New Holland, but I realised that if I made such drastic changes to the plan approved by the King I would receive none of your despatches at a time when they could reach me and it was possible that you might have instructions to give me in China, Manilla, Kamchatka or the Isle de France. This consideration led me to follow strictly my instructions. I will not put in at Cristmasaount [Christmas Sound] if I have enough water and wood and will act in such a manner so I can be in Thaïty at the date that was decided. If you consider it appropriate, My Lord, you can send me orders to the Isle de France that, on my way home, I should visit the islands or land of Sondowick, this investigation seeming to me to be a geographical issue of some importance.

You will be pleased to learn, My Lord, that there is not a single case of sickness in the two frigates and we even cured during the crossing several cases brought about by an imprudent behaviour which my advice was unable to prevent at Teneriff. We are leaving

---

[1] Sandwich Land (the South Sandwich Islands) was so named by James Cook who comments how depressed he was by the 'enexpressable horrid aspect of the Country' (Beaglehole, *Journals*, II, 633). A number of islands and geographical features were named after John Montagu, Earl of Sandwich, first Lord of the Admiralty. The word is not easy for a Frenchman, hence La Pérouse's struggle with 'Sondowick'.

St Catherine with quantities of refreshments, water and wood, in an excellent state of health, a great deal of eagerness for the success of the campaign, a deep regret that we did not have time to do everything, but I cannot conceal from you that our vessels are extremely poor sailers which causes our crossings to be inordinately lengthy.

The Vicomte de Langle and I are driven by the same enthusiasm. This letter is written from both of us and we are both anxious to meet your expectations. The greatest unity reigns among our scientists, and among these I must include the officers of the two ships who carry out astronomical observations, work on charts and drawings and in every way contribute to the success of the voyage.

I have the honour to be....

Langle to the Minister
Holograph. Unpublished.[1]

(from Concepción)

My Lord,

As Mr de la Pérouse has given you an account of his navigation and of his plans I shall limit myself to the *Astrolabe* which is a better sailer than the *Boussole*, and which behaves well during bad weather; it has made no water, either from its upperworks or from his hold since we left Brest so that our spare sails and ammunition have not been damaged. The Chevalier de Monti who is my first officer is an an excellent seaman, honourable and firm. He is careful and rigorously accurate; he has done so far all that is needed to justify the warrant of *capitaine de vaisseau* which you have been kind enough to promise him on his return.

Mr de Vaujuas, a man of considerable intelligence who manoeuvres, observes and calculates very well, is a man of considerable promise. I make no distinction between him and Mr de la Borde Marchainville; both are quite capable of commanding the king's ships.

Mr D'Aigremont, a man who shows firmness and strength, also deserves praise, but he is not as capable as the other two. No one can have a greater intention of being successful than Mr de Flassan: he is good at manoeuvring and calculations and I hope he will gain experience during the campaign. Mr de la Borde Bouterviliers, to

---

[1] AN.M. 3JJ 386, 111:2, No. 28.

whom Mr de la Pérouse gave a warrant of *enseigne* at the same time as to Mr de Flassan, is an officer well trained in manoeuvring, observing and calculating; he has moreover an excellent physique and a fine disposition.

Mr de Blondelas, *lieutenant de frégate*, has an excellent sense of judgment and all the intelligence, the patience, the accuracy and the good will that make up a good seagoing officer; all his spare time is devoted to drawing maps of roadsteads and sketches of the shore and landscape; his collection of drawings will be most interesting.

Mr de Lauriston, *Garde de la marine*, is the most zealous and eager person I have met. Since I am no longer in a position to tutor him in astronomy, I have handed him over during the stay at Concepción to Mr Dagelet who holds the greatest hopes from his excellent attitude and application.

Mr de la Martinière, an untiring and passionate botanist, spends all his time during our stops looking for material; his collection would be more interesting if Mr Prévot, the uncle, a natural history artist, had helped him. I have been forced to use my authority to compel the latter to carry out his engagements.

Father Receveur carries out his duties as chaplain with great uprightness. He carries out researches in natural history in roadsteads, and in astronomy while at sea. He is a gracious and intelligent person.

Mr du Frene deals with natural history in the roadsteads. He is a weak man who finds it difficult to get used to the sea and is continuing on the voyage with a deal of reluctance. He would have landed here if the officer in charge of Concepción had consented to give him a passport to go to Spain and thence to France.[1]

Friendly relations have reigned on board so far and between the officers and crews of the *Boussole* and *Astrolabe*. I have every reason to hope that they will not be broken and I shall leave no stone unturned to ensure that they are not. I am equally certain that all of us on board the *Astrolabe* will do all in our power to contribute to the success of the campaign, being fully motivated, My Lord, to deserve the King's favour and the continuation of all the kindnesses with which you have honoured me.

My crew is happier and stronger than when we sailed from Brest. I

---

[1] Dufresne's request to be allowed to leave the expedition at this point is held in the Archives Nationales, Marine, ref. B4 319.

shall neglect nothing that can maintain them in that condition. So far I have had only two cases of sickness and have no loss to report.

My surgeon, Mr Lavaux, fulfils his duty with complete punctuality and seems to be a man of skill.

Mr de Lesseps who joined my ship as a Russian interpreter is a very gentle person, very upright and well brought up; I hope he will be of value to us in Kamschatka.

I have presented a medal to my boatswain Lamare who has done everything to deserve it since our departure from Brest.

With respect, I remain
  My Lord,
    Your most humble and obedient servant,
      De Langle.
On board the *Astrolabe*, this 14 March 1786.

La Pérouse to the Minister
Copy with alterations in La Pérouse's handwriting. Two copies exist. Unpublished.[1]

(Mr de la Pérouse. Undated, must be from Concepción 14 March 1786)

My Lord,

I had the honour to advise you upon my departure from the island of St Catherine of my intention of limiting my searches in the Atlantic Sea to the Isle Grande of La Roche, in order to round Cape Horn during the fine season and not to risk incurring any damage or tiring my crew, which would have been very prejudicial to the success of my campaign.

I set sail on 19 November. I was ready to leave on the 16th but contrary winds held me back three days within the bay. I had taken water and wood for more than four months; my crew enjoyed the best of health, there was not a single case of sickness in either ship. I firstly set my course for the east-south-east in order to place myself to the east of the island that was the object of my search; I did not delude myself about the extreme difficulty I would have in sailing up west, the north-west to south-west winds being almost as constant in this part of the world as the trade winds are between

---

[1] AN.M. 3JJ 386, 111:2, No. 32. No. 35 is the second copy of this letter which was of considerable importance as it outlined the proposed itinerary and was no doubt forwarded back to France by a different ship in case the first copy got lost.

the tropics: but whatever the circumstances I was obliged to progress much further west to approach the coast of Patagonia which I needed to sound before passing by the Strait of Le Maire; furthermore I thought that since the latitude of Isle Grande was not satisfactorily determined it was more likely that I would come upon it by tacking between the 44 and 45th degrees of latitude than if I followed a straight line along 44 degrees 30 minutes of latitude, as I might have done by sailing from west to east. On seventh December I was in the latitude of Isle Grande in 44 degrees 38 minutes and 34 degrees of longitude, Paris meridian, according to our latest observations of distances made on the previous day.

We could see a great deal of seaweed going by and we had been surrounded by birds for several days, but of the petrel and albatross type which never go near land except in the breeding season.[1]

These weak indications of land nevertheless kept our hopes alive and consoled us for the frightful seas in which we were sailing, but I felt some concern when I thought that I still had thirty-five degrees to sail on west as far as the Strait of Le Maire which it was very important for me to reach before the end of January. Convinced however that the leading quality of a navigator is stubbornness I covered fifteen degrees along the parallel of Isle Grande as my orders required me to; I encountered several gales and constantly sailed in dreadful seas; finally on 27 December I set a course to round Cape Horn, quite sure that the island of La Roche does not exist in this latitude, and that weeds and petrels are no evidence of the near presence of land since I saw seaweed and birds on the coast of Patagonia which I reached on 22 January within sight of the Cape of Virgins that forms the northern entrance to the Strait of Magellan.

It is fortunate that the month of January was as fine as December had been awful, and taking advantage of the winds which vary from south-west to north-west I completed a voyage in 25 days that could have held me back much longer.

I coasted Tierra del Fuego at a distance of three or four leagues as

---

[1] Such remarks are common in the journals of early navigators. Sailing in largely uncharted waters and being often unable to determine their longitude with any accuracy, they relied on floating seaweed, branches or tree trunks, even fruit, and the presence of land-based birds near the ship to assess the nearness of land. At night, listening for the sound of breakers was an important alternative on which their lives often depended.

far as the Strait of Le Maire which I passed through on 25 January, 69 days after my departure from the island of St Catherine.

My first intention had been to put in at this strait, and I had given the *Astrolabe* Good Success Bay as a rendezvous in case of separation, but I had been considering for quite a time a new plan of campaign, about which however I could not make a decision until I had rounded Cape Horn: this plan was to go this year to the Northwest Coast of America.

I knew, My Lord, that you would have given me such an order in my instructions if you had not feared that I would not have enough time to complete such a long journey before the bad time of the year, for this plan of campaign carries with it many advantages, the first and no doubt the deciding one being to reach that coast before the English who were foreshadowing, prior to my departure, an expedition under the command of Captain Gore[1] and although nothing is known about his mission I have no doubt that it was intended to set up some establishment behind Hudson's Bay and I think it is good to get there before him.

The second advantage is that all the English navigators after making for the latitude of Juan Fernandes Island on leaving the straits of Le Maire or Magellan have sailed west, and my campaign would give the appearance of being copied on theirs. Instead of that I will follow a new track and cross longitudes on which I can come upon several unknown islands.

And the third is to survey more rapidly all the places that have been indicated to me, by spending two years in the northern hemisphere and two in the southern hemisphere, returning to France after four years of navigation. I append to my letter, My Lord, my new plan of campaign. It meets in its entirety the purpose of my instructions, and you will recall that you were good enough to allow me to carry out the King's wishes in the manner I would consider most appropriate for the success and glory of my campaign.

Having adopted this new plan, I gave up all thoughts of obtaining wood and water in the Strait of Le Maire, and gave to the *Astrolabe* as a rendezvous in case of separation the island of Juan

---

[1] John Gore (1730?-1790) had brought back James Cook's expedition in October 1780, following which he had been appointed post-captain and given the vacancy at Greenwich Hospital left vacant by Cook's death. He did not lead the rumoured expedition to Hudson's Bay. See Dunmore, *Who's Who*, pp. 119-21.

Fernandes where I intended to get these two items, and made without delay for Easter Island and later for the coast of New Albion.[1]

We rounded Cape Horn so easily that I am convinced this navigation is no more difficult than any other. It is an old preconception that should disappear, and Cape Horn differs in no way from all the capes that are situated in high latitudes.

On 9 February I was athwart the Strait of Magellan, sailing, as I have already had the honour of indicating, for Juan Fernandes Island. According to my reckoning I had sailed over the so-called land of Drake, but I beg you, My Lord, to allow me, to digress a little on this question.

Since my departure from Europe all my thoughts had been directed towards the routes followed by the old navigators. Their journals are so poorly written that one needs in a way to guess at their meaning, and the Parisian geographers are so ignorant that they have been unable to shed the light of critical wisdom over journals that badly need it.

They have consequently drawn islands which do not exist and which have vanished like ghosts in front of the new navigators,

In 1578 Admiral Drake, leaving the Strait of Magellan, was assailed after five days of navigation in the great western ocean by frightful gales that lasted nearly a month. It is difficult to follow his various tracks, but finally he came upon an island in 57 degrees of southern latitude; he put in there and, there is no question about it, found natives who had canoes. He only saw natives after having sailed 20 leagues to the north because the first island where he put in was Diego Ramires, and the 2nd Tierra del Fuego. He saw a great many birds. Then sailing north for twenty leagues he saw more bird-covered islands where there was wood and antiscorbutic plants. How can one fail to recognise in this description Tierra del Fuego itself, where Drake called, and in all probability Ramires Island situated in much the same latitude as Drake's Island? At this period Tierra del Fuego was not even known; Le Maire and Chouten found the strait that bears their name only in 1616, and always in the belief that there were in the southern hemisphere as there are in the northern hemisphere lands stretching out to the

---

[1] New Albion was a term used to describe the coast of north-west America and was bestowed by Francis Drake in 1579.

neighbourhood of the poles, they thought that at most the southern part of America was cut through by channels and that like Magellan they had found a second one. All these erroneous notions were likely to lead Admiral Drake into error, who had been driven by currents 12 or 15 degrees to the east of his dead reckoning, as has happened since to a hundred navigators in the same regions, and this probability becomes a certainty when one thinks that one of the vessels from this squadron which sailed north while his commander tacked south returned into the very Strait of Magellan which they had just left, proof that they had not gone very far to the west and that Admiral Drake had not gone beyond the longitude of America. One can add that it goes beyond all reasonableness to believe that an island out at sea in this latitude could be wooded when there is not even a ligneous plant on the Malouine Islands situated in 53 degrees, that there are no inhabitants on these islands, not even on Staten Island which is separated from the mainland by no more than a channel of five leagues, and that finally the description given by Admiral Drake of the natives, their canoes, the trees and the plants is so appropriate to the Pécherets[1] and generally to all the details we have on Tierra del Fuego that I am today wondering how it could be that this so-called island has not been removed from every map.

As I had made all these reflections I wasted very little time on looking for these islands, and rapidly sailed north, but a few days before we reached Juan Fernandes, having requested Mr Clonard to give me a report on the food we had left on board, I realised we had very little bread and flour because, like the *Astrolabe*, I had been compelled to leave a hundred quarters at Brest because we lacked space to accommodate them on board. Furthermore worms had attacked our biscuits; they did not make it uneatable but reduced its quantity to the point where I believe a fifth had been lost. These considerations forced me to put into Concepción instead of Juan Fernandes Island; I know that this part of Chile abounds in grain which is cheaper than anywhere in Europe; this country in addition has an abundance of every type of comestible at the most moderate prices.

[1] This name, associated with the French *pêcheur* (fisherman) but not derived from it, was widely used in French accounts of the time to refer to the natives of southern Patagonia. Its first use has been ascribed to Louis de Bougainville: 'Bougainville came into contact with a people he calls the Pêcherais', Taillemitte, *Bougainville*, I, p. 53.

APPENDIX I

I have the honour, My Lord, to give you an account in a private letter of the courteous and friendly manner with which we were received, following the orders given by the King of Spain that were carried out in a way to warrant our utmost gratitude.

We dropped anchor at Concepción on 24 February; there we caulked and repaired our vessels, loaded a four months' supply of food, in the form of flour, wheat, dried meat, vegetables, &c. We have all the wood and water we need and we are leaving today 15 March for our eventual destination, with the intention of following the plan of campaign that is enclosed unless unforeseen circumstances prevent it.

You will learn with pleasure, My Lord, that there is not a single sick case in either ship, an unparalleled situation after such a long navigation, but our food, apart from the biscuit, was of such quality, the wishes of Mr de Langle and mine in this respect have been followed so scrupulously that we have succeeded in preserving the sailors' health much better than in the best province of France.

I have the honour to be...

Note written in La Pérouse's own hand: 'The plan of campaign is on the back of this sheet.' *The text of the plan is:*

New plan of campaign containing everything laid down in my instructions but which will be carried out at different times.

I shall leave from Concepción on 15 March [1786] and must hope to arrive on 15 April at Easter Island whence I shall set sail for the north, passing to the east of the Sandwich Islands and endeavour to find some new islands in these waters which are unfrequented. I shall reach the coast of America in 36 degrees of latitude north towards the end of June. I shall spend the end of June and all July and August exploring the entire extent of the American coast from 36 degrees and 60 degrees, and on 1 September I shall sail along the Aleutian Islands, spending the whole month in determining their position and endeavouring to land if weather permits.

On 1 October I shall make for the island of Tinian. I shall endeavour to find some new islands during this crossing. I do not expect to reach the island of Tinian before 1st December because I shall devote some time to a search for land. I shall leave there on the 20th of the same month with the easterly monsoon. I shall visit the Carolines and arrive on 10 February [1787] at Manila whence I shall leave for Formosa at the end of March with the start of the south-west monsoon. I shall take advantage of the summer to visit the coast of Tartary, the land of Jesso &c, and will endeavour to arrive in Kamtchatka on 25 July. I shall leave on 10 August of the same year and shall sail without delay for Endeavour Strait which I hope to reach on 10 October. I shall pass through this strait or some other one if I find one and will sail along and visit New Holland until 1 March 1788.

By then I expect to have sailed through Queen Charlotte Channel in New Zealand. I shall then sail north to the Society Islands or the Friendly Islands according to circumstances, and will visit during the months of June, July, August and September the land of Surville [the Solomon Islands], New Caledonia and the Hebrides, and will seek a new pass along New Guinea to get to the Isle de France; if I am forced to go again through Endeavour Strait I shall visit the Gulf of Carpentaria. I shall arrive at the Isle de France in December 1788. I will leave in January 1789 and go to Bouvet's Cape Circumcision and will generally do all I can in the Atlantic Sea to find the various islands mentioned by the old navigators.

I shall arrive in France during the month of July 1789 after 4 years of navigation and more that 3 years under sail.

La Pérouse to the Minister
Holograph. Unpublished.[1]

(Mr de la Pérouse, Concepción, 14 March 1786)
My Lord,
I had the honour of advising you in my letter No. 1 that I obtained four months of supplies at Concepción, per list herewith. They were provided by Mr Thomas Delphin, a merchant to whom I gave in settlement a bill of exchange at sight for an amount of one thousand five hundred and ninety-seven piastres signed by lieutenants of the two frigates and countersigned by Mr de Langle and me because the food was shared equally between the two frigates. We drew this bill on Mr Fournier, director-general of the navy's food supplies.

With respect, I have the honour to be,
My Lord,
Your most humble and obedient servant,
Lapérouse
Concepción, Chile, 14 March 1786.

[1] AN.M. 3JJ 386, 111:2, no. 33.

APPENDIX I

La Pérouse to the Minister
Copy, incomplete. Unpublished.[1]
(Mr de la Pérouse, undated, from Concepción, 14 March 1786)
My Lord,
   I had the honour in my previous letter of explaining the reasons for my stay at Concepción. I should have very much liked not to have to do so so that, if we are fortunate enough to succeed in our enterprise, we cannot be criticised for having had more assistance than the English navigators but, My Lord, you may recall that Captain Cook obtained food supplies at the Cape of Good Hope, and that he did not travel only with the comestibles they provided him with in England. Be this as it may, the King does not require of me a feat of strength but a navigation of value to geography and natural history &c, and one especially which shows that the sailors' health can be preserved even in the longest navigations.
   The plan of campaign I had the honour of sending you in my letter No. 1 and which contains practically the entire itinerary set down in the King's instructions you gave me can, I believe, stand comparison with any other....
   Such a plan is not beyond my zeal or that of Mr de Langle and all our officers and scientists who, with the exception of Mr Dufresne, about whom I have the honour of enclosing a special letter, are full of the greatest good will.
   On 24 February, as I have already had the honour of stating, we dropped anchor in the port of Concepción. Letters from the King of Spain's Minister had announced our presence. Pilots were sent out when we were two leagues from the shore. Soon after, Mr de Postigo, captain in the Spanish navy, came on board, having been sent by the commander of Concepción, and offered me on his behalf all the assistance in his power; a great abundance of fruit, vegetables, cattle and sheep were sent and I received from Mr Quexada, commander ad interim of this part of Chile such a kind letter that he could not have found better terms to address his own compatriots; he advised me that Mr Higuins, brigadier of the King of Spain's armies and commander of southern Chile, was at the frontier at the head of all the Spanish troops, signing a peace treaty with the Indians of the Cordillera, that he had sent him a messenger to let him know of my arrival

[1] AN.M. 3JJ 386, 111:2, no. 34.

and that he thought this most important business would be settled before I left.

The town of Concepción lies three leagues from the sea. We had anchored two musket shots away from a village called Talcaguano which consists at most of fifty houses, but thanks to the Spanish we could get there everything we could need.

Since I had to devote my first care to repairing my ship I was not able to go immediately to see the commander, so that he came first; he came aboard with the leading officers of this colony and expressed such a burning desire to give me all the help he could that I was unable to find words to express my gratitude.

A couple of days later I repaid his visit with Mr de Langle and several officers; he travelled a league to meet me, and he sent a detachment of dragoons to escort us, having in addition had the kindness to send half a company to Talcaguano which was placed at our disposal as were their horses.

A few days later Mr Higuins, whom I have already had the honour of mentioning to you, arrived, having made peace with the natives in a manner most advantageous to the kingdom of Chile; he gave me the most interesting details on these people, which I shall have the honour of giving you on my return to France. This Mr Higuins is the one officer in the whole of America in whom the natives place the utmost trust; they will deal only with him. The manners of this worthy soldier are marked by a frankness and uprightness that would be impressive in any country, and I had not spent more than two hours in his company before I began to share with all the inhabitants of this part of America the esteem and the wide respect everyone has for him.[1]

Mr Higuins compounded the courtesy I had been constantly shown since my arrival, and behaved similarly towards Mr de Langle and all our officers. We decided that we would offer a reception to the whole town before we left and thus make our farewells. A large tent was erected ashore and we offered a modest meal to a hundred and fifty men and women from Concepción who had come to Talcaguano.

This meal was followed by a ball and a small display of fireworks and finally a paper balloon, large enough to provide a show.

The next day the same tent was used to provide a large meal for

---

[1] On Ambrosio O'Higgins, see note to the main text.

the two crews. We all ate at the same table, with the Vicomte de Langle and I at the head, each officer down to the lowest sailors placed according to the rank he held on board. Our plates were wooden bowls. We ate an ox and drank a barrel of wine to the King's health. Never in my life have I felt happier; every sailor's face expressed their joy; they were leaving in a spirit of extreme jollity, a thousand times happier and more satisfied than the day we sailed from Brest; I shall repeat these Saturnalia annually if circumstances allow it.

I have the honour to be....

La Pérouse to the Minister
Copy. Part is included in the Milet-Mureau edition.[1]
(Mr de la Pérouse, from Monterey, 17 September 1786)
My Lord,
I have already had the honour of giving you a special report on the officers, scientists and artists of the two vessels; I have more praise to give them today; each one is carrying out his duties with the utmost zeal and punctuality. We are all as well as we were on leaving France, and our misfortunes have not depressed our spirits. I have been forced to replace the officers we had the misfortune to lose, and have appointed Messrs Colinet and Blondelas acting *capitaines de brûlot*, and Mr D'Arbaud acting *enseigne*. To Mr Laws de Lauriston, a young man of high merit I gave the warrant of *enseigne* I was to give him only on First July 1787; I dated it the First of August 1786, in accordance with the authorisation you gave me. And finally [I gave] to Mr Broudou, *volontaire*,[2] the remaining warrant of *lieutenant de frégate* which I also dated first August.

We were received by the Spanish at Monterey like ships of their own nation. Every assistance possible was given us. The religious in charge of the missions sent us a very large quantity of all kinds of provisions, and I gave them in the King's name for their Indians a great number of small items taken on at Brest for this purpose and which will prove of the greatest value to them.

You know, My Lord, that Monterey is not a colony. It is simply an outpost with some twenty Spanish whom the King's piety

---

[1] AN.M. 3JJ 386, 111:2, No. 39; Milet-Mureau, *Voyage*, IV, pp. 153–4.
[2] Frédéric Broudou was La Pérouse's brother-in-law who was making amends for a fairly unruly youth.

maintains for the protection of the missions that are working most successfully to convert the natives, and one will not be able to accuse this new system of the cruelties that tarnished the century of Cristoply Colomb and the reign of Isabella and Ferdinand.

Our biscuit has spoilt a little, but our grain, our flour, our wine &c have kept beyond our expectations, and have played no small part in maintaining us in good health. Our vessels are in excellent condition, but are very poor sailers.

With deepest respect, I have the honour to be,
My Lord,
Your most humble and obedient servant,
Lapérouse
Monterey, 17 September 1786.

La Pérouse to the Minister
Holograph, with copy attached. Published in Milet-Mureau's edition with mentions of the king omitted.[1]

(Mr de la Pérouse, from Monterey, 19 September 1786)
My Lord,

I have already had the honour of notifying you that, as I follow in every particular the King's orders in respect of the various places I am to visit and survey, I have felt it necessary to make use of the latitude given me of altering the plan of instructions and begin by the northwest coast of America. I venture to say that my arrangements have met with the greatest success. In the space of fourteen months we have rounded Cape Horn and gone up to the end of America as far as Mount St Hélie [Elias]; we have explored this coast most carefully and arrived at Monterey on 15 September. We had been preceded by the King of Spain's orders, and we could not have received a better welcome in our own colonies. I must also inform you, My Lord, that we called at the various South Sea islands that had aroused the King's curiosity and sailed for five hundred leagues east to west along the parallel of the Sandowick [Sandwich] Islands in order to clarify several geographical questions of considerable interest. I anchored for only 24 hours at the island of Mowee and passed through a new channel which the English had not been in a position to visit, but all our charts, the complete narrative of my voyage in a fair copy for publication (if

[1] AN.M. 3JJ 386, 111:2, No. 40; Milet-Mureau, *Voyage*, IV, pp. 156–8.

that is the King's intention), all these different matters can only be sent to you from China where I shall arrive at the end of February. I shall leave in April in order to carry out the King's instructions. I shall be in Kamchatka during the first days of August and in the Aleutian Islands at the end of the same month. I have felt it advisable to put off the exploration of these islands until after my stay in Kamchatka in order to be aware of what the Russians have not done and add something to their discoveries.

From the Aleutian Islands I shall sail without delay for the southern hemisphere to carry out my orders. I make bold to claim that no other plan of voyage has been so vast. We have already spent a year under sail and seen nevertheless during our brief calls new and most interesting things. You will be happy to tell the King, My Lord, that so far not one drop of Indian blood has been shed, nor is there a sick man in the *Boussole*. The *Astrolabe* has lost one servant who died of diseased lungs and he would not have been able to survive this affliction in France.[1] We should undeniably be the most fortunate of navigators were it not for the great tragedy we suffered. I spare my feelings the sorrow of retracing it here, and beg you to accept the extract of my journal I am sending you with the request, My Lord, that you will forward a copy to the families of the officers who so unhappily perished. I have lost during this affair the only relative I had in the navy.[2] He was without doubt among all those who have sailed with me the one who showed the greatest aptitude for his profession; he was like a son to me, and I have never been so deeply affected. Messrs de Laborde, de Pierreverd and de Flassan were also very distinguished officers; as for Mr Descures, his foolishness and his pretentious vanity, pitting himself against rocks and currents that are beyond all human efforts, were the cause of our misfortunes. I must however, My Lord, let you know that he has a sister at Alençon who is without means and lived on her brother's savings, Mr de Montarnal has an uncle who is a major in the Henaud regiment and brothers in the service: I make bold, My Lord, to recommend them to your good offices.

Our troubles compelled me to use the warrant of *lieutenant de frégate* which I had left, in favour of Mr Broudou, my wife's

[1] Jean Le Fol, Vaujuas's servant, who died on 11 August 1786.
[2] In view of the presence on board of his brother-in-law, this reference is to a blood relative, namely Pierre-Armand-Léopold Guirald de Montarnal who was (fairly distantly) related to La Pérouse's mother.

brother, who joined as a volunteer and who has given me great satisfaction. I dated this warrant First August 1786. I also gave Mr Darbaud a certificate of acting *enseigne*; he is a young man whose outstanding talent will no doubt guarantee him a place at the Académie des Sciences when we return.

All the officers, scientists and artists are enjoying the best of health and carry out their duties to perfection.

With deepest respect, I have the honour to be,
My Lord,
Your most humble and obedient servant,
Lapérouse
Monterey, Northern California, 19 September 1786.

De Langle to the Minister
Holograph. Published in Milet-Mureau's edition without the final paragraph but with the post-script.[1]

(Mr de Langle, from Monterey, 22 September 1786)
My Lord,

I could add nothing to the details the Count de la Pérouse will have given you of our navigation as I have not lost his frigate from sight for one moment since our departure from Brest. Meant to follow his destiny I have shared his misfortunes. Messrs de la Borde Marchainville, Boutervilliers and Flassan perished on 13 July 1786; an excess of courage and humanity led to their loss. More eager to deserve the King's graces than to acquire them they ended their career when they were in a position to render distinguished services. The first two especially, animated by the zeal, perseverance and sense of enquiry needed to complete campaigns of the kind we have begun, had all the talents required to get out of the most difficult situations. To sum up, My Lord, I have lost in them two friends whose advice has often been a great help. This misfortune has not diminished the zeal of my remaining five officers, their duties becoming ever more strenuous in bays where the sea does not discourage them. The friendly relations they maintain between them, the lively interest they take in the success of the campaign guarantee the safety of my frigate, and the curiosity that drives them means that they give no thought to their return to France.

[1] AN.M. 3JJ 386, 111:2, No. 44; Milet-Mureau, *Voyage*, IV, pp. 158–62.

APPENDIX I

Mr de Monti, an excellent seaman, is a model of constancy, wisdom, foresight and firmness.

Mr de Vaujuas adds to these qualities an education and an intelligence that are above average.

Mr Daigremont who is now well experienced in matters related to the sea is brave and enterprising. He does not disprove the hopes that can arise out of a lively and rakish youth; he is approaching a maturity that will soon enable him to give distinguished services because he has judgment and character.

Mr de Blondelas, a very patient, wise and conscientious officer, knows his seaman's craft to perfection. He devotes his spare time to drawing maps and very appealing and unusual sketches. On 13 July the Count de la Pérouse told him to take up the functions of a *capitaine de brûlot*; I beg you, Mr Lord, to be good enough to grant him this rank which I believe to be well deserved.

Mr de Lauriston, whom the Count de la Pérouse promoted to the rank of *enseigne*, is a distinguished individual who has acquired a considerable experience of the seaman's craft; moreover he is untiring in making observations and I depend entirely on him in this regard. As curious and as eager as his comrades in respect of discoveries he gives no more thought than they do to his return to France.

I have also cause to praise the social qualities of Mr de Lesseps, Mr de la Martinière, Father Receveur and Mr Dufresne. Mr Prévot, a painter, does not share their gracious manners or their zeal. These five and similar persons employed on the expedition are the most fortunate of all: they are at leisure while at sea and can only carry out their functions during periods of call and when the weather is fine.

The loss of the four best soldiers and of three excellent sailors from my crew has not discouraged those who remain. I accordingly announced after the events of 13 July that there would be a bonus of two months' pay offered by His Majesty.

The man François La Mare, my boatswain, is a person of outstanding merit to whom I presented a medal in accordance with His Majesty's intentions; if he continues to behave as he has done so far I will give him the warrant of permanent appointment that has been sent for him.

My boatswain certainly deserves the two favours His Majesty has made available for him, but seeing that they would give rise to some jealousy I have felt it necessary to promise to Mathurin Léon, my senior pilot, to Robert Le Gal, my chief carpenter, to Jean François Paul, my chief caulker, that I would ask you, My Lord, with my most earnest entreaties, to be given the right to bring forward the date of their appointment, and to speed the appointment of Jean Grosset who, although younger than the others, is not inferior to them in ability and intelligence; I believe that I owe the good atmosphere that reigns on board to these undertakings, and I attribute the happiness and good will that exist to their good example.

The man Gaulin, master-at-arms and acting master gunner, is also a distinguished individual; his modest pay and the means I have of increasing it enable me to reward him.

The chronometer No. 18 has been remarkably regular. I believe that in consequence the longitudes of all the land we have visited since our departure from Concepción have been determined with the utmost accuracy.

Timekeeper No. 27, less regular than No. 18, is as satisfactory as I could have hoped and in line with what Mr Bertoud had stated. We always give preference to the circles developed by Mr de Borda over sextants when we determine longitudes by lunar distances. There is always a great similarity between the results obtained by Messrs de Vaujuas, de Lauriston and me by means of these instruments which, a few minor manufacturing defects aside, are I believe the best there are for determining longitudes at sea. Fr Receveur and 4 of my pilots are also well versed in these types of observations.

Among the latter is a Mr Brossar, a relative of Lécuyer, an officer attached to the Duke of Orleans. As I am keen on his education, I do not wish him to leave the category of pilots before we return to the Isle de France, but I believe that he will then be in a position to carry out the functions of *lieutenant de frégate*; he is currently a second pilot, he is intelligent and well mannered and deserves being taken up and leave the poverty in which he was born and which runs quite counter to his behaviour and his bearing.

Don Bertrand Josep Martinez,[1] commanding the King of Spain's

---

[1] The 'Bertrand' is incorrect. The officer in question is Esteban [=Stephen] José Martinez (1742–98) who was to sail from San Blas with the *Princesa* and the *San Carlos* on 8 March 1788 for the Northwest Coast and the Aleutian Islands; his expedition eventually gave rise to the Nootka Incident which almost led to war between Spain and England. *Who's Who*, pp. 174–5.

APPENDIX I

frigate *La Princessa* from St Blas, was at anchor in Monterey Bay when we arrived; he untiringly saw to our needs and gave us all the help he could. He asked me, My Lord, to beg you to recommend him to his minister: I should be delighted to have an opportunity to contribute to his promotion.

I close, My Lord, with my assurance of the keen interest I shall not cease to bear towards the success of the expedition because I believe that the renown of the nation depends on it and that you are most eager that it should succeed. My attachment for the Count de la Pérouse has linked me to his fate and I shall always rejoice when I have an opportunity of contributing to his successs.

Respectfully, I am,
    My Lord,
        Your most humble and obedient servant,
            de Langle.

Post-script

I am leaving here without any sick. The attention of Mr Lavaux, my chief surgeon, has not been able to save Mr de Vaujuas's servant who was suffering from a wasting disease when we sailed from Brest, which caused his death on 11 August 1786. The wheat and corn taken on at Brest have kept perfectly. Mills we have had built which two men can operate when the wind is weak give us 20 pounds of ground grain an hour. We adapted with it the mills used by the Bailly de Suffren[1] during his last campaign; I gave one to the religious at Monterey mission.

Lamanon to the Minister
Holograph. Published in the Milet-Mureau edition.[2]
         (Mr de Lamanon, from the China Seas, 1 January 1787)
My Lord,
    I should have liked after a voyage of ten thousand leagues to be able to give you a report on our discoveries in natural history and

---

[1] Pierre-André de Suffren Saint-Tropez (1729–88) of whom La Pérouse subsequently wrote in a report on Manila sent from Petropavlovsk and dated 10 September 1787: 'the Bailly de Suffren's army proved of the utmost value to the colony of Manila because it kept the entire enemy forces occupied and prevented them from thinking about any distant campaign' (AN.N. 3JJ:386, 111:69 and 387 111:10). The reference is to Suffren's vigorous campaigns in Indian waters in 1782–3. Suffren who was at Brest when the *Astrolabe* and the *Boussole* were being prepared for the expedition was most helpful to La Pérouse.
[2] AN.M. 3JJ 386, 111:2, No. 49; Milet-Mureau, *Voyage*, IV, pp. 163–4.

of my personal work. But all the matters I deal with are so interlinked that I would have had to send you whole volumes. In my field I have neglected nothing that could advance our aims; my studies range from the sand that sticks to the sounding lead to the mountains I have been able to reach. I am taking collections of fishes, shells and insects, and numerous descriptions of animals, and I hope to add considerably to the known number of structured beings. I am in turn attracted by the natural history of the sea, the land and the atmosphere. We may not be the first circumnavigators to concern themselves only with the advancement of the sciences, but the English will no longer be the only ones. After a beneficial peace, it only remained, My Lord, for us to bring into being this glorious undertaking of value to everyone.

At the beginning of the previous century, our neighbours[1] discovered a new world while looking for gold; in ours the French have determined the shape and dimensions of the earth by their calculations, the English have eliminated the error of a passage through the northern seas in which they had themselves believed, they began a general survey of the globe in which we are today taking part under your auspices and which the next generations will one day complete; but what will always make this voyage stand out, what will be the glory of the French nation in the eyes of philosophers, of our contemporaries and of posterity, will be that we have frequented people believed to be uncivilised without shedding a single drop of blood. It is true that the campaign is not over, but our leader's feelings are known and I can see the support he receives. In a moment of unrest and danger arising out of a misunderstanding, he called out, 'Take up your muskets, but do not load them'; peace was fully restored through his caution, everything settled down as a result of his prudence; to the merit of being an able navigator, a fighting man, a good writer, Mr de la Pérouse joins another which is much dearer to his heart, that of being at the ends of the world the worthy representative of his prince's humaneness and of his country's virtues. Our voyage will prove to the world that the Frenchman is good and that natural man is not evil.

I have taken a few memoirs from my journals which I am forwarding to the royal academy of sciences. I beg you, Monsieur

---

[1] The reference is to the Spanish.

le Maréchal, to hand them over to the Marquis de Condorcet, permanent secretary of the academy and my correspondent. At the same time I have taken the liberty of placing under this cover a few letters, sure that in this way they will arrive more safely.
With respect, I have the honour to be,
  My Lord,
    Your most humble and obedient servant,
      Chev^{er} de Lamanon.
In the China Seas, 1 January 1787.

La Pérouse to the Minister
Holograph with copy attached. Published in the Milet-Mureau edition.[1]
  (from Mr de la Pérouse in Macao roadstead, 3 January 1787)
My Lord,
All the enclosed maps were drawn up by Mr Bernisét, a very bright and precise young man. Although all the officers took part in the astronomical observations it was fair to show them under Mr Dagelet's name, as he was in charge. It is in addition not enough that they should deserve the confidence of navigators, they must inspire it, and the name of a professional astronomer, a member of the Academy of sciences, is very appropriate to meet this requirement.

Mr Dagelet and all the officers have also taken bearings, but Mr Berniset is the main one who worked on this without interruption; furthermore he recorded them, discussed them and rejected those which did not link up with the others, and so I have had to accept trigonometric operations as being part of geography which is an opinion I did not share when I sailed. He is fully conversant with all the aspects of mathematics that are required for his functions, paints, draws and prepares charts with the utmost facility, and I am convinced that his talents would make him invaluable to an army general who appointed him as his aide-de-camp in wartime. He can also be most useful in the navy and I am very anxious to find him a position when he returns.

The *Astrolabe* has always carried out the same astronomical and trigonometrical operations as the *Boussole*; the Vicomte de Langle himself made observations of distances and angles, with Messrs de

[1] AN.M. 3JJ 386, 111:2, no. 50; Milet-Mureau, *Voyage*, IV, pp. 165–7.

Vaujuas and de Lauriston, and he had among his staff Mr de Blondelas, *lieutenant de frégate*, who filled exactly the same role as Mr Bernisét. I would have had the honour of sending you the *Astrolabe*'s maps had I not, when I compared them with ours, found that they were so much like ours that despatching them seemed to serve no purpose, but the similarity of the results achieved in both vessels is evidence of the exactness of our work.

I have the honour of sending you, My Lord, two drawings by Mr Blondelas that are not inferior to the four by Mr Duché. This painter depicts costumes with the greatest accuracy. His drawing of Easter Island will give a much more accurate impression of the monuments than Mr Hoges's[1] engraving, and since they had aroused the King's interest I instructed Mr Bernisét to make an accurate plan. Moreover I endeavoured in my narrative to complete the portrait of these islanders who will not be much visited by Europeans because their island offers no resources. Mr Duché's other three drawings are also very lifelike. This is only a sample of his activity; there are still twenty left in this artist's portfolio.

Young Mr Prévost drew all the birds, fishes and shells; I felt that I owed it to his zeal to send you three of his drawings of birds. I would very much like to be able to be equally complimentary about his uncle, an artist who embarked in the *Astrolabe* with Mr de la Martinière, doctor and botanist, but his unmanageable character makes him almost useless for his area of responsibility, and as he is over fifty it is difficult to correct him; I do not believe that he has drawn 15 plants since our departure, claiming that he does not have convenient quarters.

I append, My Lord, to our own charts those of the three presidios of Northern Caliphornia which I obtained at Monterey with two maps of the west coast of America from the port of St Blas up to that of St Francis in 38 degrees north. They have been carefully drawn up by the Spanish in recent times and will enable the French depository of journals to supply an excellent map as far as Mt St Hélie by combining it with the one we have made.

---

[1] William Hodges (1744–97), appointed by the Admiralty as artist on board the *Resolution* on Cook's second voyage. His 'Monuments of Easter Island' is currently at the National Maritime Museum, Greenwich. A number of engravings were made from it which sometimes romanticised and 'improved' the original.

APPENDIX I

The Spanish map of the Great Ocean which I have the honour of sending you and on which I have traced my route from Monterey to China is detestable; I am adding it to the others merely to prove that knowledge of this vast sea has not advanced over two centuries because the Manila galleons always follow the same track and do not diverge from it by ten leagues.

With deepest respect, I have the honour to be,
My Lord,
    Your most humble and obedient servant,
        Lapérouse.

On board the *Boussole*, 3 January 1787, day of our arrival in Macao roadstead.

A pencilled note at the bottom of the page states: 'These 2 maps were not found in the collection of maps brought by Dufresne.'

La Pérouse to the Minister
Holograph. Unpublished.[1]

(Mr de la Pérouse, 3 January 1787, appended: Mr de Monneron's memoir on military surveys)
My Lord,

I have the honour of sending you the military observations of Mr de Moneron. Our calls have given him few opportunities so far of putting forward any plan of substance, but those that remain to be done will be more interesting. We are to visit this year Manila, Formosa and the entire coast of China from Tartary to Katchatka, and from the reports which I have already had the honour of giving you on Mr de Moneron you must be quite sure, My Lord, that his zeal will lead him to see everything and his knowledge place him in a position of noting everything that can be of interest to the government.

Respectfully, I have the honour to be,
My Lord,
    Your most humble and obedient servant,
        Lapérouse
Macao, 3 January 1787.

---

[1] AN.M. 3JJ 386, 111:2, No. 52.

La Pérouse to the Minister
Holograph, with copy attached. Published in the Milet-Mureau edition.[1]

(Mr de la Pérouse, appended: memoir No. 1 on otter furs)
My Lord,
I have the honour of enclosing the complete narrative of my voyage up to Macao, with the table of route followed each day. I append the drawings of the coast we sailed along, of Port des François of which we took possession, of the various islands we visited, as well as Necker Island and French Frigate Shoal which we discovered and where we came so close to being lost. I have traced the route of the two frigates on the general map enclosed; it goes across several islands that do not exist and which unnecessarily fill on world maps spaces where there has never been any land.

Our map of the northwest coast of America is certainly the most exact ever drawn, and it only lacks details that time and a long series of navigations will complete.

We identified the entrance to the archipelago of St Lazarus (if one can retain this name), determined its true position in latitude and longitude as well as its width from east to west and its depth for twenty leagues approximately to the north. The season which was already well advanced, the shortness of the days and the rest of our plan of campaign did not allow me to penetrate into the back of this labyrinth, which would have required two or three months on account of the precautions one must take for such an undertaking which, although it would have satisfied curiosity, would never have produced results of interest to navigation or of value to France; I would nevertheless not have hesitated to complete this survey if I had found myself at the entrance to this archipelago during June, but at the end of August, approaching the equinox, with darkness lasting twelve hours and almost continuous fog, the enterprise was I venture to say impossible and I would have put the remainder of the voyage at risk without any benefit for geography.

I am confident, My Lord, that you will notice that in eighteen months we have spent fifteen at sea and three only on our various calls, and our cares have been so successful that we have been affected neither by illness nor by scurvy, but although at the point

---

[1] AN.N. 3JJ 386, 111:2, no. 53. There are two copies of this letter in the same file, nos 54 and 55. Milet-Mureau, *Voyage*, IV, pp. 167–71.

when I have the honour of writing to you we have covered ten thousand leagues we have hardly completed a third of our campaign, and I dare not hope that we shall be equally fortunate during the remainder of the voyage, if however one can use the word fortunate at the dreadful tragedy we encountered at Port des François, of which I had the honour of informing you in my letters from Monterey. Since the precautions I had taken could not prevent it, it proves only too well that we cannot control our destiny.

I have scrupulously avoided changing the names which Captain Cook bestowed on the various headlands he discovered, but you will notice, My Lord, that we saw the coast of America from much nearer than that celebrated navigator, and so we were permitted to name ports, bays, islands and inlets he did not even suspect. Usage has allowed me to choose these names from among those whom I most fondly recollect and I felt that I should pay this kind of homage only to my friends and to Messrs de Sartine and Neker whom out of respect I cannot call thus, but they treated me with friendliness in the days of their highest good fortune and they rightly deserve the country's esteem. As grateful, My Lord, for your extreme kindness towards me as Captain Cook was to Lord Sandowick I have been most eager to make a discovery of sufficient importance to give it your name. Since I am only at the start of my campaign it is likely that I shall find on the coast of Tartary or in the southern hemisphere a feature of equal importance to geography as the Sandowick Islands. I will consider the world too small if my hopes prove vain.

It is my earnest wish, My Lord, that your duties will leave you time to glance at the various chapters of my narrative, so that you may inform the King of the punctilious manner in which I have endeavoured to carry out every aspect of my instructions. I have called at Easter Island, sought the so-called islands to the east of the Sandowicks which do not exist, seen Mowhee Island in the Sandowicks where Captain Cook had not landed, the Northwest coast of America from Mt St Hélie as far as Nootka, but from Nootka to Monterai I sought only the points which Captain Cook had not been able to survey and which remained as dotted lines on his chart.

I have obtained on the Spanish settlements the added information which my special instructions asked me for. I have the honour of sending you herewith a memoir on this subject.

I crossed the Great Ocean along a parallel one hundred and sixty leagues from that of other navigators. I discovered Neker Island and French Frigates Shoal. I proved by my route that the islands of La Gorta, Deserte, La Mira, Des Jardins do not exist and as requested visited one of the islands to the north of the Marianas whence I sailed to China. I shall leave at the beginning of the season to navigate between the coast of that vast empire, of Korea and Tartary and the islands of Japan and the Kurils. I shall then put into Kamtchatka, and leaving it I shall visit the Aleutian Islands as well as those shown to the east of Japan whose existence is more than doubtful.

After that all that will be left to me will be to sail towards the southern hemisphere, not overlooking the fact however that I have left to the north of the Line the Caroline Islands which I am required to investigate. It is only from Kamtchatka, My Lord, that I will be able to advise you of the later plan of my voyage, because I will be in a position to plan it in full only when I know the precise date of my departure from Siberian waters and am still unaware of the amount of time I shall have to devote to my navigation on the coasts of Tartary. The south-west monsoon one meets south of the Line from the first days of November does not allow me at this stage to make plans which the slightest delay would upset, but if I can foresee crossing Endeavour Strait before this monsoon sets in, my first navigation will be around New Holland. Otherwise I shall begin by Cook's Entry in New Zealand,[1] the south of New Caledonia, the Arsacides and the Carolines; then sailing through the Moluccas with the northern monsoon I will survey New Holland whence I shall make for the Isle de France.

This programme is very extensive, but it is beyond the zeal of none of the people engaged in this expedition. The most difficult thing is to complete this task in four years and it may be impossible for our vessels, our rigging and our provisions to last any longer. Whatever happens I beg you, My Lord, to assure His Majesty that I shall make all endeavours to carry out my instructions in full, but I cannot allocate much time to our various places of call and this long period at sea does not appeal to our botanists and our mineral-

---

[1] There is no such place. It is evident from other references that La Pérouse fully intended, until his orders were countermanded in Kamchatka, to put in at Queen Charlotte Sound, which is reached by way of Cook Strait, between the North and the South islands.

ogists who can only exercise on land their talents for their various scientific fields of endeavour.

With the deepest respect, I have the honour to be,
My Lord,
Your most humble and obedient servant,
Lapérouse.
Macao roadstead, 3 January 1787, day of arrival.

La Pérouse to Fleurieu
Copy. Published in Milet-Mureau's edition.[1]

(Macao roadstead, 3 January 1787)

I am forwarding the plan of Monterey we prepared. I had the opportunity to make the acquaintance at Monterey of officers of the small naval forces of St Blas who certainly are not lacking in education and who appeared to me to be quite able to draw accurate charts....

You will see that I amended my plan of navigation as experience and reflection dictated. That is the only way a plan as vast as ours can be carried out.

For instance, I sailed direct from the Sandwich Islands[2] to Mt St Elias because if I had begun by Monterey and then gone up north I would have faced the continual opposition of the north-west winds; whereas with these very winds I was able to sail down the coast of America and follow it as I wished. But the fog is a constantly recurring obstacle which makes one waste a great deal of time which one must sacrifice in the name of prudence: I do not think one can depend on more than three days of clear weather in a month. The currents are very strong and also require a great deal of caution: they caused at Port des Français[3] the misfortune my letters told you about and which I will always look back on with sorrow.

I do not know whether you will regret my not having explored more throroughly the St Lazarus archipelago, if however one should continue to use such a name which is certainly not something I support: but note that I only discovered its entrance at the

[1] AN.M. 3JJ 389, 22; Milet-Mureau, *Voyage*, IV, pp. 209–13.
[2] It is worth noting in passing that the copyist, whoever he was, knew the correct spelling of 'Sandwich' and was able to put right La Pérouse's usual 'Sondowick'.
[3] A similar comment applies here: La Pérouse uses the old word-ending *-ois*, but the modern form *-ais* was coming into general use by this time. Both spellings were pronounced in the same way by the eighteenth century, certainly among the younger generation.

end of August, that the days were becoming very short, that the fog never ceased and that we found by Cape Hector currents of more than six miles an hour. It was consequently impossible to sail up between all these islands in the space of two or three months, and the season ends at the beginning of September. To be complete such an exploration would require an expedition whose sole object it would be and which lasted not less than two or three years. Nothing is more time consuming that the exploration of a coast littered with islands, broken by several gulfs, which frequent fogs and eternally violent and uncertain currents allow one to approach only with prudence and precautions. Be that as it may, I have no doubt that the voyage of Admiral Fuentes, at least as it has been told to us, is greatly exaggerated, if it is not a dream:[1] one does not cover such a distance as he did in such a short time; and I would be very tempted to believe that Admiral Fuentes and his captain Bernarda are imaginary characters, and the tale of the voyage credited to them a fable. It is nevertheless true that from Cross Sound to Cape Fleurieu the great Spanish navigator Maurelle, Captain Cook and I coasted along nothing but islands distant forty or forty-five leagues from the mainland; and my opinion is based on the trend of the mainland shore that I sighted once more at Cape Fleurieu. Most of these islands are extensive, and as they overlap, this arrangement gives the impression of a continuous coastline. I several times suspected that the lands I could see were not all on the same level, but this suspicion became a certainty when, having rounded Cape Hector, I advanced twenty leagues to the north. All these details assume that you have before you the charts and maps I am sending and that you are following my route while reading my narrative....

You will sense that not a great many details can be expected from us: in order to travel, in the space of four years, to all the places listed in my instructions, we cannot waste a single day. But our navigation will prove that a crew's health does not have to suffer as a consequence of the longest of periods spent at sea: we are arriving at Macao without a single case of scurvy, and yet during the

---

[1] It was more than a dream: it was a fabrication. Bartholomew Fuentes, or de Fonte, was the subject of an article in a short-lived periodical, *The Monthly Miscellany, or Memoirs for the Curious*, published in 1708. No evidence exists for his voyage of 1640 to an archipelago named St Lazarus. In his journal, La Pérouse further inveighs against this imaginary expedition.

APPENDIX I

eighteen months the campaign has already lasted fifteen were spent in a toilsome navigation which required us to pass through vastly different climates....

I am writing to you in haste, without any methodical arrangement; I am casting my thoughts on the paper as they come to me. I am anchored five miles from this place, with which I have not yet communicated; and as they told me that a ship was leaving tomorrow for Europe I am rushing off all my despatches. I am enclosing my narrative and charts and maps with the letter I am writing to the minister: I will send him copies at the first opportunity, so that if some misfortune should overtake us on the coast of Tartary navigators will not lose the benefit of at least the beginning of our campaign. You will be pleased to note as you read through my various chapters that, although the uncivilised people we called on have done us some little harm we have been fortunate enough never to have to do them any in return. You know better than anyone that I am expressly ordered to treat this as a final recourse, and you know also that I strongly share that view.

We bought close to a thousand otter skins on the coast of North America, but most of them were in shread and rotting. I felt that I should engage in these transactions with a scrupulousness and a delicacy of approach which none of the navigators who have landed on this coast has shown. Furs were dealt with only by Mr Dufresne: I gave him the task of carrying out this trade and he fulfilled this delicate duties with as much zeal as wisdom. I shall send an account to the minister, just as a supercargo would to his employers, and I shall append the receipts of all those who received payment. I did not want a single fur to be reserved for the officers, the scientists and artists, or for me. Profits from this campaign must go to the sailors; and the glory, if there is any, will be the share that goes to the officers who directed the expedition and their helpers. I must admit, my dear friend, that a thousand *écus* would not have persuaded me to carry out this campaign, but I did not hesitate to undertake it out of a sense of duty and of gratitude for the trust that was placed in my zeal, more, undoubtedly, than in my talents.

Lapérouse.

Langle to the Minister
Holograph. Published in Milet-Mureau's edition with some modifications.[1]

(Mr de Langle, from Macao, 18 January 1787)

My Lord,

The navigation of the *Astrolabe* has been very satisfactory during the crossing from Monterai to Macao. I did not lose one man and not even one case of sickness. The frigate will be able to continue the campaign as soon as the rigging and sails have been repaired.

The eagerness and good will of my crew have not diminished for one moment and we will all continue with great pleasure to contribute to the Count de la Pérouse's success.

Mr de Monti's firmness, wisdom and foresight are contributing to everyone's happiness and I have the utmost confidence in his talents.

I admitted Mr de Vaujuas to the grade of Chevalier of St Louis[2] on 1 January 1786; I have not come across such an accomplished naval officer since I have been in the King's service.

Mr Daigremont is a man of spirit, good judgment and firmness; he is made to inspire the utmost confidence at sea; he is practising to take readings and will succeed.

Mr de Blondelas, a very good naval officer, displays exemplary wisdom and firmness; he devotes his spare time to the charting of bays and produces most attractive and lifelike sketches.

Mr de Lauriston's zeal in adding to his knowledge of his craft has not diminished for a moment; he is now an excellent naval officer and in a position to make great progress in astronomy; I have passed over to him everything that relates to this subject.

I owe the constancy with which the *Astrolabe* has stayed with the *Boussole* in fog and darkness to these five officers' talents and the good relations they maintain between them. They take such an interest in the safety and the preservation of the vessel and the success of the campaign that I find myself today the one who has the least to do.

[1] AN.M. 3JJ 386, 111:2, No. 56; Milet-Mureau, *Voyage*, IV, pp. 179–82.
[2] The decoration being in effect admission to an order, the bestowal of the cross of St Louis involved a formal reception into the order by someone who was already a member. The same procedure is followed today for most French decorations, including the award of the Legion of Honour which replaced the order of St Louis after the Revolution.

APPENDIX I

All my hopes would be fulfilled if they were to receive at the Isle de France the rewards you felt, My Lord, that they could claim upon their return, namely the warrant of *capitaine de vaisseau* for Mr de Monti, the Cross of St Louis for Mr Daigremont and the warrant of *capitaine de brûlot* for Mr de Blondelas. I think, My Lord, that Mr de Vaujuas who was the senior when you appointed him lieutenant and who was born without means could have a claim to the pension of 800 *livres* granted to the late Mr Descures.

I believe also that Mr de Lauriston deserves to rank as an *enseigne* from 13 July 1786, date when Mr de la Pérouse gave him his warrant. I cannot, My Lord, praise adequately the amiability and all the good qualities of Mr de Leseps. Father Receveur carries out his duties as chaplain with decorum; he is friendly and intelligent; while at sea he deals with meteorological and astronomical observations and when we are at anchor, with matters related to natural history.

Mr de la Martinière deals with botany in an assiduous manner; in this he is very poorly assisted by Mr Prévot who has made it his rule not to paint on board ship; he is a considerable burden and all the more useless to the expedition in that his nephew who has now more talent than he has is full of keenness and good will and draws everywhere.

Mr Dufrene made himself useful in the trade in otter skins; he has done a great deal for their preservation and their sale; as he wishes to return to France and I now consider him of very little use, Mr de la Pérouse has allowed him to go back. I have a great deal of praise, My Lord, for Mr Laveaux, my senior surgeon, and of Mr Guilloux, his assistant; by their foresight they have contributed a great deal to my crew's good health. Fortunately they have had considerable free time so far, and during our periods at anchor they have devoted this to acquire knowledge in botany and natural history and to gather specimens for the King's collection.

I wish to call on your kindness, My Lord, on behalf of Mr Brossar, a Breton gentleman who after 40 months of service as a volunteer in several royal ships joined as assistant pilot on the *Astrolabe*. Since 13 July 1786 he has carried out with much zeal and intelligence the duties of second pilot. I beg you, My Lord, to be kind enough to send him at the Isle de France a warrant of appointment as *lieutenant de frégate*. Allow me also, My Lord, to recommend to you my masters, pilot, gunner, carpenter, sailmaker and caulker. They are all former servants of the King who have shown proof of their intelligence and firmness, and they greatly contribute to creating the happiness that reigns on my ship and to the good relations that exist among all individuals. I beg you, My Lord, to grant them appointments. I am not mentioning my boatswain because I have given him the medal and will give him his warrant of appointment if he continues to behave with the firmness and distinction he has displayed until now.

The Chevalier de Bellegarde has transferred from the storeship *Marquis de Castries* to the *Astrolabe*. He is a man whom Mr de Richeri has highly praised to me and I shall have the honour of telling you more about him when I know him better. He is a *garde de la marine*.

With respect, I am,
  My Lord,
    Your most humble and obedient servant,
      de Langle.
At Macao, 18 January 1787.

La Pérouse to the Minister
Holograph. Published in Milet-Mureau's edition with modifications.[1]

(from Macao, 18 January 1787)
My Lord,

I owe you a report on each of the officers and passengers of the expedition, and as I have much that is favourable to report this is a pleasant duty.

The Vicomte de Langle is an excellent officer with a great ability for his profession and a firm and unshakeable character. He has been so careful to follow me that we have never once been beyond hailing distance, except when I instructed him to sail away ahead, his frigate being a much better sailer than mine.

The departure of Mr Monges has not affected in any way the astronomical observations being made on board the *Astrolabe* because the Vicomte de Langle is as good a naval astronomer as was the professor; he has been most ably assisted by Mr de Vaujuas, a highly trained officer, and he has trained in the work of observation Mr de Lauriston, a young man who under every heading has

---

[1] AN.M. 3JJ 386, 111:2, No. 57; Milet-Murau, *Voyage*, IV, pp. 176–8.

shown himself well educated, of good character, zealous and devoted to all his duties.

I authorised the Vicomte de Langle to report to you himself on the abilities, character and behaviour of each of his officers and passengers, I know that he is quite unprejudiced or given to any favouritism, and so the truth will reach you without any concealment.

Mr de Clonard, my first officer, is an officer of outstanding merit in whom are combined the skills of his profession, accuracy, zeal, honour and a desire for success which make him one of the most estimable men I have ever met. In accordance with your instructions, I gave him, My Lord, his warrant of *capitaine de vaisseau* on 1 January 1787 so that he can acquire seniority from that date and take up his position among other captains as advised by the letter I had the honour to reeive from you, written in Versailles on 23 June 1785.

Mr Boutin is a man of spirit, talent, untiring energy, firmness and coolness in difficult situations, whom I can never praise highly enough. I owe to this quality the preservation of the small boat that went through the breakers in the pass of Port des François on the day when our unfortunate companions were drowned. I would on that same day have made use of the permission you gave me in your letter of 26 June to put forward or put back the King's rewards, and the Cross of St Louis was certainly due to the officer to whom I owe the survival of six other people and who had himself escaped from such extreme danger, but we were all so distressed that I felt I should put off this award until 1 January 1787, at the same time as Mr Vaujuas's for which you had laid down that date. I thereby put forward by six months the date decided for Mr Boutin.

If it was less painful for me, My Lord, to remind you of the losses we sustained, I would have the honour to point out that almost all the rewards you kindly granted to the division's officers have now lapsed. Mr Descures had a pension of 800 *livres*, Mr de Pierreverd one of 500 with the warrant of lieutenant, Mr de Marchainville the Cross of St Louis, Messrs de Bouterviliers, de Flassan and de Montarnal a warrant of *enseigne* each. Your letter, My Lord, allowed me only to make promises to Messrs de Monti, Daigremont, Colinet and Blondelas: if they all became a reality when we reach the Isle de France, we should all have proof that our actions have deserved your approbation, but the special promotion of the vicomte de Langle would for me be a priceless favour; he will be 47 years of age when he returns to France and if at that date he has not reached the rank of *brigadier* it would not be possible for him to attain the higher ranks at a time when his outstanding ability would enable him to be of the greatest use.

Messrs Colinet, St Céran, Darbaud, Mouton and Broudou, to whom I gave the two warrants of *lieutenant de frégate*, are full of zeal, and energy and carry out their functions excellently well. Their duties vary greatly, as each boat is always under the command of an officer. Their number would have been insufficient after our misfortunes without the two replacements I made.

With deepest respect, I have the honour to be,
My Lord,
    Your most humble and obedient servant,
        Lapérouse.

La Pérouse to the Minister
Holograph. Unpublished.[1]

(from Macao, 26 January 1787)
My Lord,

I have instructed Mr du Fresne to hand you a packet I intend for the Queen if you consider it worthy of her and of being presented by you. It contains four very fine otter furs, and two lengths of material made by the Indians of Port des François.

The parcel of samples which my instructions required me to send to France so that furs from that part of America might be known contains, with fours length of material, eight otter skins which after examination could be forwarded to the collections of the King and the Comte d'Artois.[2]

Knowing your scrupulousness, My Lord, I have not dared to send you any otter skin, your work not allowing you time for a natural history collection, but only a minute is needed for you to

---

[1] AN.M. 3JJ 386, 111:2, No. 60.
[2] Louis XVI's brother, later King Charles X (1757–1836).

cast your eyes on the two lengths of material which I have asked Mr du Fresne to have the honour of giving you.

With deepest respect, I have the honour to be,
My Lord,
Your most humble and obedient servant,
Lapérouse.

Macao, 26 January 1787.

La Pérouse to the Minister
Holograph. Unpublished.[1]

(Mr de la Pérouse, from Macao, 26 January 1787)
My Lord,

I had the honour in all my despatches to give you an account of our deals in furs.

I appended the list Mr du Fresne had drawn up, as well as the price he hoped to get, but all our hopes were dashed by events and the two hundred and six thousand nine hundred and nineteen *livres*, Mr du Fresne's asking price per his estimate enclosed herewith, was reduced to nine thousand five hundred piastres or a little less than fifty thousand *livres*, and the deal was concluded by Mr Vieillart, French Consul,[2] as you will see from the enclosed letter. I wished, I wanted even, that the King's man should be aware of this affair, as my feelings and those of the two staffs would have suffered from the slightest lack of authenticity. I have had the honour of notifying you of all the officers' refusal to share in this, even when these otter skins were thought to be worth four times what they sold for. They even made a point of keeping for themselves not even enough material to make a button for their uniform, and I make bold enough to give my word of honour that no trader has ever been more faithfully dealt with in a matter of business in which he was directly involved; Mr du Fresne himself, whom I had appointed supercargo for the two ships, and to whom I had granted a commission of three per cent on the sale price,

[1] AN.M. 3JJ 386, 111:2, No. 61.
[2] Philippe Vieillard had been in China since 1776 and had been eager to return home for some years. His ability, like those of other French officials in Canton, was in doubt and La Pérouse condemned them en bloc. There can be no doubt that, by the time the expedition reached China, Vieillard had allowed things to slide and had lost all interest in his work. It had never been easy to trade with the Chinese authorities and the fur trade was entering a difficult period, but Vieillard's indifference certainly made matters worse.

refused to accept it and wanted even the value of his endeavours to be distributed. I will have the honour of sending you at the earliest opportunity a statement of this distribution, signed by all those receiving any payment, but I shall borrow from this total a small sum for the requirements of the two frigates, which the King will take into account on our return.

I believe that I have the honour of pointing out at this stage, My Lord, that although the expenditure incurred by the two frigates is modest we do use fairly large sums, but these are merely advance drawings on the pay or the keep of [the complement of] both frigates who, in accordance with your decrees, are entitled to request payment once two months are due, and so our expenses are in no way out of the ordinary and consist only of the campaign's normal outgoings which can be paid out equally well during the voyage or on our return. Moreover our accounts are quite in order; we took two thousand piastres on each frigate when we sailed from France and Mr Vieillart had two thousand piastres loaned to us in the King's name when we reached China. Mr de Clonard, my first officer, will render an account of how all these sums were used to the Council of Marine at Brest, and I am sure that in this matter as all others his accuracy will be praised.

Although the memoir on otter skins, which I had the honour of sending you,[1] concerns only America and the ease with which one can obtain this article of trade on that coast, the circumstances surrounding the sale we concluded in China will prove, My Lord, that all the speculations carried out in Europe would end in ruin. Nine hundred and ninety-eight furs, three hundred of them whole and of the best quality, fetched a mere fifty thousand *livres*, and at the price reported in Cook's third voyage this same quantity would have produced at least a hundred thousand, but an English company from Bengal made six voyages to Cook's River and Williams Sound this year, and although these expeditions resulted in very few furs, the mere rumour of these voyages brought about an extraordinary drop in prices. I will find out in Kamtchatka whether Russian trade in northern China has suffered as much as the Canton trade, but even if it has survived these shocks on account of the great distance between Canton and Kiakia, it seems impossible to me that it will not be upset by the forthcoming arrival of the

---

[1] 'Rapport sur le commerce des loutres', AN.M. 3 JJ 386, 111:61.

APPENDIX I

Spanish with ten thousand skins, and I have not noticed that the Chinese make any difference between furs from the north and those of California, of which we had a small quantity; all they look at is the size of the skins.

With deepest respect, I have the honour to be,
My Lord,
Your most humble and obedient servant,
Lapérouse.
Macao, 26 January 1787.

La Pérouse to the Minister
Holograph. Published in the Milet-Mureau edition.[1]
(Mr de la Pérouse, Macao, 2 February 1787)
My Lord,

I have often mentioned our furs to you; I have even added that they had been sold. The accounts I had the honour of sending you were reliable, but not the buyers, and at the point of delivery they raised arguments that led to the deal being broken. I intended bringing them back to France, where I am sure they would have found a more certain and more profitable market than in China, but realising that my return to Europe is still quite some time away I accepted the kind offer made to me by Mr Stekinstron, manager of the Swedish Company, who agreed to store them in his premises, and promised to sell them on behalf of our sailors and to remit the proceeds to the Isle de France where I propose to share them out among the crews of the two ships, unless you issue different instructions, and since we will not arrive at the Isle de France for another two years we shall have time to receive your final instructions on this matter and carry out your wishes. As I am sailing for Manila tomorrow morning, My Lord, I think I can assure you that this is the final chapter of the otter skin saga.

I cannot fail to advise you, Mr Lord, that in China at the present time the French nation does not have a single individual who inspired enough confidence in me to be entrusted with this small consignment. The Company's two supercargos are insane or at least were, because Mr Thérien has blown his brains out and Mr

[1] AN.M. 3JJ 386, 111:2, No. 62; Milet-Mureau, *Voyage*, IV, pp. 182–4.

Dumoulin, who has remained in Macao,[1] has carried out several insane actions which in Europe would have led to his being confined; he nevertheless remains in charge of fairly important assets because no one has felt adequately empowered to depose him and I was much further removed than the King's employees to meddle in such unsatisfactory matters that were quite alien to me and might give rise to lengthy lawsuits in Europe, because the Company affairs which have gone very badly in China are likely to make the directors very litigious.

And so while every country, even Denmark and Sweden, has top-flight men in Macao, the French are privileged not to have a single person worthy of being a village magistrate. I shall permit myself some comments on this, My Lord, which I will have the honour of sending you from Manila whence I will sail during the first days of April to continue my campaign in northern Asia.

I have not had the honour of advising you in my previous despatches that I found in Macao roadstead the King's storeship *Marquis de Castries*, commanded by Mr de Richeri, *enseigne* of the King's ships. Since that vessel has been sent by Messrs de Cossigni and Dentrecasteau,[2] they will let you know about his mission, but I felt I could take it on myself to discharge Mr de Bellegarde, *garde de la marine*, and appoint him to the *Astrolabe* in replacement of the three officers from that frigate who were lost on the coast of America.

With deepest respect, I have the honour to be,
My Lord,
your most humble and obedient servant,
Lapérouse.
Macao, 2 February 1787.

---

[1] François Thérien had arrived in China at the age of 21 and committed suicide after fourteen years, despairing of ever being allowed to return to France. Dumoulin is Fougueux des Moulins who arrived in 1783; he was neither incompetent nor unco-operative with La Pérouse but suffered from nervous depression; he later expressed the wish to join a Franciscan order.

[2] It was Joseph-Antoine Bruny d'Entrecasteaux (1737–93) who was to lead the great expedition of 1791–93 to look for La Pérouse and his ships. Richeri was Joseph de Richery (1757–98) and Cossigni was David Charpentier de Cossigny (1740–1809).

APPENDIX I

La Pérouse to the Minister
Copy. Unpublished.[1]
My Lord,
  I have the honour of enclosing a list of the provisions supplied to us at Manila, which added to the cost of the repairs of the vessels in the port makes a total of four thousand eight hundred and seventy-nine piastres in respect of which I have given the civil administrator of Manila a bill of exchange drawn on the Treasurer General of the Navy, signed by Messrs de Clonard and Monti and countersigned by Mr de Langle and me. You will notice, My Lord, that we took at Manila a year's supply of food for the two frigates, and that we have been most careful with our expenditure.
  With deepest respect, I have the honour to be,
    My Lord,
      Your most humble and obedient servant,
        Lapérouse.
Manila, 7 April 1787.

La Pérouse to the Minister
Copy. Altered version in the Milet-Mureau edition.[2]
                (Mr de la Pérouse, 7 April 1787, from Manila)
My Lord,
  The despatches I have asked Mr Dufresne to have the honour of handing over to you, duplicates of which will reach you through Mr Vieillard, former consul in China, contain all the details of our navigation from Brest to Macao with the maps of the various islands, ports and bays where we put into in the South sea and on the Northwest coast of America; I have enclosed with these maps drawings by Messrs Blondelas, Duché and Prévots to show you their ability and their zeal. If your functions have allowed you, My Lord, to cast an eye on our narrative, I dare flatter myself that you will have seen that we have neglected nothing that would render our voyage interesting and useful. Our map of the Northwest of America from Mt St Elias to Monteray will prove widely satisfactory to navigators. Our misfortunes in Baye des François, far from lessening our zeal, convinced us even more of the obligations we have undertaken towards the King and the nation, and we do not

[1] AN.M. 3JJ 386, 111:2, No. 63.
[2] AN.M. 3JJ 386, 111:2, No. 65; Milet-Mureau, *Voyage*, IV, pp. 184–7.

cease to regret that it is no longer possible to hope to find a new continent, but only a few islands of minor importance that will not add anything to our knowledge and our commerce. These same despatches brought by Mr Dufresne will have informed you, My Lord, that after disposing of our furs, I planned to sail to Manila to obtain food supplies, inspect our rigging, repair our rudder and place us in a position to continue our voyage by passing through the strait of Formosa and coasting along the west coast of Japan and Tartary. I do not know, My Lord, whether you recollect that this is the section of our navigation which the King considered the most difficult when I had the honour of being presented to him, and that if we are fortunate enough to explore these coasts with the same care as the coast of America I am sure His Majesty will not fail to realise that his vessels will have been the first to undertake this navigation, which is subject to the most violent storms in quite narrow seas that are completely unknown, shrouded in fog and in all likelihood reef-strewn. All these difficulties are present in our imagination merely in order to arouse our caution and increase our eagerness.

I left Macao on 6 February and did not reach Cavitte in Manila Bay until the 28th. The detail of our navigation is fairly interesting from the point of view of navigation and will make up an extra chapter in our narrative

I preferred the port of Cavitte in Manila Bay because we are close to an arsenal and within reach of every kind of assistance. This we have received and we owe it to the orders issued by the Governor and even more to the friendly help given by Mr Gonsolez Carvagual, Civil Administrator of the Philippines, that we are sailing from Cavitte as well supplied with fresh food as when we left Brest. I shall have the honour of sending you from Kamtchatka, My Lord, in accordance with your instructions, a detailed memoir on Manila, on that colony's resources, on its administration, on the New Company and the character of its administrators who are far from having adopted towards the French the feelings of the Madrid Cabinet, with the exception of the Civil Administrator who has constantly given us signs of the utmost good will, and personally called twenty times a day on all our suppliers because, knowing the dilatoriness of his compatriots, he was concerned lest we wasted a single day.

I am leaving on 8 April although the north-west monsoon has

not turned, but I will be in a position to benefit from the first wind changes to go north, and before sailing I had the satisfaction of seeing the arrival in Manila Bay of the King's frigate the *Suptile*, commanded by Mr de la Croix de Castries whom Mr Dantre Casteau [D'Entrecasteaux] had sent in part to let me know of his negotiations in China so that they did not run counter to ours if our instructions required us to sail on the northern coasts of that empire.

Mr Dantre Casteau will give you an account, My Lord, of the native rebellion in Formosa and the steps he felt he should take to offer his assistance to the Chinese to put it down; it was not accepted and I admit that I would have felt saddened to see the French navy allied to the most iniquitous and the most oppressive government there is on earth.[1] I can today with a clear conscience express my good wishes for the Formosans and even more so because the French are going towards that land before the English whose activities and thirst for conquest in India are beyond words; I do not believe in the least that this ambition comes from the Court at London, but the English government finds itself drawn in this manner by undertakings that may not have its support, and it no doubt owes several of its Asian provinces to the concerns of the governors and councils of Madras and Calcutta. Already new English forts are rising in the Strait of Malacca; the dissensions in Cochinchina[2] are also attracting the greed of traders in Bombay and Calcutta. Asia is not sufficient for their vast speculations and the slightest hopes give rise to expeditions costing several millions.

You know, My Lord, that they have this year sent six vessels to trade in otter skins, and that their company is well structured to leave to private trade in eastern waters all the opportunity for growth it can find.

---

[1] La Pérouse's liberal ideas, like those of the eighteenth-century French *philosophes*, are strongly expressed on a number of occasions. Uprisings by the Formosan people against their Chinese conquerors were frequent and continued until the island was taken over by the Japanese in 1895, whereupon the Taiwanese began to rebel against their new rulers.

[2] Cochin China (the southern part of Vietnam) was ruled by the Annamese, but had rebelled on a number of occasions. When La Pérouse wrote this comment, a claimant to the throne of Annam was ruling over Cochin China and in 1787 he signed a treaty with the French which led to his conquering Annam and the northern kingdom of Tonkin; these military operations spread over 25 years; in time French influence spread over the whole of Indochina.

I am replying to Mr Dantre Casteau that my navigation on the coasts of China will not alarm that government, that I shall never fly my ensign and will carefully avoid everything that could cause him some anxiety, and I am adding that, although a good Frenchman, I am in this campaign a cosmopolitan, remaining outside all the politics of Asia.

You sent me, My Lord, prior to my departure from Brest a memoir on Formosa by Mr Vieillart, and I saw with some surprise at Macao that this same Mr Vieillart had no knowledge of that country, could answer none of my questions and had simply copied a manuscript which is in the hands of every European in Macao. Although, My Lord, it is quite outside my mission to report to you on French employees in Canton, I would feel that I would not be justifying the trust you have placed in me if I did not make you aware that Messrs Vieillart, Costar, de Guignes and du Moulin should never have been entrusted with the interests of a great power, and I had to approach Mr Stokinstrom, Head of the Swedish Company, for all our transactions.[1]

I shall have the honour of sending you a private letter on this matter.

With deepest respect, I am,
    My Lord,
        Your most humble and obedient servant,
        Lapérouse.
Manilla, 7 April 1787.

La Pérouse to the Minister
Copy. Published in the Milet-Mureau edition.[2]
                (Mr de la Pérouse, from Manila, 7 April 1787)
My Lord,

Mr de la Croix de Castries's arrival in Manila has been one of the happiest events of this campaign.[3] As I have already had the honour of advising you, he has been kind enough to take our furs to France, and has shown the greatest willingness to make up for the losses we have sustained since our departure from France by trans-

---

[1] Stockenstrom's report to La Pérouse on his transactions and the low price of furs is held in Archives Nationales, Marine, 3JJ 386, 111:2, no. 64.

[2] AN.M. 3JJ 386, 111:2, No.66; Milet-Mureau, *Voyage*, IV, pp. 188–9.

[3] Anne-Jean-Jacques-Scipion de la Croix de Vargnas, vicomte de Castries (1756–1825).

ferring to each of our frigates four men from his crew with one officer; in accordance with this arrangement Mr Guyet de la Villeneuve, *enseigne de vaisseau*, transferred to the *Boussole* and Mr Gobien, *garde de la marine*, to the *Astrolabe*. This was very necessary because we had the misfortune three days ago to lose Mr D'Aigremont, *lieutenant de vaisseau*, on the *Astrolabe*, who died of dysentery, and Mr de St Céran's health has so deteriorated that I am forced to send him to the Isle de France to recover, all the surgeons agreeing that it was impossible for him to continue on the campaign, and so we have eight officers fewer than when we sailed from Europe, seven of them having died and the other leaving us with little hope for him. However we have lost from natural causes only one officer and a servant in two years. Both had joined the *Astrolabe* whose crew however has enjoyed even better health than the *Boussole*'s.

   Respectfully, I have the honour to be,
     My Lord,
        Your most humble and obedient servant,
           Lapérouse.
Manila, 7 April 1787.

From La Pérouse to Fleurieu
Copy. Published in the Milet-Mureau edition.[1]

              (Mr de la Pérouse, from Manila, 8 april 1787)
   I shall give you no details about my campaign, my dear friend: you have my letters to the Minister before you, and I flatter myself that you will have read my narrative with some interest. You will have noticed that we are certainly the first navigators who, in the same year, have reached Mt St Elias after visiting Easter Island and the Sandwich Islands and endeavoured to clear up certain geographical matters. Our maps, our charts, our journals, our tables &c, everything will prove to you that we have neglected nothing to ensure the accuracy of our various undertakings.

   What remains to be done this year is even more difficult, and all the information we have been able to obtain in China on this part of that empire's coast merely convinces us that the currents are extremely violent in the straits, that there are numerous shoals and that the fog almost never lifts.

---

[1] AN.M. 3JJ 387, 22, No. 9; Milet-Mureau, *Voyage*, IV, pp. 213–14.

But as I know that everything can be overcome with determination and patience, these obstacles merely increase my eagerness and I have the greatest faith in my star.
Lapérouse.

La Pérouse to the Minister
Copy. Modified version published in the Milet-Mureau edition.[1]

(Mr de la Pérouse, Avatska, 10 September 1787)

My Lord,

I dare to hope you will be pleased to place before the King the account of our navigation from Manila to Kamtchatka. His Majesty's frigates have followed an entirely new route, passed between Korea and Japan, followed the coast of Tartary up to the neighbourhood of the Segalien River, surveyed the Oku Yesso and the Yesso of the Japanese, and found a new strait leading out of the Sea of Tartary, which I felt you would be good enough to allow me to name after you; we verified and linked up our discoveries with those of the Dutch which most geographers were beginning to discount and which the Russians have found more convenient to eliminate from their maps, and we finally emerged north of Company Land from where we sailed to Kamtchatka. Our frigates anchored in Avatska Bay on 7 September after a voyage lasting one hundred and fifty days, one hundred and forty of them under sail, and there has not been one case of sickness in the two frigates although we were constantly sailing in the thickest of fogs, forced to drop and weigh anchor time and again, involving hardships which may seldom have been paralleled during the voyages of Captain Cook. The care we have taken to preserve our crews' health has been so far crowned with even more success than enjoyed by that famous navigator, because in the 26 months since our departure from Europe no one has died on board the *Boussole* and we do not have a single sick in either vessel.

I recall, My Lord, that when you were kind enough to present me to the King, His Majesty commented that this part of our navigation would be difficult and interesting since it could not be less important for geography to know the boundaries of the continent on which we live than those of the southern lands or of North America. We have been fortunate enough to present geographers

---

[1] AN.M. 3JJ 386, 111:2, Nos 70-1; Milet-Mureau, *Voyage*, IV, pp. 189-92.

with two islands as large as the British Isles and to settle finally what is possibly the last geographical problem still to be solved on the globe. Only from now do I dare to compare our voyage with Captain Cook's, and if death had not put an end to that great man's work it is probable that he would not have left the exploration of eastern Tartary to those who come after him. If your duties permit you, My Lord, to cast an eye on the various chapters of my narrative, you will find with navigational details all the observations I have had the opportunity of making on the people we visited, and the soil and products of their country. Generally, I have neglected nothing that might be of interest to the government relating to trade, without losing sight of the need to keep occupied the scientists, who may be waiting for our return to publish new analyses, but I do beg you, My Lord, to accept my apologies for the errors of style, grammar and spelling in my narrative; it is moreover horribly copied because I have no secretary on board, but the Chevalier de Fleurieu is in a position to make it worthy of being submitted to you and to be presented to the King. I add to my narrative all the maps, charts and tables of latitude and longitude necessary, as well as Messrs Duché and Blondelas's drawings, the accuracy of which I can guarantee.

I have the honour of sending you, My Lord, two memoirs, one on Manila and the other on Formosa relating to the political part of my instructions. They are of a very summary nature because I know how valuable is your time and they contain only what I felt I could not put in my narrative which the King may order to be published.[1] I would not have dared to entrust them to the mail service, but I believe, My Lord, that you will agree with the course I am adopting of sending to France Mr Le Cepts [Lesseps], our Russian interpreter. I felt that Mr Le Ceps's pay and his keep until the time of our return to France would cost approximately the same as his voyage from Kamtchatka to Paris, and I had qualms about dragging into the southern hemisphere a young man destined to follow a consular career who would waste a time that is precious for his training. I have therefore given him my despatches and I

---

[1] 'Mémoire sur Manille', AN.M. 3 JJ, 386, 111:69, and 387, 111:10; 'Mémoire sur Formose', AN.M. 3 JJ, 386, 111:72, and 387, 111:11. It would, as La Pérouse points out, have been unwise to send these from Manila, where they might have been opened, read and then resealed by the Spanish authorities.

hope that, by the time he has the honour to being with you, His Majesty's frigates will be in New Zealand.

I have the honour, My Lord, of sending you a private letter on the subsequent plan of my campaign which will last almost four years during which we shall have spent at least 38 months under sail, which may be unparalleled among navigators.

With deepest respect, I have the honour to be,
My Lord,
Your most humble and obedient servant,
Lapérouse.

La Pérouse to Fleurieu
Copy. Published in the Milet-Mureau edition.[1]

(From Avatscha Bay, 10 September 1787)
My dear friend, I am going to give you my news in no particular order, but I shall try to leave out nothing I have to tell you.

The Minister should have received from Mr Dufresne details of our campaign since our departure from France up to our arrival at Macao; and I am giving Mr de Lesseps the rest of our narrative from Macao to Kamtschatka....

I hope you will be pleased with the section of our voyage from Manila to Kamtschatka: it was the newest, the most interesting and certainly the most difficult on account of the constant fog around these places, in the latitudes we followed. The fogs are such that I had to devote one hundred and fifty days to explore the part of this coast which Captain King, in the third volume of Cook's third Voyage, assumed could be surveyed in the space of two months. However I stayed only three days in Ternai Bay, two in De Langle Bay and five in De Castries Bay, so I did not waste any time. Even so I neglected to sail round Chicha Island by going through Sangaar Strait. I would even have liked to anchor off the northern point of Japan, and might have chanced sending a boat ashore, although such a step would have required careful consideration because it is likely my boat would have been arrested, and such an event, which might be viewed as almost without importance if it had been a merchant vessel, could be regarded as an insult to the national flag when the boat belongs to a royal ship. The recourse of capturing

---

[1] AN.M. 3JJ 386, 111:2, No. 72 (in draft form); Milet-Mureau, *Voyage*, IV, pp. 214–19.

and burning some sampans is a feeble compensation with a nation that would count a hundred Japanese as being of no importance compared with a single European they wanted to make an example of. Be that as it may, I was not in a position to send a boat to the coast of Japan, and I cannot judge at present what I would have done if I had been.

I would find it difficult to describe the hardships of this part of the campaign, when I did not undress once and there were not four nights when I was not forced to spend several hours on the bridge. Imagine six days of fog, with only two or three hours when it cleared, in very narrow seas, completely unknown, and where the imagination, led by the information one had, visualised dangers and currents that did not exist. From the time of our landfall on the coast of eastern Tartary up to the strait we discovered between the islands of Tchoka and Chicha we did not pass a single feature without setting it down and you can rest assured that no creek, harbour or river has escaped us. Remain assured too that there are many maps of European coasts that are less accurate than those we will give upon our return, because the map enclosed in this despatch is so to speak no more than a sketch, admittedly very carefully done, but where there may be a few errors, of ten or twelve minutes of longitudes, in the position given for some features.

We have then finally cleared up the famous question of the lands of Jesso, Oku-Jesso, the strait of Tessoy, &c, which so worried geographers.

I have neglected nothing that can give a true impression of the people who live on these islands and on the mainland.

The Russians had found it more convenient to erase these two large islands from their maps, although they are ten times larger than their Kurils which are only barren rocks with a population not exceeding three thousand. Fog did prevent us from surveying the Kurils north of Marikan as far as Lopatka Point but I plan to carry out this work when I leave Avatscha Bay, although I do not consider it important: the English having determined the position of Paramousir and Marikan, the islands situated between these two points can hardly have been shown on the map with any significant degree of inaccuracy.

You will notice that our work in this part ties up marvellously with that of the Dutch whose navigation is possibly the most precise made at the time of the *Kastricum*'s voyage. You will find among the maps I am sending the Minister the one you gave me of Captain Uriès's discoveries: he did not suspect the presence of a sea behind the land he was coasting along, and even less a strait north of the village of Acqueis in front of which he had anchored. It may be deduced from his account that the people of Chicha and those of Tchoka are absolutely the same, since having left Acqueis and reached Aniva he did not suspect that he was no longer on the same island.

Another advantage deriving from the Dutch campaign is that it gives us the width of Tchoka Island up to Cape Patience and beyond, because the Dutch longitudes, taken close to the latitude of Cape Nabo, are roughly correct.

On your map, which I am sending the Minister, I have shown the strait we discovered, in the middle of the Dutch's mountains, and traced our route within sight of Staten Island, from Uries Strait and Company Land.

You will surely note when you read my account with the map before you that I could have followed the coast of Korea up to the 42nd degree, which would have been much easier and possibly more brilliant than what I did; but I felt that it was more important to determine accurately a feature of Japan that would give us the width of the Sea of Tartary and even of the island from Cape Nabo. I am sure you will agree with the course I adopted; however, you will regret that circumstances did not allow me to follow more of the coast of Japan, and so do I; but do not forget, my dear friend, when you look at my voyage, the continual fog which did not permit me to complete in a month the work one could do in three days under the fine sky of the Tropics; do not forget, finally, that without the lucky storm which gave us forty-eight hours of northerlies in the Gulf of Tartary, we would not have reached Kamtschatka this year.

Once more, although we did not accomplish everything, I am convinced that one could hardly do more, and that our campaign can stand comparison with those of the English, which was not the case when I returned from the coast of America because we had been forced to sail along it too quickly, and that furthermore several expeditions would not be enough to survey it in detail even from Cross Sound to the port of San Francisco. Imagine, at every league, inlets of an immeasurable depth, the back of which is out of sight, currents like those of Le Four and the Raz on our Breton coasts, and a fog that almost never lets up: you will conclude that a whole season would hardly be enough to examine every feature along twenty leagues of such a coast; and I could not guarantee to provide after six months of work a correct and detailed account of the land comprised between Cross Sound and Port Bucarelli, and much less as far as Cape Hector, which would require several years. So I was forced to limit myself to determining the latitude and longitude of the main capes, to find and trace the real trend of the coast from one point to the next, to determine the geographical position of the islands that lie several leagues away from the mainland. The immense plan of our voyage did not allow me to attempt any other work. Captain Cook may have done less on this coast – not of course that I want to disparage in any way the merit of that celebrated navigator – but, faced with contrary winds, squeezed as I was by time restrictions that prevented him from developing his discoveries, he sailed much further from the coast than conditions allowed me to do; and when he approached it near Cook's River and Williams Sound, it was in the ill-founded hope, which he never gave up, of getting away to the north and to make for his favourite target, a passage to the gulf of Baffin or Davis. His exploration of Williams Sound still leaves much to be desired; but, I repeat, these explorations require a great deal more time than he or I could give to our investigations.

I obtained while in Manila the journal of the voyage carried out by the Spanish pilot, the famous D. Francisco Antonio Maurelle, on the Northwest Coast of America.[1] And so, by adding this journal to that of the first Spanish campaign made in those waters, which Mr Barrington published in his *Miscellanies*,[2] an extract of which in translation had been included among the material you were good enough to collect for my information, we shall have all Maurelle's secrets. I left that navigator in Manila, commanding one of the new company's vessels intended for trading between Cavite and Canton. I am sending you a very detailed plan of Port Bucarelli

---

[1] A French translation, 'Relation du voyage..de la Princesse..178 et 1781', is held among papers relating to the La Pérouse expedition (AN.M. 3JJ, 111:4, no. 5) as well as the text in Spanish (ibid. No. 6).

[2] Daines Barrington, lawyer, antiquary, one-time patron of John Reinhold Forster, an influential advocate of a search for the North-West Passage, published a translation of Maurelle's 'Journal of a Voyage in 1775' in his *Miscellanies*, London, 1781. The account obtained by La Pérouse was reprinted in French in the Milet-Mureau edition of 1797.

and the surrounding islands, which I obtained in Manila. In their second campaign the Spanish penetrated as far as Williams Sound, and believing themselves to be on the coast of Kamtschatka they were afraid of an imminent Russian attack. I am not sending you their general map, because to be frank it would do more harm than good to geography. Did they want to mislead us, or more likely were they mistaken? Be that as it may, they saw land only near Port Bucarelli and at the entrance to Prince William harbour.

You will find among the maps nine drawings by Mr Duché: they are most lifelike. Mr Blondela adds to this packet a view of the harbour of St Peter and St Paul which is not drawn from the same angle as the one in Captain Cook's third Voyage, and a set of drawings of the various seagoing craft used by the various peoples we visited. This collection is very interesting and deserves the honour of being engraved.

I shall sail from Avatscha on 1 October. We have been received there with the greatest display of friendship, but the ship from Okhotsk has presumably been lost in the crossing, and the governor of Kamtschatka, in spite of the utmost good will, could not let us have a single case of flour. This shortage will force me to put into Guaham to try to get some.

Here is the subsequent plan of my campaign, subject always to circumstances and unforeseen happenings.

You know that I have already transposed part of the original plan outlined in my instructions, because I had been authorised to do so. I thought that it was quicker to begin by the northern hemisphere and end with the southern one, as I was to complete my voyage with a call at the Isle de France which is south of the Line. I must also admit to you that I was somewhat afraid of being preceded by the English who, prior to my departure, had announced a plan for a new voyage of discovery; I was concerned over the coast of Tartary, &c which was the only really new region I had to survey, and on no account did I want to be forstalled there.

Upon leaving Avatscha I will sail for the Kurils and determine the position of these islands up to Boussole Strait. I will make for the 37th parallel to look for land reported to have been discovered in that latitude by the Spanish in 1610. I will go to the islands north of the Marians and the archipelago itself as far as Guaham where I shall call to obtain some provisions. I shall spend only five days at Guaham and from there will make for the Carolines if I can hope to

APPENDIX I

reach from these islands Cape Choiseul on Surville's Land of the Arsacides and pass through Bougainville Strait; I will then sail south where I should find westerly winds &c.

If, on the contrary, the information I obtain at Guaham and what I find during the crossing persuade me that, by exploring the Carolines I would drop too far to leeward to reach New Zealand by 1 February 1788, I will give up the Carolines, which are of little importance, and sail to New Zealand from Guaham by following as easterly a course as I can. I will survey everything on the way and this route, which is totally new, should bring me to unknown islands that possibly will be of greater value than the Carolines. Either plan will permit me to arrive at Queen Charlotte Strait around 1 February. From there, I shall devote six months to sailing through the Friendly Islands, to obtain refreshments, along the south-west coast of New Caledonia, Mendana's Santa Crux Island, the southern coast of the Arsacides and of the Louisiades up to New Guinea; and I shall seek in this region another strait than that of the Endeavour. I will spend the months of August and September and part of October exploring the Gulf of Carpentaria and the west coast of New Holland, but arranging my operations so as to veer back north to the tropic and reach the Isle de France at the end of November.

I will leave the Isle de France around 25 December 1788. I will sail towards Cape Circumcision whence I will make for France without calling anywhere or with a call at the Cape of Good Hope, depending on circumstances; and I hope to arrive at Brest in June 1789, forty-six or forty-seven months after my departure from that port.

That is my new plan into which you can see I cannot include the south coast of New Holland or Van Diemen's Land from which I could only get to the Isle de France, on account of the westerlies, if I sailed all the way round. Such a route, which would be much longer, does not seem practicable: the condition of my rigging, the condition of my vessels even, prevent my attempting it.

I have not mentioned the Society Islands because they are so well known that they can offer nothing of any further interest; possibly it is to the credit of the expedition's leader, and certainly to the benefit of the crews, if one can go round the world without calling at O-Taïty. You know anyhow that the Society Islands, the Friendly Islands, Mendana's and others, which are already well known were only mentioned in my instructions to the extent that I might need supplies and was given the option of calling at islands where I could obtain refreshments, but could do so or not. I shall not forget however that you advised me, as something that would be of importance to the improvement of geographical knowledge, to determine the correct position of some of the features seen by Carteret, so that one could have an accurate basis on which to correct errors of reckoning along that navigator's entire route, as he had no marine chronometers and moreover seems to have carried out only a few astronomical observations.

The same Antonio Maurelle I have already mentioned, Spain's Cook, although in my opinion very inferior to the English Cook,[1] made a third voyage from Manila to North America at the beginning of 1781, during which he attempted to reach the high southern latitudes to work up later towards the east by means of the westerlies found near New Zealand; but he could not carry out this plan, being short of food, and had to make his way north towards the Marianas whence he took the regular route of the galleons to sail to San Blas. I am sending you the journal of this third voyage which I have been able to obtain, in which Maurelle claims he made numerous discoveries because he is not aware of any made by modern navigators. At first I wanted to keep this journal to check whether Maurelle had in fact found some new island in the neighbourhood of the Friendly Islands, an area where the natives report that there are quite a number they know about but Europeans do not; but after examining it I realised that it would only mislead me if I tried to use it: it is an almost shapeless chaos, a badly written account, in which longitudes are drawn from most uncertain reckonings and fairly badly observed latitudes.[2]

I obtained an excellent map of Manila and a few other interesting maps. Be assured that I did not get these without a great deal of trouble or without having to make some sacrifices, because as you know the Spanish are far from communicative: yet they have more to receive than to give. Other maritime nations have quickly made

---

[1] La Pérouse shows again here how he used James Cook as a yardstick for evaluating the work of other navigators, including himself.

[2] Francisco Antonio Maurelle (sometimes found as Mourelle) (1754–1820) had sailed with Bodega y Quadra from San Blas to the northwest coast of America in 1775; in 1780 he sailed to Manila in the *Princesa* and endeavoured to return to America by the central and southern Pacific; he made a number of discoveries, but was forced eventually to go to Guam and follow the more traditional Spanish route back to San Blas where he arrived in September 1781. See *Who's Who*, pp 176–7.

APPENDIX I

Europe aware of the things they so mysteriously wanted to conceal from us. I had the opprtunity in Manila of confirming my opinion of their anxious and pointless caution. The island governor has a map that covers the area from Manila to Kamtschatka. A single glance showed me that it is nothing more than Bellin's French map, drawn on the largest scale – and you know the work of our hydrographer and the errors of this map which is possibly less accurate than all his others: the governor allowed me to look at it only for a minute, and even then not closely, so worried was he that my powers of retention might be good enough for me to draw it from memory. I must admit that I found his fear so childish that, overlooking for a moment his serious mien, I could not prevent myself from telling him that I would shortly be in a position to know much more than he did and than all his maps could ever teach me.

If you take a moment to summarise the length of my stays in each port, from 1 August 1785 when I left Brest to 7 September 1787 when I arrived in Kamtschatka, you will see that in this time I spent only five months and thirteen days on our various calls and that some twenty-one months have been spent at sea, and you will be pleased to learn that, in spite of the hardships and privations that are unavoidable on such a long navigation, not a single man died on my frigate, and not one is sick. The *Astrolabe* lost one officer; but the illness that carried him off, the result of his imprudence, is due in no way to the hardships and dangers of the campaign. The crew's health on that frigate is furthermore as excellent as that of my own men. You can rest assured that the care Captain Cook took of his crews was no greater, no more thorough than the attention Mr de Langle and I give to the preservation of the valuable men who share our work; and if, at the end of our voyage, we find ourselves in as good a position as we have been until now, we will prove, as did Cook, that with care and an enlightened diet, one can preserve sailors from scurvy and the other illnesses that seem inseparable from long voyages. But one should not conclude from these experiences that the same would apply to ships of the line, to crews of eight hundred, a thousand, twelve hundred men who are often recruited among convalescents leaving hospital, and who cannot be fed as one feeds a crew of a hundred men selected for a special expedition, with top quality flour from Moissac,[1] wine

---

[1] A small market town in south-west France.

from Cahors and Teneriffe at six hundred *livres* a barrel, and all the antiscorbutics that pharmacists and physicians have been able to bring together &c. Notice too that the lack of space on large vessels, due to the number of men, does not allow to give each one a sizeable hammock, and that officers are not numerous enough for their supervision to extend to matters that may seem of minor significance, such as requiring the sailors to change their linen at regular intervals and in their presence, to protect these good people from the laziness which is natural to man when it is a question of personal cleanliness, a laziness he overcomes when it is a matter of overcoming hardships and facing danger. To these numerous and constant attentions, I have taken good care, regardless of cost, to put into ports where I can be certain of obtaining food supplies of an excellent quality, such as Concepción in Chile, Monterey in California, Macao, Manila &c. I considered that one of the tests being carried out during this campaign was to see if men, well fed, well cared for, can bear the hardships of a very long navigation, under all kinds of climates, in every latitude, in fog, or a burning sky, &c; and so far I can answer in the affirmative: but my campaign is still far from over. May the constancy of our cares and our zeal always meet with the same success!

Lapérouse

La Pérouse to the Minister
Copy. Reprinted in part in the Milet-Mureau edition; but see the following extract from another version of the same letter.[1]

(Letter from Mr de la Pérouse, in Kamtchatka, Awatska Bay)
I propose to leave from Awatska Bay on 1 October. I will sail for an exploration of the northern Kurils as far as Boussole Strait, from where I will make for the 37th degree to look for the land claimed to have been discovered by the Spanish in 1610; from the 37th parallel I will sail for the archipelago to the north of the Marianas and will follow that chain of islands as far as Guam, where I will put in for only five days to load fruit and a few oxen to preserve our crews from scurvy for the remainder of our lengthy navigation; from Guam I shall make for the Carolines if the information I get there allows me to be sure of reaching Cape Choiseul on the Land of the Arsacides and sail through the same strait as Mr de Bougainville

---

[1] Based on AN.M. 3JJ 386, 111:2, No. 73; Milet-Mureau, *Voyage*, pp. 192–5.

APPENDIX I

to proceed south and, with the help of the westerlies, arrive in Queen Charlotte Channel in New Zealand around 20 June 1788.[1] If on the other hand my own observations and my various investigations show that such a route is not possible, I will give up the Carolines which would force me to sail 150 leagues to leeward of the Marianas and make straight for New Zealand by sailing as far east as possible. It is likely that I would find along that route, which is quite new, various islands of even greater interest than the Carolines and about which certainly less is known. From Queen Charlotte Channel I will go up to the Friendly Islands and carry out absolutely everything required of me in my instructions relative to the south of New Caledonia, Mindanue's Santa Cruz Island, the southern coast of the Land of the Arsacides, Bougainville's Louisiades, checking whether it is attached to New Guinea or not, and pass at the end of July by a different strait from that of the Endeavour, if however such a one exists. During the months of August, September and part of October, I will visit the Gulf of Carpentaria and the coast of New Holland, but in such a way that I can sail north towards the tropics and arrive at the Isle de France at the beginning of December 1788. I will leave from there very promptly to investigate the so-called Cape Circumcision of Bouvet and arrive in France, with or without a call at the Cape of Good Hope depending on circumstances, in June 1789, 46 months after my departure from France having spent 37 or 38 of them under sail, a longer navigation than any of Captain Cook's, and if we are fortunate enough to carry out this plan I will make bold enough to say that our campaign will be next in importance after that great man's.

I flatter myself that the King will approve and will even be glad that during this lengthy voyage I will not have needed to call at these eternal Society Islands about which a great deal has already been written in various European kingdoms, and I must admit that I am glad not to have to mention either Taïty or Queen Oberea:[2] I have taken particular care to keep away from the routes of the navigators who preceded me and I think I can guarantee that, when I return, there will not be one place of importance to explore on the globe.

[1] This a copyist's error for January. The following letter, with the introductory section omitted by Milet-Mureau is correct.
[2] He would not have had much new to report about Oberea (or Purea): she had died some years earlier.

The political part of my instructions required me to discover whether the Spanish had a settlement in the Carolines. I think I can say that, on the contrary, they no longer know where to place them on their maps; their inertia is such that they do not even know the shoals surrounding the island of Luçon and have only a few presidios in the southern Philippines, rather like those of Oray and Ceuta: they feebly made up for their losses in the south by the conquest of the Babujames Islands to the north and in sight of the island of Luçon: 30 men commanded by Mr Tobias, formerly commander in the Marianas, added this feather to the Spanish cap with little difficulty and added, without any compensatory benefit, eight or ten thousand piastres to the cost of sovereignty.

These islands are in a manner of speaking only large rocks inhabited by a very small number of natives who live on roots and bananas: and therefore this conquest will not change the balance of power in Europe.

I have the honour &c.

La Pérouse to the Minister
Copy. The latter part of this letter repeats almost word for word the programme outlined in the previous one, except that the date of the expected arrival in New Zealand is given as 20 January 1788.

(Mr de la Pérouse. Awatska, in Kamtschatka,
21 September 1787)

My Lord,

I have had the honour of sending you by Messrs Dufresne and Le Ceps [Lesseps] the narrative of my campaign from my departure from Brest to my arrival in Kamtchatka. I still have to advise you of the later plan for my navigation, since I made use of the right you had given me of making to the programme outlined for me changes that would seem appropriate while staying within my instructions as often as I could. I felt it necessary to begin by the northern hemisphere and end by the southern one where is situated the Isle de France where I consider my voyage ends.

I dare flatter myself, My Lord, that so far I have fully and completely met all the hopes you placed in me, and I was so excellently assisted by the Vicomte de Langle that, if you consider the campaign to have any value, he must share the credit, and the vessels, in spite of fog, have sailed so close to each other, and the

APPENDIX I

degree of harmony was so great that one could almost say that the expedition had only one ship and one captain.

I propose to sail from the Bay of Awatska on 1 October to explore the northern Kurils as far as Boussole Strait, whence I shall make for the 37th parallel to seek the land claimed to have been discovered by the Spanish in 1610. I do not believe in the existence of this land which is very close to the ordinary track of the galleons, and all the information I have obtained leads me to think that the Spanish know nothing about it....

La Pérouse to the Minister
Holograph. Unpublished.[1]

(From Mr de la Pérouse, from St Paul and St Peter, 23 September 1787)

My Lord,

I have the honour of enclosing the narrative of my voyage from Macao to Kamtchatka, with the tables of latitude and longitude drawn up on the two frigates by the Vicomte de Langle and me.

Mr du Fresne whom I sent from Macao must have had the honour of handing you the account of the beginning of this same voyage from my departure from France to China; and so you have before you, My Lord, all the details of this half of the campaign, which I consider as the most difficult and the most interesting as we are the first navigators to sail east of Japan and to explore the coast of Eastern Tartary and of Oku Yesso.

The drawings by Messrs Duché and Blondelas leave nothing to be desired in respect of accuracy and the maps by this same Mr Blondelas and Mr Berniset have been prepared with the utmost precision. I would like to have carried out my function of historian with the same degree of success, but the art of writing is a craft like any other, and I did not learn it.

Moreover, the account I have the honour of sending you is full of spelling mistakes and even more of errors of punctuation. I beg you, My Lord, to read it only after a good copyist has made it worthy of its being submitted to you.[2]

[1] AN.M. 3JJ 386, 111:2, No. 75.
[2] La Pérouse is constantly worried about his inadequate spelling and syntax; the subtext of this concern is his dissatisfaction with the lack of secretarial help on the expedition.

With deepest respect, I have the honour to be,
My Lord,
Your most hmble and obedient servant,
Lapérouse.
In Kamtchatka, harbour of St Peter and St Paul, 23 September 1787.

La Pérouse to the Minister

Holograph. Published with alterations in Milet-Mureau's edition.[1]
(From Mr de la Pérouse, St Paul and St Peter, 25 September 1787)
My Lord,
The Vicomte de Langle has shown me the reports he has had the honour of giving you on his officers, and I join him in requesting the favours he seeks on their behalf, pointing out, My Lord, that the misfortune we suffered on the Northwest Coast of America have nullified most of those you had been kind enough to allocate to the staff of the two frigates. Messrs Descures and de Pierreverd each had a pension which could be given to Messrs de Vaujuas and Boutin, officers of equal merit and equally distinguished by their talents, their zeal and their good will. Messrs de Bellegarde and Gobien, *gardes de la marine* whom we have associated with our work and who expressed a great keenness in Macao and Manila to replace the officers we had the misfortune of losing, will have well deserved upon their arrival at the Isle de France the warrants of *enseigne* that had been given to Messrs de Boutervilier, de Flassan and de Montarnal. Messrs de Blondelas and Colinet, *lieutenants de frégate* to whom you have allowed me to give the hope of obtaining a warrant of *capitaine de brûlot* on their return, have already deserved this favour by their good behaviour, which I beg you, My Lord, to send me for them at the Isle de France, together with Mr de Monti's warrant and a letter of commendation to Mr de Clonard who, having been raised to the rank of *capitaine de vaisseau*, has nothing further to expect, but he has continued to serve as lieutenant and to see to a thousand minor details with a zeal and an attention that are worthy of the greatest praise, and if I was not afraid of showing some partiality because he is my close friend, I would make bold enough to assure you, My Lord, that the King could not have in his service a better officer or a man of greater honour and virtue.

---

[1] AN.M. 3JJ 386, 111:2, No. 76; Milet-Mureau, *Voyage*, IV, pp. 195–7.

APPENDIX I

I have also a great deal of praise for Mr Guiyet de la Villeneuve who transferred at Manila from Mr de la Croix de Castries's frigate to mine to take the place of Mr de St Céran whose extremely poor state of health forced me to send back to the Isle de France, and for Messrs Mouton and Broudou to whom I gave the warrants of *lieutenant de frégate* which you were good enough to give me in blank before my departure.

Mr de Langle has handed over the work of astronomy to Mr de Lauriston, a young man who is full of ability, zeal and merit; he has become a pupil who has no further need of a master. Mr Darbaud has also been a valuable assistant to Mr Dagelet, and I am sure that there is probably no youth of his age in France of equal attainments.

I should give you, My Lord, a particular report on the scientists, but I can only judge their behaviour on board, and in general they belong to a class of people who are so full of self-esteem and vanity that they are very difficult to lead in long campaigns of this kind. I have succeeded nevertheless in getting them to tolerate each other, and that is no small task; I make a complete exception for Mr Dagelet who is doing the same work as we do and probably better: among a hundred good and pleasing qualities, I can only fault him for having a very delicate health.

As for the Vicomte de Langle, he is above all praise, and I wish for the good of the service and the state that he may attain to higher rank before the years and exhaustion lessen his abilities.

Mr Rollin, a medical doctor and my senior surgeon, is a man who is as noteworthy for his knowledge as chief surgeon Andresson[1] who died during Cook's voyage. He preserved us from scurvy and all other illnesses. You authorised me, My Lord, to promise him a pension on his return if the mortality rate did not exceed three per cent in my frigate, and in the 26 months since we left not one person has died of natural death on the *Boussole* and we do not have a single case of sickness.

Mr de Langle is also very pleased with Mr Lavau, his chief surgeon. He has lost only one servant, whose lungs were affected, and Mr Daigremont who poisoned himself by attempting to cure

---

[1] William Anderson, surgeon's mate in the *Resolution* on Cook's second voyage, and surgeon in the same vessel on the third expedition, died on 3 august 1776. He was not only professionally competent, but knowledgeable in the field of natural history. He kept a journal, part of which, however, is missing.

himself from dysentery with flamed brandy. The *Astrolabe*'s assistant commissary has also died, as the result of a broken skull caused by fragments from the explosion of a musket he was holding.[1]
Respectfully, I have the honour to be,
My Lord,
Your most humble and obedient servant,
Lapérouse.

De Langle to the Minister
Holograph. Extracts published in the Milet-Mureau edition.[2]

(De Langle, 25 September 1787.)
My Lord,
The Comte de la Pérouse has allowed me the honour of sending you advices concerning the *Astrolabe*. The fog which has fairly constantly surrounded us since our departure from Manila has had a serious effect on her rigging. I hope that with the spares I still have on board I will succeed in taking her at least as far as the Isle de France at the date set in our plan. The frigate for the rest is in good condition. I have always sailed within hailing distance of the *Boussole* in spite of the fog because the Comte de la Pérouse has considered it his duty to keep me with him and my officers feel it is a matter of self-esteem not to separate. I would like to be able to add to the praises which I have already had the honour of sending you about their ability, the patience with which they await the end of the campaign and their desire of making new discoveries. Mr de Monti is an model in this respect, which affects all the persons on board the *Astrolabe*. I owe to his wisdom and attentions the good order which reigns on board, and the preservation of the food and of all the King's property; his poor sight has forced him to give up taking readings, but he has the skill required to supervise them well and I consider him an excellent naval officer.

Mr de Vaujuas is the most accomplished officer I have ever met. I believe him fit to lead an expedition.

Mr de Blondelas is a very good naval officer. He has used most ably the material we gave him to prepare the maps drawn on board the *Astrolabe* and he personally surveyed the bays we have visited.

[1] Jean-Marie Kermel, died on 7 September 1787. The servant mentioned in the same paragraph is Jean Le Fol.
[2] AN.M. 3JJ 386, 111:2, No. 77; Milet-Mureau, *Voyage*, IV, pp. 198–9.

APPENDIX I

He has a passion for drawing and spends all his leisure time on it; he displays judgment and spirit and is made to inspire a great deal of confidence.

Mr de Lauriston continues to progress in astronomy; although devoted to this science he does not neglect the working of a vessel which he knows excellently well, and is so painstaking that one can refer to him about every matter relating to it; he is moreover a person of considerable character.

Mr de Bellegarde has talent and much wisdom. He is progressing in astronomy and is made to inspire confidence.

Mr de Gobien is keen, and displays spirit and firmness of mind. He is keen on his trade, progresses perfectly every day and will undoubtedly become a distinguished officer.

Mr de Lamartinière, botanist, fulfils his functions with much zeal and has gathered a very large collection of plants which he is carefully preserving.

Father Receveur concerns himself eagerly with everything that relates to natural history; he is very assiduous in meteorological observations and is also having some success with astronomy.

Mr Prévot, the uncle, has done nothing since our departure from Manila and has not even once expressed a desire to go ashore.

Mr de Leseps, vice-consul at St Petersburg, rendered us essential services during our stay in Awatska Bay. He is a young man who is full of charm; I look with pleasure upon his success and I would have been very happy to keep him on board.

Messrs Lavo and Guilloux, surgeons, take infinite pains for the preservation of the crew's good health. Their skill could not save the commissary's assistant in whose hands a musket exploded, fragments of which pierced his skull. Mr Lavo showed ability in drawing up several large vocabularies of the languages of the people we visited, and sees, together with his colleague, to a collection of plants and items relating to natural history.

I have petty officers who are men of considerable intelligence and very concerned with preserving the King's property; their exemplary behaviour and the lack of eagerness they display for a return to France contribute greatly to the excellent discipline and the happiness that reign on board my ship.

My wishes would be gratified to the full, My Lord, if I received at the Isle de France the warrant of *capitaine de vaisseau* which you were kind enough to promise Mr de Monti on his return from the campaign.

The grant to Mr de Vaujuas, who was born without any wealth, of the pension allowed to the late Mr des Cures.

The warrant of *capitaine de brûlot* for Mr Blondelas.

For Mr de Lauriston the certainty of receiving the Cross of St Louis some years before the date fixed by the regulations, and for Messrs de Bellegarde and de Gobien an appointment as *enseigne*. The former is aged 22, the other 21.

I also have to beg you, Mr Lord, to send me a warrant of *lieutenant de frégate* for Mr Brossar, acting as second pilot on my ship, and of permanent appointment for the masters, carpenter, pilot, caulker and sailmaker. I do not mention the boatswain, because you were kind enough to grant him an appointment and the medal.

I am able to reward the master-at-arms who is also acting as master gunner, while waiting for his promotion to officer, which I believe he has well deserved, and which he will be in a position to request at the conclusion of the campaign.

My role in adding to the renown of my country, and to the Comte de la Pérouse's success leads me to tell you, My Lord, how glad we are of the success of the perilous and difficult navigation we have brought to a satisfactory conclusion thanks to the untiring vigilance of our leader, to his prudence and his talents; I shall always make it my duty to assist him as a worthy servant of the King and out of gratitude for all the signs of friendship he has shown towards me at all times. I am also aware, My Lord, that you take an interest in the success of the campaign. Nothing can make me forget the kindnesses you have honoured me with, and I hold it dear to my heart to deserve their continuation.

With respect, I remain,
My Lord,
Your most humble and obedient servant,
de Langle.

Awatska Bay, 25 September 1787.

APPENDIX I

Monneron to the Minister
Holograph. Unpublished[1]
(Mr Monneron, on board the *Boussole* in Awatska Bay,
27 September 1787)

My Lord,
   The only observations of any importance which the circumstances in which I have found myself since our departure from China have enabled me to make concern the Philippines archipelago in general, and the island of Luçon and Manila in particular; but the nature of these observations and the reflexions that accompany them, as I feel confident you will agree if one day you read them, does not make it appropriate for them to be sent by a foreign route.
   Respectfully, I am,
   My Lord,
      Your most humble and obedient servant,
         Monneron.

La Pérouse to the Minister
Holograph. Published in the Milet-Mureau edition.[2]
(Mr de la Pérouse, Awatska, 27 September 1787)

My Lord,
   Mr Le Ceps [Lesseps] whom I have asked to have the honour of presenting you my despatches is a young man whose conduct has been exemplary throughout the campaign, and sending him to France represents a real sacrifice considering my friendship with him, but as he is in all likelihood destined to take up one day his father's position in Russia, I felt that a land journey across that vast empire would enable him to acquire some knowledge of value to our trade and likely to increase our links with that kingdom whose products are so necessary for our navy.
   I gained the impression that Mr Le Ceps speaks Russian as fluently as French; the services he rendered us in Kamtchatka are invaluable and if the reversion of his father's post was the reward for his voyage around the world by land and by sea, I would

---

[1] AN.M. 3JJ 386, 111:2, No. 78.
[2] AN.M. 3JJ 386, 111:2, No. 79; Milet-Mureau, *Voyage*, IV, p. 199.

consider that favour, My Lord, as evidence that you approve of our actions.

With respect, I have the honour to be,
My Lord,
Your most humble and obedient servant,
Lapérouse.
Kamtchatka, 27 September 1787.

La Pérouse to Fleurieu
Copy. Extracts published in the Milet-Mureau edition.[1]

(Mr de la Pérouse, from Avatscha, 25 September 1787)
I am sending you, my dear friend, a memoir by Mr Rollin, senior surgeon on the *Boussole*; you will read it and you will surely decide that it should form part of the collection of memoirs and other works our scientists are each preparing. This Mr Rollin is a man of great merit who has not lost a single man in twenty-six months, does not have one case of sickness and is constantly seeing to our food supplies, their preservation, their improvement and generally to preventive medicine, which I much prefer to curative medicine....

We found here Mr de Lisle de la Croyère's tomb; I placed a correct inscription on it. I do not know whether they are aware in France that this scientist had married in Russia and left a posterity which enjoys the consideration due to its father's memory. His grandson is an adviser with the Siberian mines and for this receives a fairly substantial salary.
Lapérouse.

La Pérouse to Fleurieu
Copy. Extracts published in the Milet-Mureau edition.[2]

(Mr de la Pérouse, from Avatscha, 28 September 1787.)
I am again writing, my dear friend, to advise you of the receipt of the despatches arrived by way of Okhotsk the day before we were due to sail. I am treated with a kindness and a distinction which neither my services, nor my good will can ever repay....

The orders I have received change nothing to the later plan I had made for my campaign; only I will put in at Botany Bay, on the east

---

[1] AN.M. 3JJ 389, ref. 22; Milet-Mureau, *Voyage*, IV, pp. 229–30.
[2] AN.M. 3JJ 389, ref. 22; Milet-Mureau, *Voyage*, IV, pp. 231–2.

APPENDIX I

coast of New Holland. I would have missed these useful objectives if I had begun by the southern hemisphere; but the greatest advantage I find in the decision I made is the certainty that I have not been forestalled on the coast of Tartary &c by any English vessel. I know that all those sent out from India sailed east of Japan: the largest was lost on Copper Island near Bering Island; only two men escaped, to whom I have spoken and who are being sent overland to St Petersburg.

The vessel being built at Okhotsk which Russia intends to use for discoveries in these seas has hardly been started, and it is possible that it may not be able to sail for three or four years....

Farewell: I am leaving tomorrow in very good health, as is all my crew. We would make another six voyages around the world, if such an undertaking was useful or merely pleasing to our fatherland.

Lapérouse.

La Pérouse to the Minister
Holograph. Unpublished.[1]
(Mr de la Pérouse, Awatska, 28 september 1787)
My Lord,

The despatches you sent me in Camtchatka arrived on 26 September, and I was due to leave two days later to continue my navigation towards the south; all my letters were sealed and handed over to Mr Lesseps. I have the honour of forwarding them to you just as they were, as your subsequent orders do not require any other change in my plan of navigation than setting course for New Holland instead of going to New Zealand where I wrongly expected to find the English settlement.

I undertake, My Lord, to give you a good account of this new colony, which will in all probability languish for quite some time on account of its remoteness.[2]

I will then follow the plan I have had the honour of sending you, I will carry out other searches if I have time, and will treat like reefs the routes of previous navigators.

[1] AN.M. 3JJ 386, 111:2, No. 80.
[2] The loss of this report which La Pérouse no doubt worked on as soon as he left Botany Bay or may even have begun while still in the English colony is one of the more unfortunate consequences of the shipwreck; a French overview of the brand-new penal settlement in Port Jackson would have made an important contribution to our knowledge of the beginnings of the colony.

The remarkable favour you have deigned, My Lord, to bestow on me is a hundred times above my talents and my work. I would not have dared to hope for it, but I feel that I must justify it, and that motive is a thousand times more powerful for me than the hope of future rewards: it is a debt of honour to be repaid.[1]
With deepest respect, I have the honour to be,
My Lord,
Your most humble and obedient servant,
Lapérouse.

De Langle to the Minister
Holograph. Unpublished.[2]
(Mr de Langle, St Paul and St Peter, 29 September 1787)
My Lord,
The flattering letters you made me the honour of sending me at Awatska Bay when announcing the pension of 1000 *l.* which the King has graciously granted me are evidence of His Majesty's satisfaction that can only be due to your kindness, My Lord, and to the favourable account Mr de la Pérouse will have given you. I most sincerely share his pleasure at his promotion; all my officers who have the same active interest in his success as I have will redouble their efforts to ensure the two vessels can finish the campaign without becoming separated. Allow me to beg you again to take note of the recommendations contained in the letter I had the honour of sending you before receiving yours, My Lord, and to thank you for the warrants of appointment which you have allowed me to announce to the petty officers of the *Astrolabe*.
Respectfully, I remain,
My Lord,
Your most humble and obedient servant,
de Langle.
Awatska Bay, 29 September 1787.

---

[1] La Pérouse was promoted *chef d'escadre* (commodore) on 2 November 1786 and learnt of this while in Petropavlovsk.
[2] AN.M. 3JJ 386, 111:2, No. 81.

APPENDIX I

La Pérouse to the Minister
Holograph. Unpublished.[1]

(Mr de la Pérouse, Awatska, 29 September 1787)
My Lord,
 I have already had the honour of informing you from Macao that the English had sent six vessels from India to the Northwest Coast of America; two were back in China before my departure, they had encountered setbacks which ruined their enterprise, they brought back no more than three or four hundred skins for which the Chinese offered only four thousand piastres; Mr Strange, these two ships' supercargo, told me that rather than sell them at such a low price he would send them to Europe. I had no knowledge of the fate of the other vessels, all I knew was that all had sailed out of sight and east of Japan and did not bother with discoveries. I learned when I reached the harbour of St Peter and St Paul that one of these four ships had put into Awatska Bay, being under the orders of Captain Peters and that he had proposed on behalf of his owners a commercial arrangement with the governor who showed me the letter the English owner had written to him on this matter, in which he suggests to the Kamchatka administration sending a ship every year with a cargo of items from China and India to be paid for in furs and roubles. The governor did not dare reply to such a proposal which he sent on to the Cabinet in St Petersburg.[2]

 This proposal, My Lord, seems to me to be evidence more of the large means the English dispose of than of their judgment; because the entire Kamchatka peninsula does not consume a hundredth of such a cargo, and furs are much dearer than at Canton since the normal price of an otter skin is 30 piastres. It appears that the trade at Kiakia has not yet felt the effects of the immense quantity of otters entering China by the south. This empire is so vast that trends are communicated but slowly. The Russian Court would moreover display little awareness of its own interests if it agreed with such a proposal, but what will nullify it even more is the loss of that same vessel on the Copper Islands; only two men were saved, with whom I spoke and who are returning to Europe overland; I had them clothed and equipped from tip to toe. I do not

[1] AN.M. 3JJ 386, 111:2, No. 88.
[2] For the Peters expedition, see La Pérouse's journal, Ch. XVIII. Its failure and other unsuccessful ventures turned English traders away from the idea of commercial enterprises in the north-west Pacific.

know the fate of the other three vessels, and some concern was being expressed about them when I left Macao.

The Russian voyage of discovery in the northern seas will not leave for three or four years; the vessels are hardly begun in Okotsk and you will have time, My Lord, to organise a second circumnavigation before they have left harbour.[1]

Mr Coslof, Governor of Okotsk which includes the whole Camtchatka peninsula, received us with such courtesies that his attitude could not be bettered. He totally refused to be paid for the cattle he provided for us, and the few gifts we were able to make him accept are poor compensation, all the more because we could not decline his.

With respect, I have the honour to be,
My Lord,
Your most humble and obedient servant,
Lapérouse.

Kamtchatka, 29 September 1787, two hours prior to my departure.

La Pérouse to the Minister
Holograph. Unpublished.[2]

(Undated letter)

My Lord,

I have the honour of enclosing the two memoirs on Formosa and Manila which I informed you about in my previous letter. As I have referred to these two islands in fair detail in my narrative you

---

[1] For an overview of Russian plans for the exploration of the Pacific, including the influence the voyages of Cook and La Pérouse had on Russian thinking, see Glynn Barratt's four-volume *Russia and the South Pacific 1696–1840*, Vancouver, 1981–92, and especially his introductory section I, 1–44. These studies followed on his *Russia in Pacific waters 1715–1825: A Survey of the Origins of Russia's Naval Presence in the North and South Pacific*, Vancouver, 1981, and *Russian Shadows on the British Northwest Coast of North America 1810–1890*, Vancouver, 1983. The particular expedition being contemplated at Okhostsk during La Pérouse's call in Kamchatka was being organised by Joseph Billings and Gavriil Sarychev (see Dunmore, *Who's Who*, pp. 26, 220); an account was written by the latter and first published at St Petersburg in 1802 as *Travels of Naval Captain Sarychev in the Northeastern Part of Siberia, the Arctic Ocean and the Eastern Ocean*.

[2] AN.M. 3JJ 386, 111:2, No. 83.

will find in these memoirs only what I felt I should not include in a work which the King may order to be published.

With deepest respect, I have the honour to be,
My Lord,
Your most humble and obedient servant,
Lapérouse.

La Pérouse to the Minister
Copy. Published in the Milet-Mureau edition.[1]

(Mr de la Pérouse, from Botany Bay, 5 February 1788)

My Lord,

I feel confident that when this letter reaches you, you will have received the journal of my navigation from Manila to Kamtchatka which I had the honour of sending you with Mr Leseps who left the harbour of St Peter and St Paul for Paris on 1 October 1787. That part of the campaign, undoubtedly the most difficult through seas that were quite new for navigators, was nevertheless the only one where we did not meet with any misfortune, and the most frightful disaster was awaiting us in the southern hemisphere: I could only repeat here, My Lord, what you will read in greater detail in my Journal. Messrs de Langle and de Lamanon with ten others fell victim to their humaneness and if they had allowed themselves to fire on the islanders before they surrounded them, our longboats would not have been destroyed and the King would not have lost one of His navy's best officers.

Although this event considerably reduced the two frigates' complements I did not feel that I ought to alter my subsequent programme of navigation, but I was forced to explore more quickly various interesting islands in the South Sea in order to have time to build two longboats in Botany Bay and to complete the various sections of my instructions before the monsoon change that would make such an exploration impossible.

We reached New Holland without a single case of sickness in either vessel; eighteen of the twenty wounded we had when we left Mahouna had completely recovered, and Mr Lavau, senior surgeon of the *Astrolabe*, who had been trepanned, and one other sailor from that frigate leave us with no anxiety about their state of health.

Mr de Monty who was first officer with the Vicomte de Langle

[1] AN.M. 3JJ 386, 111:2, No. 84; Milet-Mureau, *Voyage,*, IV, pp. 200–202.

retained the command of the *Astrolabe* until they arrived at Botany Bay; he is such a good officer I did not feel I should make any changes in the staff until our first port of call when I could not overlook Mr de Clonard's rightful claim, who has been replaced on my frigate by Mr de Monty whose zeal and ability are above all praise and whose good conduct guarantees him the warrant of *capitaine de vaisseau* which, My Lord, you had the goodness to promise him if he received favourable reports.

The English forestalled us in Botany Bay by a mere 5 days. To the most marked courtesy they joined every offer of assistance that was in their power, and we had to regret to see them leave as soon as we arrived, for Port Jackson, fifteen miles north of Botany Bay. Commodore Philip rightly preferred that port and left us masters and alone in this bay where our longboats are already on the stocks and I expect them to be launched at the end of this month.

We are only ten miles by land from the English and consequently in a position to communicate frequently with each other. As it is possible that Commodore Philip may organise expeditions into the South Sea I felt I should give him the latitude and longitude of Mahouna Island so that his vessels may beware of the perfidious welcome the natives of that island might give them if he came upon it in the course of his navigation.

With deepest respect, I have the honour to be,
My Lord,
Your most humble and obedient servant,
Lapérouse.
On board the *Boussole* in Botany Bay, 5 February 1788.

La Pérouse to Fleurieu
Holograph. Published in the Milet-Mureau edition.[1]

(Lapérouse, from Botany Bay, 7 February 1788)
It seems that I will only ever have misfortunes to tell you about, my dear Fleurieu, and my extreme prudence is continually negated by events that are impossible to foresee, but about which I have always had, in a way, a secret feeling of foreboding, and I must admit that I must blame myself for having, on that fateful day of 11 December, given in, almost in spite of myself, to the insistence, and I would even say the stubbornness, of the Vicomte de Langle

---

[1] AN.M. 3JJ 389, 22, 164–7; Milet-Mureau, *Voyage*, IV, pp. 233–4.

APPENDIX I

who claimed that fresh water was the best antiscorbutic, and that his crew would be totally affected by this sickness [scurvy] before we reached New Holland. We nevertheless got there without any sick, and I am quite convinced personally that good quality water, whether it is fresh or not, is equally healthy.

You will read in my journal, my dear Fleurieu, the detailed account of this unhappy event: my feelings are too deeply affected by it for it not to be a torture to have to repeat it. You will surely find it incredible that a man of great common sense, of perfect judgment, preferred to a known and large bay, where water was excellent, a place that was uncertain, where his longboats were grounded, surrounded by eighteen hundred to two thousand Indians who cut them to pieces after killing all those who did not have time to take refuge in the boats that had remained afloat at the foot of the reefs, while our vessels were quietly bartering for food, two leagues from the shore, where admittedly we were far from foreseeing the likelihood of such an accident.

Some thirty Indians were killed ashore by the men of our longboats during that fatal day, and had I not controlled the fury of our crews who wanted to sink all the canoes that were trading safely alongside us, I could have let five hundred others be killed: but I felt that such barbarous behaviour would not undo our misfortune, and one can only allow harm to be done when it is absolutely necessary.

All I found near the length of coast where Massacre Village is situated is a bad coral ground, with a swell driving us towards the land, where I am certain our cables would not have resisted for two hours, which would have placed our frigates in the greatest peril without even being able to get within gunshot of this infernal little bay, and I did not feel that the pleasure of burning five or six huts was a sufficient reason to lead our frigates into such imminent danger. I would however have tried if I had had any hope of recovering our longboats, but the savages had broken them up and dragged their carcasses onto the beach.

You will support me, my dear friend, in my decision that such a misfortune should not cause me to alter the later plan of the voyage, but it did prevent me from exploring thoroughly the Navigators archipelago which I believe to be larger, more populated and better provided with food than that of the Society, including O-Taïty, and ten times larger than all the Friendly Islands put together. We saw their Vavao archipelago, which the Spanish pilot

Morel had seen but without determining a longitude that approximated anything like reality, which would have added a new element of confusion were it not for our determinations, or rather those of Captain Cook who described the Hapaee group so well that it is impossible not to identify it as Morel's Galvès Islands.

You will find in my journal that I saw Pilstard Island, Norfolk Island and finally reached Botany Bay without a single sick in either vessel: minor symptoms of scurvy disappeared when we used the fresh foodstuffs I obtained at the Navigators Islands. I swear to you, my dear friend, that sea air is not the cause of this sickness, but rather the foul air in the tween-decks when it is not renewed, or even more the poor quality of the food. It is not possible that biscuit that is worm-eaten, and like a honeycomb, meat the substance of which is totally corroded, it is impossible, I repeat, for such foodstuffs in the long run to make up for daily losses in the body. It follows from this that the decomposition of the humours, the blood &c. and equally extracts of cochlearoa and all remedies contained in flasks are nonsense; and that fresh food alone, either animal or vegetable, can cure scurvy so radically that our crews, fed for a month on pigs bought in the Navigators Islands arrived at Botany Bay in a healthier condition than when they sailed from Brest, and yet they had only spent 24 hours ashore on the island of Mahouna. And I consider malt,[1] spruce-beer, wine, sauerkraut &c to be antiscorbutics only because these substances, liquid or solid, suffer very little deterioration, and finally are a suitable food for man; they are however insufficient to cure scurvy, but I believe they must slow its advance, and in every respect one cannot recommend them too highly, and I regard as medical subtleties the fixed air &c of English and French doctors: one could swallow bottlefuls of it that it would not do as much good to sailors as good slices of roast beef, turtles, fish, fruit or herbs.

My theory on scurvy therefore comes down to these aphorisms, which are not from Hippocrates: any type of food suitable for man and able to make up for daily losses; external air introduced as often as possible in the tween-decks and the hold; the damp caused by fog

---

[1] By this word, which he leaves in English, La Pérouse adds the explanation *'drèche'*- draff. Spruce beer and sauerkraut are similarly left in English, with no suggested French equivalent, even though the word *'choucroute'* for the latter would have been known in northern France at the time.

regularly countered by fumigations and even with braseros; finally, cleanliness and frequent inspections of the sailors' clothes.

I have no confidence in Captain Cook's observation on the corruption of water in barrels. I believe that what is in good condition when loaded on board, after passing through the two or three stages of decomposition known by every sailor, which cause it to smell for a few days, thereafter keeps in excellent condition, and as light as distilled water, because all the heterogeneous matter has precipitated and remains as a sediment at the bottom of the barrels, and at the moment I am writing to you, although we are quite close to a fairly good watering place, I am drinking water from Port des Français (coast of America) which is excellent. It is however this erroneous opinion, which I have never shared, that caused our misfortunes at the island of Mahouna: how can one go against a captain when he assures you that all his crew will suffer from scurvy within a fortnight if he does not have any fresh water, and who gives me the most unreliable account of a bay which he alone had visited the day before without calculating that the low tide would leave him grounded?

Mr Dagelet is writing to you about his observations, and so I shall not mention them. It is enough for me to say that combining our two methods, observations of distances and the chronometers, so completely solved the problem of longitude that we sailed with less error in longitude than we had in latitude ten years ago when observations were made with wooden octants, and four times less than when we used the arrow or the ninety degree quadrant.

Mr de Langle's death led to no changes in the *Astrolabe* as far as astronomical observations are concerned. For close on a year, young Lauriston has been in sole charge of them; he is an officer of the highest merit who could challenge even our astronomers when it comes to accuracy; I also know that his observation register is kept in most orderly fashion.

As the English have established their settlement at Port Jackson, they have left this entire bay to us. I have had a very good retrenchment set up here in order to store our [new] longboats in safety, which are well advanced and will be usable by the end of the month. These precautions were needed against the Indians of New Holland who, although very weak and not numerous, are, like all savages, very ill-natured and would set fire to our boats if they had means of doing so. They threw spears at us one minute after receiving our presents and signs of friendship. I am a hundred times more angry against the philosophers who so praise them as against the savages themselves. Lamanon, whom they murdered, was telling me the day before he died, that these men were worth more than us. A rigid follower of the King's instructions, I have always behaved towards them with the utmost moderation; but I would not undertake another campaign of this kind without asking for different orders, and a navigator leaving Europe must consider them as enemies, very weak ones, to be honest, whom it would be dishonourable to eliminate, but whom one has the duty of forestalling if a feeling of suspicion allows it in all fairness.

I advised you, my dear friend, in the letters I wrote from Kamtschatka of my subsequent plan of campaign which I needed to follow if I was to arrive in Europe in June 1789. Neither our stocks of food, nor our rigging, nor even our vessels, would allow me to lengthen my voyage which will be, I believe, the most considerable ever undertaken by a navigator, at least in respect of the length of the route. I still have some very interesting things to do, and illnatured people to visit: I cannot guarantee that I will not fire a few guns at them, because I am quite convinced that only fear can put a stop to their evil intentions.

I shall sail on 15 March from Botany Bay, and I shall not waste my time until December, by when I hope to arrive at the Isle de France.

I append to my letter the table of my route from Kamchatka, a copy of which I am sending a duplicate to the minister at the end of my journal. Do your utmost, my dear friend, to get hold of this journal, because I can foresee that the King and even the Maréchal de Castries will only have time to read extracts from it selected by you; I also beg you to correct the errors of style &c &c and to make it worthy of the public should the Court order it to be printed.

You will also find at the end of the journal the plan of six of the Navigators Islands. The islanders gave us the names of ten of them, and I believe that, in order to complete this archipelago one needs to add to it the island of the Beautiful Nation of Quiros and the two of Cocos and Traitors, but I am not completely certain of this: the last two are very small and of little importance, but it would not surprise me greatly if the islands of Mahouna, Oyolava and Pola had a total population of four hundred thousand people. Mahouna is very much smaller than the other two and yet we bought there in

APPENDIX I

the space of 24 hours five hundred pigs and an enormous quantity of fruits.

I would have liked to include with the plan of the Navigators Islands one of the Friendly archipelago, plus the islands of Vavao, Latte &c, but much to my regret it is not finished and cannot be before we sail. Failing the plan, you will find the longitudes and latitudes of these islands in the tables; they are shown there more accurately than in my journal: although it is an historical narrative it has been written as things happened and with longitudes which had not yet undergone a final check, after which quite often they were corrected.

Mr de Clonard is now in command of the *Astrolabe*; Mr de Monti has taken his place in the *Boussole*: they are officers of the highest quality. We have lost one of even greater merit in Mr de Langle; he was gifted with great quality, and the only fault I knew in him was stubbornness, and he was so fixed in his attitude that one had to fall out with him if one wanted to continue to disagree: he tore from me rather than obtained the permission that caused his loss. I would never have given in if the report he gave me on the bay where he died had been accurate; and I shall never understand how a man who was as prudent and as enlightened as he was could have been so grossly mistaken.

You see, my dear friend, how affected I still am by that event; I come back to it time and again in spite of myself...

Lapérouse.

La Pérouse to the Minister
Copy. Extract published in the Milet-Mureau edition.[1]
(Lapérouse, from Botany Bay, 7 February 1788)
...I shall go up to the Friendly Islands, and will do exactly what my instructions require me to do with respect to the southern part of New Caledonia, Mendana's Island of Santa Cruz, the south coast of Surville's Arsacides, and Bougainville's land of Louisiades, endeavouring to assess if the latter forms part of New Guinea or not. I shall pass, towards the end of July, between New Guinea and New Holland, by another channel than the Endeavour's, if such exists. In

---

[1] AN.M. BB4:992; Milet-Mureau, *Voyage*, IV, pp. 202–203.

September and part of October I shall visit the Gulf of Carpentaria and the entire west coast of New Holland as far as Van Diemen's Land; but in such a way as to enable me to go back north in good time to reach the Isle de France in December.

    Lapérouse.

APPENDIX II

The Muster Rolls

The Muster Rolls are held at the Archives Nationales under reference Marine C6 885. First established as at the date of departure on 1 August 1785, they were subsequently brought up to date by order of Claret de Fleurieu in his role as Minister of Marine, so that payments and pensions due to the dependants might be calculated. Fleurieu's instructions of 4 January 1791, issued prior to the despatch of the D'Entrecasteaux expedition and forwarded to the Comte d'Hector, Director of the port of Brest, and Redon, the civilian administrator, reflect official concern for the men's families during the long months of uncertainty and strain:

> Although, gentlemen, a strong belief can be held that the frigates *Astrolabe* and *Boussole* have been totally lost, we should not give up hope that the men are alive. However, as it is only fair to assist the families, I have sought the King's instructions, and His Majesty has decided that the campaign shall be considered as having ended on 31 October 1788, since Mr de la Pérouse sent advices from Botany Bay that he would reach the Isle de France during that month Consequently, I hereby authorise Mr Redon to pay what was owed at that date, both to the officers and crews of the two frigates, and to the scientists and artists who were on board. To simplify the work of the Bureau des Armements and prevent so far as is possible any errors it might make because of inadequate information I hereby enclose what my Office has been able to gather.

Another set of muster rolls was later drawn up as at 1 January 1789, which is held at the Archives Nationale under the reference Marine C6 956. This was prepared partly in case the D'Entrecasteaux expedition came upon the wreckage of the frigates and found some survivors, but above all because the new, republican government considered that the campaign ought not to be considered at an end until the ships sent out to look for it came back to France. This decision, overriding Fleurieu who had now been dismissed, was motivated by a desire to help the ordinary seamen's wives who were going through a period of hardship and is recorded in a letter dated 3 November 1793:

> To Citizen Sané, Head of the Navy's Civil Administration:
> The National Convention, citizen, acting upon a report from its Navy and Finance committees and in response to a petition presented by the wives of seamen belonging to the crews of the frigates *Astrolabe* and *Boussole* commanded by the C. la Peyrouse that the provisions of the law of 4 May be applied to them equally with the wife of that expedition's commander, having decreed that the assistance granted until 31 October 1788 to the families of the sailors who embarked on these ships shall continue to be paid from 1 January 1789 until the ships sent to look for these crews have returned, I request you to have this intention recorded and to send me a detailed statement of costs, with a request for the appropriate funds, so that I may issue instructions to ensure this decree is carried out. [signed] Dalbarade [Minister of Marine].

In the following list, the first rank shown is the one held at the time of departure. Promotions or rises are indicated in italics in cases where the officer or sailor concerned could not have been notified because they occurred just before or after the expedition left Botany Bay on the final leg of the voyage.

Ranks are not translated. The British naval structure was not strictly comparable to that of the French royal navy and modern titles do not always correspond to earlier ones. It may be more useful to note the rates of pay per annum, which were not shown in respect of the officers on the muster rolls, but indicate the range of grades. They were, in French pounds (*livres*), beginning with the more junior rank: *volontaires*, 180 to 360, depending on the length of service; *garde de la marine*, 360; *garde du pavillon*, 432; *enseigne de vaisseau*, 800; *lieutenant de frégate*, 840; *lieutenant de vaisseau*, 1,600; *capitaine de vaisseau*, 3,000 to 3,600 according to seniority; *chef d'escadre*, 6,000. The latter rank, which was formally granted to La Pérouse from 2 November 1786, can be equated to that of commodore, the commander of a squadron at sea, and just below that of rear-admiral. The rank of *sous-lieutenant* appears among promotions granted during the voyage; it did not exist at the time of the expedition's departure from Brest, but was created by a decree of

APPENDIX II

1786 and replaced the two ranks of *enseigne de vaisseau* and *lieutenant de frégate*.

Supplementary payments were made to certain officers and men who had special responsibilities or were given additional functions. Rises granted to the men during the voyage make it clear that special skills were recognised, and are an indication that a number of men took the opportunities offered during a lengthy voyage to study and improve themselves.

Most of the men came from Brittany, many of them from Brest and Morlaix or from nearby villages. With few exceptions, the others came from Normandy, especially Caen or Le Havre. We find the occasional recruit from Orleans or Dieppe, but these are rare cases among the sailors. This meant that many of them would have known each other before they joined the expedition, probably from their early boyhood; some, like the two Hamons, were obviously related, possibly brothers, and there must have been a number of cousins among them. Among those who were not primarily sailors, such as the carpenters, armourers, cooks and servants, one finds a wider range of birthplaces – Paris, Tours, the provinces of Béarn and Dauphiné – but even among these the Bretons and Normans predominate. The place of origin of the soldiers does not appear in the roll, but it is likely that the pattern there would have been different from that of the seamen. Although there were a number of Bretons and Normands among the officers, we get a much wider range: although Langle was a Breton, La Pérouse came from south-west France. As far as the scientists were concerned, they had few links with Brittany or Normandy.

The following lists incorporate the information provided by the three main rolls: the one established as at the date of sailing from Brest, the 'Fleurieu update' and the final update. Some minor errors or omissions have been rectified. The lists include additional information such as transfers, changes of duties, payments to be made to family members back home, and deaths.

The sailors' requests for special provision to be made for their families are of interest. Men leaving on long voyages were naturally concerned that their dependants should be looked after, and the practice became generalised from the end of the seventeenth century. It applied both to merchantmen and naval vessels. There was, however, no legislation covering this practice until the early nineteenth century. It was regarded as a favour, not an obligation imposed on shipowners or the naval authorities – this no doubt created problems at times, but it had the advantage that creditors could not lay a claim by way of lien on a man's advance or back pay. A number of the men who sailed with La Pérouse were supporting their parents, some a sister, others a wife. The request could be expressed in cash or in kind: a sailor could ask his wife or other relative to be paid a stipulated sum of money at some specified interval, or he could ask for food, 'a loaf', to be supplied weekly. Whether this shows concern about rising food prices or about a wife who was likely to drop into the nearest tavern on her way back from the naval pay office, can only be speculated on.

A general roll call and inspection took place on 12 July 1785 at Brest after the frigates were towed away from the quayside, as La Pérouse was hoping to sail on the 15th. At this point the men were paid an advance of six months' pay. Messages to the shore and the handing over of some of this money to their families with the connivence of the few ship's boats still maintaining contact with the authorities on land were still possible, but going ashore for a last joyful fling was not allowed. One man did sneak ashore and stayed away until the 16th, by which time he had spent all his advance pay; he was placed under arrest, deleted from the roll and sent back to Brest to serve a prison sentence for aggravated desertion.

Deaths are indicated by an asterisk next to the name of the person concerned; this naturally indicates a death occuring before 10 March 1788 when the expedition sailed from Botany Bay, since all those who left with the two frigates on that day were lost in the shipwrecks or not long after. A small dagger indicates that the person concerned did not sail from Botany Bay with the two frigates for some other reason, such as a transfer or, for instance in the case of De Lesseps, leaving the expedition on an earlier occasion.

LA BOUSSOLE

Jean-François de Galaup de LA PEROUSE, Capitaine de Vaisseau, acting Chef de Division from 18 June 1785, Chef d'Escadre from 2 November 1786, promotion notified by despatches received in Kamchatka 28 September 1787.

APPENDIX II

SENIOR OFFICERS

Robert Sutton de CLONARD, Lieutenant de Vaisseau, Administrative Officer, Capitaine de Vaisseau from 1 January 1787. Supplement payable as flag captain from 2 November 1786, 120 *livres*. Transferred to the *Astrolabe* 12 December 1787.
*Charles-Gabriel Morel D'ESCURES, Lieutenant de Vaisseau. Drowned at Port des Français, Northwest Coast of America at 7.30 a.m. on 13 July 1786.
Charles-Marie Fantin de BOUTIN, Enseigne de Vaisseau. Lieutenant de Vaisseau from 1 May 1786. *Major de vaisseau from 14 April 1788.*
*Ferdinand-Marc-Antoine Bernier de PIERREVERT, Enseigne de Vaisseau. Lieutenant de Vaisseau from 1 May 1786. Drowned at Port des Français, Nortwest Coast of America at 7.30 a.m. on 13 July 1786.
– COLINET (or COLLINET), Lieutenant de frégate. Sous-Lieutenant from 1 May 1786. His Christian names are not recorded.

Additions

Anne-Georges-Augustin de MONTY, Lieutenant de Vaisseau. *Capitaine de Vaisseau from 14 April 1788.* Transferred from the *Astrolabe* on 12 December 1787.
Pierre-Louis GUYET DE LA VILLENEUVE, Lieutenant de Vaisseau. Joined from the *Subtile* at Cavite, Philippines, on 8 April 1787.
Jérôme LAPRISE-MOUTON, joined from the *Northumberland*. Senior Pilot at 70 *livres* plus 5*l.* with supplement as clerk of 20*l.* Promoted from petty officer to Lieutenant de Frégate on 1 May 1786, a rank later equated to that of Sous-Lieutenant. Supplementary pay 20 *livres* as officer in charge of food supplies.

GARDES DE LA MARINE

*Pierre-Armand-Léopold Guirald de MONTARNAL. Drowned at 7.30 a.m. on 13 July 1786 at Port des Français, Northwest Coast of America.
†Henri-Marie-Anne-Jean-Baptiste MEL DE SAINT-CERAN. Promoted Lieutenant de Vaisseau from 1 July 1786. Discharged at Cavite, Philippines, and joined the *Vénus* on 16 April 1787.

VOLONTAIRES

Frédéric BROUDOU. Promoted to Lieutenant de Frégate July 1786, Sous-Lieutenant 1 August 1786.
Roux D'ARBAUD. Promoted Elève de la Marine, first class, on 1 January 1786. Lieutenant de Vaisseau from 14 April 1788.

SURGEONS

Claude-Nicolas ROLLIN, Senior Surgeon at 100 *livres* per month, plus 1 *sol* per man or 5*l*. 15*s*. for 115 men.
Jacques-Joseph LE COR (or LE CORRE), Assistant Surgeon at 40 *livres*.

CHAPLAIN

Jean-André MONGEZ, Regular Canon at Ste Geneviève. Chaplain at 600 *livres* with a supplementary pay of 600 *livres* for scientific work.

SCIENTISTS, ENGINEERS AND ARTISTS

★Jean-Honoré-Robert de Paul de LAMANON, Physicist, mineralogist and meteorologist, paid 3,000 *livres* in Paris for the first year of the campaign and due to receive 9,000 on his return. Killed at Tutuila, Samoan Islands, 11 December 1787.
Joseph LEPAUTE DAGELET, member of the Académie Royale des Sciences, Astronomer, 3,000 *livres* a year.
Gérault-Sébastien BERNIZET, Surveyor-Geographer, 1,200 *livres* a year.
Gaspard DUCHE DE VANCY, Portrait and Landscape Artist, 1,500 *livres* a year.
Jean-Louis-Robert PREVOST, known as 'The Younger', Botanical Draughtsman, 1,200 *livres* a year.
Paul-Mérault de MONNERON, Captain in the Royal Engineering Corps, embarked as Chief Engineer, 3,000 *livres* a year.
Jean-Nicolas COLLIGNON, Gardener and Botanist, 50 *livres* plus 5 as a half-ration, taken on as supernumerary.
Pierre GUERY, Armourer and Watchmaker, 60 *livres* plus 5 as half-ration, taken on as supernumerary.
Special duties:
Rollin: natural sciences and botany.
Mongez: physicist.

APPENDIX II

PETTY OFFICERS

Jacques DARRIS, from Brest, Chief Petty Officer, 70 *livres* plus 5 as half-ration. Requests that his wife be given half his pay.
Etienne LORMIER, from Brest, Assistant Chief Petty Officer, 60 *livres* plus 5 as half-ration. Increased to 70*l.* from 1 August 1786. Gives 30*l.* a month to his wife.
Vincent LE FUR, from Brest, Boatswain, 50 *livres* plus 5 as half-ration. Requests that his family be paid 25*l.* a month or to give 2 loaves a week to his wife. His wife received 34 loaves in 1787 to a value of 26*l.* 15*s.* 6*d.* and 57 loaves in 1788. *Increased to 70 livres from 1 August 1788.*
François TAYER, from Dinan, Boatswain's Mate, 36 *livres* plus 5. Gives half his pay to his wife or his sister. *Increased to 50 livres from 1 August 1788.*
François ROPARS, from Recouvrance, Boatswain's Mate, 33 *livres* plus 5. *Increased to 55 from 1 August 1788.*

PILOTS

*Jean-Baptiste LEMAITRE, from St Malo. Second Pilot at 35*l.* plus 5. Drowned at Port des Français, Northwest Coast of America, at 7.30 a.m. on 13 July 1786.
Eutrope FAURE, from Saintes. Assistant Pilot at 24 *livres* plus 5. Joined from the *Portefaix*. *Increased to 45 livres from 1 August 1788.*
Jean QUERENNEUR, from Le Conquet. Coastal Pilot at 70 *livres* plus 5*l.* Joined from the *Clairvoyant*.

Additions and Deletions

Adrien DUMANET, from Dinan. Second Pilot at 45 *livres* plus 5*l.* *Increased to 60l. from 1 August 1788.* Transferred from the *Astrolabe* on 15 July 1786.
Jérome LAPRISE-MOUTON, transferred to the officers' list on 1 May 1786.

GUNNERS AND MARINES

*Pierre TALIN. Senior Master Gunner. Naval Quartermaster on shore at 32 *livres* with supplement of 38*l.* plus 5*l.* From 1 May 1786 Quartermaster-Sergeant at 24*l.* with supplementary pay of 10*l.* plus 22*l.* Master-at-Arms from 1 May 1787 with supplement of 6*l.*

Increased to 70*l*. plus supplement of 15*s*. 5*s*. Killed at Tutuila, Samoa, on 11 December 1787.
Edmé-François-Mathieu LIVIERE. Sergeant at 24*l*. 10*s*. plus 5*l*. Assistant Chief Gunner with supplement of 5*l*. 10*s*. Master Gunner from 1 May 1786 at 20*l*. 12*s*. 6*d*. *Promoted to assistant quartermastersergeant at 22l. from 1 August 1788.*
Antoine FLHIRE. Corporal at 16*l*. 15*s*. plus supplement of 5*l*. Assistant Master Gunner from 1 May 1786 with 16*l*. 2*s*. 6*d*.
François DIEGE. Marine. 11*l*. 5*s*. plus 5*l*. Gunner First Class from 1 May 1786 at 12*l*. 10*s*., supplement 2*l*.
*Georges FLEURY. Marine at 11*l*. 5*s*. Gunner First Class from 1 May 1786 at 12*l*. 10*s*. Drowned at Port des Français, Northwest Coast of America on 13 July 1786.
*Jean BOLLAY. Marine at 11*l*. 5*s*. Gunner First Class from 1 May 1786 at 12*l*. 10*s*. Drowned at Port des Français, Northwest Coast of America on 13 July 1786.
*Pierre LIETOT. Marine at 11*l*. 5*s*. Assistant Gunner at 9*l*. 5*s*. plus 5*l*. Gunner first Class at 12*l*. 10*s*. from 1 May 1786. Drowned at Port des Français, Northwest Coast of America on 13 July 1786.
Etienne DUTERRE. Drummer at 13*l*. 10*s*., supplement 7*l*. 10*s*. Gunner Second Class at 8*l*. 12*s*. 6*d*.; First Class from 1 June 1787 at 12*l*. 10*s*.

CARPENTERS, CAULKERS AND SAILMAKERS

Pierre CHARRON, from Rochefort. Head Carpenter at 60*l*. plus 5*l*. Joined from the *Portefaix*.
Jean-Baptiste François SOUDE, from Le Havre. Carpenter's Mate at 21*l*. *Increased to 36l. from 1 August 1788.*
André CHAUVE, from Le Croisic. Carpenter's Mate (replacing Jean-Pierre Kerneis, Master Butcher), at 21*l*. plus 5*l*. *Increased to 36l. from 1 August 1788.*
Jean GARNIER, from Rochefort. Carpenter at 18*l*. Joined from the *Portefaix*. *Increased to 30l. from 1 August 1788.*
Pierre ACHARD, from Caen. Carpenter at 17*l*. *Increased to 24l. plus 5l. from 1 August 1788.*
Pierre MESCHIN, from Rochefort. Master Caulker, joined from the *Portefaix*. 60*l*. plus 5*l*. Gives 30*l*. a month to his wife.
Claude NEVIN, from Brest. Caulker's Mate at 21*l*. *Increased to 36l. from 1 August 1788.*

APPENDIX II

Jean FAUDIL, from Brest. Caulker's Mate at 21*l*. *Increased to 36l. from 1 August 1788.*
Alexandre MOREAU, from Nantes. Caulker's Mate at 21*l*. *Increased to 36l. from 1 August 1788.*
Jacques FRANCHETEAU, from the Ile de Ré. Head Sailmaker at 60*l*. plus 5*l*. Joined from the *Thétis*. Gives his wife 4 months of his pay per year.
Laurent POINTEL, from Lambezellec. Sailmaker's Mate at 21*l*. plus 5*l*. *Increased to 36l. from 1 August 1788.*
Bertrand DANIEL, from Lorient. Sailmaker at 18*l*. *Increased to 30l. from 1 August 1788.*

ABLE SEAMEN AND HELMSMEN

Guillaume DURAND, from Morlaix, at 20*l*. *Increased to 30l. from 1 August 1788.*
Jean MASSON, from St Brieuc, at 20*l*. Gives his mother 2 months of his pay per year. *Increased to 30l. from 1 August 1788.*
Jacques POHIC, from Quimper, at 20*l*. *Increased to 30l. from 1 August 1788.*
Julien HELLEC, from Vannes, at 20*l*. Joined from the *Experiment*. Gives half his pay to his mother. *Increased to 30l. from 1 August 1788.*
François GORIN, from St Brieuc at 20*l*. *Increased to 30l. from 1 August 1788.*
Pierre BRETAUD, from Le Croisic, at 20*l*. Gives his family half his pay. *Increased to 30l. from 1 August 1788.*
Jean FRICHOUX, from Recouvrance. Able Seaman at 20*l*. Paid 60*l*. to his sister on 11 May 1786. *Increased to 30l. on 1 August 1788.*
Guillaume STEPHAN, from Brest. Able Seaman at 20*l*. *Increased to 30l. from 1 August 1788.*
Pierre-Marie LASTENNEC, from Brest. 20*l*. *Increased to 30l. from 1 August 1788.*
François LOSTIS, from Brest. 20*l*. *Increased to 30l. from 1 August 1788.*
Jean-Marie DREAN, from Vannes. 20*l*. Gives half his pay to his mother. *Increased to 30l. from 1 August 1788.*
Alain MARZIN, from Quimper. 20*l*. Gives his father 3 months of his pay per year. *Increased to 30l. from 1 August 1788.*
Pierre BONNY, from Le Croisic. 20*l*. gives half his pay to his mother. *Increased to 30l. from 1 August 1788.*

Charles LEDUC, from Orleans. 20*l.* Joined from the *Seine*. Increased to 30*l. from 1 August 1788.*
Paul-Joseph BERTHELE, from Ushant. 20*l.* Gives half his pay to his father. *Increased to 30l. from 1 August 1788.*
Jean MAGUEUR, from Le Conquet. 20*l.* Gives his father or his mother 4 months of his pay per year. *Increased to 30l. from 1 August 1788.*
Jean-François DUQUESNE, from St Malo. 20*l.* Joined from the *Héros.* In hospital 1 June whence discharged on 8 July 1785. Gives his wife 4 months of his pay per year. *Increased to 30l. from 1 August 1788.*
André-Marie LE BRICE, from Brest. 20*l. Increased to 30l. from 1 August 1788.*
Jean-Pierre CHEVREUIL. 20*l.* Requests bread and food supplies for his wife. (She received 36 loaves in 1786, value 28*l.* 7*s.*, and 28 in 1787, value 22*l.* 1*s.*) *Increased to 30l. from 1 August 1788.*

SAILORS

Jean GOHONNEC, from Morlaix. 16*l. Increased to 24 + 5 from 1 August 1788.*
Yves LE BIHAN, from Quimper. 15*l.* Gives half his pay to his father or mother. *Increased to 21l. from 1 August 1788.*
Corentin LEYER, from Quimper. 15*l.* Gives 2 months of his pay per year to his sister. *Increased to 21l. from 21 August 1788.*
Jean LUCO, from Vannes. 18*l.* Gives his wife half his pay. *Increased to 30 from 1 August 1788.*
Louis PLEMER, from Vannes. 16*l.* Gives 4 months of his pay per year to his father. *Increased to 24 from 1 August 1788.*
François GLOAHEC, from Vannes. 16*l.* Gives half his pay to his father or mother. *Increased to 24 from 1 August 1788.*
Joseph LE BARS, from Morlaix, living in Brest. 18*l.* Joined from the *Etoile. Increased to 30 from 1 August 1788.*
Jean-François PLEUREN, from Brest. 18*l.* Gives his father half his pay. *Increased to 30 from 1 August 1788.*
Jean DARRON, from St Malo. 16*l. Increased to 24 + 5 from 1 August 1788.*
Jean DONETY, from St Malo. 18*l. Increased to 30 from 1 August 1788.*
Louis LE BOT, from Ploucastel. 18*l. Increased to 30 from 1 August 1788.*

APPENDIX II

Alain ABGRAL, from Recouvrance. 18*l*. *Increased to 30 from 1 August 1788*.
Charles-Antoine CHAUVRY, from Caen. 16*l*. *Increased to 24 + 5 from 1 August 1788*.
Guillaume PICHARD, from St Malo. 15*l*. Joined 11 July 1785, landed 12th, rejoined on 18th. *Increased to 21l. from 1 August 1788*.
Hilarion-Marie NORET, from Ushant. 18*l*. Originally on the *Astrolabe*, transferred to the *Boussole* on 27 July 1785. *Increased to 30 on 1 August 1788*.

Additions

Jean MONEUR, from Morlaix. 16*l*. Transferred from the *Astrolabe* on 15 August 1785. 18*l*. from 1 August 1786, 21*l*. from 1 August 1787.
Julien ROBERT, from Paris. 14*l*. Joined from the *Subtile* in Cavite on 9 April 1787.

GUNNERS, ASSISTANTS, SOLDIERS

César-Augustin DEROZIER, fusilier at 10*l*. 5*s*., supplement 7*l*. 15*s*. Supplement as surgeon's orderly 8*l*. 10. Promoted to 1st class from 1 May 1786, 12*l*. 10.
★Michel BERRIN, fusilier at 10*l*. 5, supplement 7*l*. 15. 2nd class from 1 May 1786. Drowned on 13 July 1786 at Port des Français, Northwest Coast of America.
Henry Salomon VEBER, fusilier at 10*l*. 5., supplement 7*l*. 5. Gunner 2nd class from 1 May 1786 at 8*l*. 12*s*. 6*d*.; gunner 1st class from 1 March 1787 at 12*l*. 10*s*.
★Pierre PRIEUR, fusilier at 10*l*. 5, supplement 7*l*. 5. Gunner 2nd class from 1 May 1786 at 8*l*. 12*s*. 6*d*. Drowned on 13 July 1787 at Port des Français, Northwest Coast of America.
★Marius [Marcus] CHAUB, fusilier at 10*l*. 5., supplement 7*l*. 5. Gunner 2nd class from 1 May 1786 at 8*l*. 12*s*. 6*d*. Drowned on 13 July 1786 at Port des Français, Northwest Coast of America.
François-Joseph VAUTRIN, fusilier at 10*l*. 5*s*., supplement 9*l*. 15. Gunner 1st class from 1 May 1786 at 12*l*. 10*s*.
★André ROTH, fusilier at 10*l*. 5, supplement 4*l*. 15*s*. Gunner 1st class from 1 May 1786 at 12*l*. 10*s*. Killed at Tutuila, Samoa, on 11 December 1787.
Jean BLONDEAU, fusilier at 10*l*. 5., supplement 6*l*. 15. Gunner 2nd class from 1 May 1786 at 8*l*. 12*s*. 6*d*.

Michel MITERHOFFER, fusilier at 10l. 5s., supplement 4l. 16s. Gunner 1st class at 12l. 10s.
*Jean-Pierre FRAICHOT, fusilier at 10l. 5., supplement 4l. 15. Supplement as Captain's secretary 12l. 10s. Gunner 3rd class from May 1786, 6l. 6s. 8d. Drowned on 13 July 1786 at Port des Français, Northwest Coast of America.
Pierre GUILLEMIN, fusilier at 10l. 5., supplement at 7l. 5. Supplement as scientists' secretary 12l. 10s. Gunner 1st class from 1 May 1786 12l. 10.
Jean GILLET, fusilier at 10l. 5s., supplement 5l. 15s. Gunner 1st class from 1 May 1786 at 12l. 10.

SUPERNUMERARIES

Jean LOUVIGNY, from Brest, Senior Steward, 21l. + 5l.
Simon ROLLAND, from Nantes. Cooper at 21l. + 5l. Previously a sailor.
Joseph BONNEAU, from Aunai, Poitou. Baker at 21l. + 5.
Jean-Pierre DURAND, from Paris, Master Armourer at 40l. + 5. *Increased to 42l. + 5 from 1 May 1786.*
Jean-Marie BLEAS, from Brest. Blacksmith at 40l. + 5. *Increased to 50l. from 1 May 1786.* Gives half his pay to his mother.
Jacques QUINIOU, from Brest. Cook at 21l. + 5.
René-Marie COSQUER. Master Ship's Carpenter. Is not a sailor. At 60l. + 5. Owes Mr Pilstat, constable with the provostry 47l. 12; owes Hervé Thomas 88l. 4s. against receipt. Pays his wife 40l. per month; his wife died around the year 1789. The debt to Thomas was settled on 27 December 1791.
Others:
Jean QUERENNEUR, coastal pilot, listed with pilots,
Jacques LE CORRE, surgeon, listed with the officers,
Jean COLLIGNON, gardener-botanist, and Pierre GUERY, armourer and clockmaker, listed with scientists, engineers and artists.

SERVANTS

For Mr de la Pérouse, captain:
Pierre CARAURANT, from Guiasse in Béarn,
François BISALION, from Royan in Dauphiné. Cook.
For Mr de Clonard, lieutenant de vaisseau:

François BRETEL, at 15*l*. Transferred to the *Astrolabe* 12 December 1787.

For Mr D'Escures, lieutenant de vaisseau:
†Michel SIRON, at 15*l*., discharged 26 January 1787 into the naval storeship *Marquis de Castries*.

For Mr. Boutin, enseigne de vaisseau:
René de SAINT-MAURICE (mulatto) at 15*l*.

For Mr de Pierrevert, enseigne:
Louis DAVID, from St Brieuc, at 15*l*.

For Mr Gillet de Villeneuve, lieutenant:
BENJAMIN (black) at 16*l*. Joined from the *Subtile*, transferred to the *Boussole* on 6 April 1787 at Cavite.

JOINED IN MANILA ON 7 APRIL 1787

SOLDIERS

Jean-Charles MASSEPIN
Dominique CHAMPION
Pierre LEBIS
Jean JUGON
Pierre MOTTE

SAILORS

Six Chinese.

Total at time of departure from Brest: 109; at time of departure from Botany Bay: 108.

Losses: Drowned at Port des Français on 13 July 1786, 11 including 3 officers; killed at Tutila, Samoan Islands, on 11 December 1787, 2.

L'ASTROLABE

*Paul-Antoine de LANGLE, Capitaine de Vaisseau, *promoted to Chef de Division effective 14 April 1788*. Killed by natives at the island of Manoua, Navigators archipelago [Tutuila, Samoa] on 11 December 1787. Supplement as commanding officer 120 *livres*.

SENIOR OFFICERS

Anne-Georges-Augustin de MONTY, Lieutenant de Vaisseau. Transferred to the *Boussole* 12 December 1787. *Promoted to Capitaine de Vaisseau effective 14 April 1788.*

Jean-François Tréton de VAUJUAS, Enseigne de Vaisseau, Lieutenant de Vaisseau from 1 May 1786, *promoted Major de Vaisseau from 14 April 1788.*

★Edouard-Jean-Joseph de LA BORDE MARCHAINVILLE, Enseigne de Vaisseau, Lieutenant de Vaisseau from April 1786. Drowned at 7 a.m. on 13 July 1786 at Baie des Français, Northwest Coast of America.

★Prosper-Philippe D'AIGREMONT PEPINVAST, Lieutenant de Vaisseau 1st class April 1786. Died of dysentery at Cavite on 25 March 1787.

– BLONDELA, Lieutenant de Frégate, Sous-Lieutenant de Vaisseau from 1 May 1786.

Additions

Robert Sutton de CLONARD, Capitaine de Vaisseau, transferred from the *Boussole* on 12 December 1788, commanding officer at time of departure from Botany Bay 10 March 1788.

GARDES DE LA MARINE

★Joseph-Ignace RAXI DE FLASSAN (PLASSAN), promoted to Lieutenant de Vaisseau on 1 May 1786. Drowned at 7 a.m. on 13 July 1786 at Baie des Français, Northwest Coast of America.

★Edouard-Jean-Joseph de LA BORDE MARCHAINVILLE, promoted to Lieutenant de Vaisseau on 1 May 1786. Drowned at 7 a.m. on 13 July 1786 at Baie des Français, Northwest Coast of America.

Jean-Guillaume LAW DE LAURISTON, promoted to Lieutenant de Vaisseau from 1 August 1786.

Additions

Pierre LE GOBIEN, Elève 1st class. Joined from the *Subtile* in Cavite on 9 April 1787. *Promoted to Lieutenant de Vaisseau 2nd class from 5 March 1788.*

Gabriel-Jean DUPAC DE BELLEGARDE, Elève 1st class. Joined from the royal storeship *Marquis de Castries* on 1 January 1787. *Promoted to Lieutenant de Vaisseau effective 4 March 1788.*

APPENDIX II

SURGEONS

Simon-Pierre LAVAUX (LAVAU). Navy surgeon. 1s. per man and per day for 115 men 5 *livres* 15 *sols*.
Jean GUILLOU. Assistant surgeon.

CHAPLAIN

*Claude-François-Joseph RECEVEUR, 50 *livres* as chaplain plus 50 *livres* per month for his work as naturalist. Died at Botany Bay 17 February 1788.

SCIENTISTS, ENGINEERS AND ARTISTS

†Louis MONGE, Astronomer at 2400 *livres* per year. Discharged at Tenerife on 29 August 1785 to return to Europe.
Joseph Boissieu de la MARTINIERE, Botanist, doctor, at 2400 *livres* per year.
†– DUFRESNE, Naturalist. Discharged at Macao on 1 February 1787 and joined the *Maréchal de Ségur*.
Guillaume PREVOST. Botanical artist at 1200 *livres* per year,
†Barthélémy de LESSEPS. Joined as Russian interpreter. Discharged in Kamchatka on 29 September 1787 to carry dispatches to France across Russia.

PETTY OFFICERS

François LAMARE, from Brest, Senior Boatswain, at 70 *livres*.
François-Marie AUDIGNON, from Morlaix, Boatswain, at 60 *livres*.
Sébastien ROLLAND, from Brest, Boatswain's Mate, at 38*l*. Increased to 50*l*. *from 1 August 1788*.
Guillaume-Marie GAUDEBERT, from Brest, at 36*l*. Increased to 50*l*. *from 1 August 1788*.
Bastien TANIOU, from Recouvrance. Leading Seaman at 27*l*. + 5. *Increased to 39l. from 1 August 1788*.

PILOTS

Mathurin LEON, from Brest. Senior Pilot at 70*l*. Supplement as secretary 20*l*.
Adrien DUMANET, from Dinan. Second Pilot at 40*l*. Increased to 45*l*. on 1 May 1786. Transferred to the *Boussole* on 14 July 1786.

Jean LAINE, from St Brieuc. Assistant Pilot at 24*l. Increased to* 45 *from 1 August 1788.*
Pierre BROSSARD, from Morlaix. Assistant Pilot at 32*l. Promoted Sous-Lieutenant de Vaisseau 2nd class from 14 April 1788.* Paid to Miss Brossard on 31 January 1792 146*l.* 6*s.* 8*d.*
François QUERRE, from St Brieuc. Coastal Pilot at 70*l.* + 5.

GUNNERS AND SOLDIERS

Jean GAULIN. Sergeant, 24*l.* 10*s.* Supplement as Master-at-Arms 6*l.* Assistant Quartermaster from 1 May 1786 22*l.*
*Léonard SOULAS. Corporal at 16*l.* 5. Supplement 10*l.* 5. Drowned on 13 July 1786 at Baie des Français, Northwest Coast of America.
Jacques MOREL. Fusilier 11*l.* 5. Supplement 9*l.* 5.
Pierre CHAUVIN. Fusilier 11*l.* 15. Gunner 1st class from 1 May 1786, supplement 12*l.* 10.
*Pierre PHILIBY. Fusilier 11*l.* 15. Gunner 1st class from 1 May 1786, 12*l.* 10. Drowned on 13 July 1786 at Baie des Français, Northwest Coast of America.
Christophe GILBERT. Corporal 14*l.* Supplement 7*l.* 2*s.* 6*d.* Second Master Gunner from 1 May 1786, supplement 16*l.* 2*s.* 6*d.*
Jean-Pierre HUGUET. Drummer 13*l.* 10. Supplement 7*l.* 10. Gunner 1st class from 1 May 1786, supplement 12*l.* 10.
François SAUTOT. Fusilier 11*l.* 5. Supplement 15*l.* 5.

CARPENTERS, CAULKERS AND SAILMAKERS

Robert-Marie LEGAL, from Recouvrance. Master Carpenter at 60*l.* + 5. To be paid to his wife: 40*l.*
Jean BERNY, from Le Croisic. Second Carpenter at 36*l.* 5. *Increased to* 55*l. from 1 August 1788.* Half his pay to his wife.
François BIZIEN, from Brest. Assistant Carpenter at 24*l. Increased to* 45*l. from 1 August 1788.*
Jean LE CAM, from Brest. Assistant Carpenter at 24*l.* Reduced to sailor's rank at 18*l.* on 17 July 1785. Restored to 24*l.* on 1 August 1786. *Increased to* 36 *from 1 August 1788.* 10*l.* per month or 3 loaves per week to be paid to his wife. Owes 9*l.* 9*s.* for 12 loaves supplied in 1787; 70*l.* 17*s.* 6*d.* for 90 loaves supplied during the last 6 months of 1787; 59*l.* 1*s.* 3*d.* for 75 loaves supplied in 1788. Owes 335*l.* 8*s.* 4*d.* for 408 loaves.

Pierre FOUACHE, from Caen. Carpenter at 18*l. Increased to 30l. from 1 August 1788.*

Jean-François PAUL, from Brest. Master Caulker at 60*l.* + 5. To be paid to his wife: 30*l.* per month.

Louis MEVEL, from Brest. Master Caulker at 45 + 5. *Increased to 60 from 1 August 1788.*

Yves-Marie QUELENEC, from Brest. Master Caulker. [45 + 5?]

François LE BOUCHER, from Brest. Assistant Caulker at 26 + 5. To be paid to his wife: 20*l.* per month. *Increased to 45 from 1 August 1788.*

Jean GROSSET, from Brest. Master Sailmaker at 60*l.* + 5. To be paid to his wife: 30*l.* per month.

Olivier CREACHCADEC, from Brest. Assistant Sailmaker at 24 + 5. To be paid to his wife 12*l.* per month. *Increased to 45 from 1 August 1788.*

Yves BOURHIS, from Lorient. Assistant Sailmaker at 21*l. Increased to 33l. from 1 August 1788.*

ABLE SEAMEN, HELMSMEN AND SAILORS

Louis ALLES, from Tréguier at 20*l. Increased to 30l. from 1 August 1788.*

Pierre-Marie RIO, from Vannes, at 20*l. Increased to 30l. from 1 August 1788.*

Jean MOAL, from Morlaix at 20*l. Increased to 30l. from 1 August 1788*

Joseph LE QUELLEC, from Morlaix, at 20*l. Increased to 30l. from 1 August 1788.*

*Guillaume DUQUESNE, from Brest, at 20*l.* Increased to 21*l.* from 1 May 1786. Drowned on 13 July 1786 at Baie des Français, Northwest Coast of America.

Charles-Jacques RIOU, from Morlaix, at 20*l. Increased to 30l. from 1 August 1788.*

François LE LOCAT, from St Brieuc, at 20*l. Increased to 30l. from 1 August 1788.*

Yves-Louis GARANDEL, from Brest, at 20*l. Increased to 30 from 1 August 1788.*

Bertrand LEISSEIGNE, from Quimper, at 20*l. Increased to 30 from 1 August 1788.* Gives 6*l.* per month to his mother.

Jean LEBRIS, from Recouvrance, at 20*l. Increased to 30l. from 1 August 1788.*

Denis LE CORR, from Recouvrance, at 20*l*. *Increased to 30l. from 1 August 1788.*
Jean LE GUYADER, from Tréguier, at 15*l*. *Increased to 21l. from 1 August 1788.*
Pierre BANNIOU, from Morlaix, at 16*l*. *Increased to 24l.* + 5 *from 1 August 1788.* To be paid to his mother: 6*l*. per month
Joseph RICHEBECQ, from Morlaix, at 18*l*. *Increased to 30l. from 1 August 1788.* To be paid to his mother: 6*l*. per month.
François-Marie VAUTIGNY, from Morlaix, at 18*l*. *Increased to 30l. from 1 August 1788.*
★Yves HAMON, from Morlaix, at 16*l*. *Increased to 24l.* + 5 *from 1 August 1788.* Killed by the natives in the island of Maouna, Navigators archipelago [Tutuila, Samoa] on 11 December 1787.
Jean HAMON, from Morlaix, at 16*l*. Increased to 24*l*. + 5 on 25 July 1785.
★Gilles HENRY, from St Brieuc, at 18*l*. *Increased to* 24 + 5 *from 1 August 1787.* Drowned on 15 October 1787.
★Goulven TARRAU, from Brest, at 16*l*. Drowned at Baie des Français, Northwest Coast of America on 13 July 1787.
†Jean-Marie BASSET, at 18*l* . Joined from the *Résolution*. Increased to 21*l*. on 1 August 1786. Discharged at Macao on 19 January 1787 and joined the *Marquis de Castries.*
Pierre-Marie-Fidèle PAUGAM, from Morlaix, at 18*l*. *Increased to 30 from 1 August 1788.* Give his pay to his wife or supply her with 2 loaves per week from 3 April 1790. Owing for 170 loaves: 146*l*. 4*s*. 6*d*.
★Jean-Louis BELLEC, from Lorient at 15*l*. Joined from the *Grondin*. *Increased to 21 from 1 August 1788.* Killed by the natives at Tutuila, Samoan Islands, on 11 December 1787.
Joseph LEBLOIS, from Quimper, at 12*l*. *Increased to 18 from 1 August 1788.*
Jean-Marie LETANAFF, from St Brieuc, at 16*l*. Joined from a merchantman. *Increased to 24 from 1 August 1788.*
Michel-Guillaume Lambert NIEULE, from Dieppe. Seaman at 14*l*. For his attitude 23*l*. *Increased to 21 from 1 August 1788.*
Jean MONEUR, from Morlaix, at 15*l*. Transferred to the *Boussole* on 15 August 1785.
Louis MEZON, from Brest, ropemaker, at 15*l*. *Increased to 21 from 1 August 1788.*
Guillaume QUEDEC, from Brest, at 15*l*. *Increased to 21 from 1 August 1788.* Pay his mother 12*l*. per month.

APPENDIX II

*Jean NEDELLEC, from Morlaix, at 18*l*. *Increased to* 21*l*. *from 1 August 1788*. Joined from the port. Killed by the natives of Tutuila, Samoan Islands, on 11 December 1787.

Guillaume AUTRET, from Morlaix, at 18*l*. *Increased to 30l. from 1 August 1788.*

Claude LORGI, from Crozon, at 18*l*. *Increased to 30l. from 1 August 1788.*

Jean GOURMELON, from Brest, at 18*l*. *Increased to 30 from 1 August 1788.*

Jean BERNARD, from Brest, at 18*l*. *Increased to 30 from 1 August 1788.*

†Jean CREE, from Brest, at 18*l*. Deserted at Concepción, Chili, on 14 March 1786.

*François FORET, from Brest, at 18*l*. Increased to 21*l*. on 18 August 1786. Killed by the natives at Tutuila, Samoan Islands, on 11 December 1787.

Mathurin CAUZIAU, from Brest, at 18*l*. *Increased to 30l. from 1 August 1788.*

Guillaume RICHARD, from St Brieuc, at 17*l*. *Increased to 24 + 5 from 1 August 1788*. Joined from the apprentice gunners.

*Laurent ROBIN, from Lanveaux, at 20*l*. Increased to 21 from 1 May 1786. Pay his wife 10*l*. per month. Killed by the natives of Tutuila Island in the Samoas on 11 December 1787.

Julien MASSE, from Morlaix, at 16*l*. *Increased to 24 + 5 from 1 August 1788.*

*Jean-Thomas Hainon ANDRIEUX, from Morlaix-Roscoff, at 20*l*. Increased to 21*l*. on 1 May 1786. Joined from the barque *St-Louis*. Drowned at Baie des Français, Northwest Coast of America, on 13 July 1786.

GUNNERS, ASSISTANTS, SOLDIERS

Pierre GUIMARD. Grenadier at 6*l*. 5 + 12*l*. 10. Gunner 1st class from 1 May 1786.

Joseph FRETCH. Fusilier at 7*l*. 15. Gunner 1st class from 1 May 1786 at 12*l*. 10.

*Louis DAVID, fusilier at 7*l*. 15. Gunner 1st class from 1 May 1786 at 12*l*. 10. Killed by the natives of Tutuila Island, Samoan archipelago, on 11 December 1787.

†Louis SPAN. Fusilier at 7*l*. 15. Supplement 5*l*. 15. Deserted at Concepción, Chili, on 14 March 1786.

Chrétien THOMAS. Fusilier at 7*l*. 15. Supplement 5*l*. 15. Gunner 1st class from 1 May 1786 at 12*l*. 10.
*Joseph RAYES. Fusilier at 7*l*. 15. Gunner 1st class from 1 May 1786 at 12*l*. 10. Transferred to the *Boussole* on 14 July 1786 as fusilier at 10*l*. 5*s*., gunner 1st class at 12*l*. 10*s*. Killed at Maouna (Tutuila, Samoan Islands) 11 December 1787.
Jean-Baptiste PLINIER. Fusilier at 9*l*. 15. Gunner 1st class from 1 May 1786 at 12*l*. 10.
Considérant LENDEBERT. Fusilier at 7*l*. 15. Gunner 1st class from 1 May 1786 at 12*l*. 10.
Jean-Gautier PLEUMEUR. Fusilier at 10*l*. Supplement 7*l*. 15. Gunner 1st class from 1 May 1786. at 12*l*. 10.
*Julien LE PENN. Fusilier at 10*l*. 5. Supplement 7*l*. 15. Gunner 1st class from 1 May 1786 at 12*l*. 10. Drowned at Baie des Français, Northwest Coast of America, on 13 July 1786.
François BIGNON. Fusilier at 10*l*. 5. Supplement 4*l*. 5. Gunner 1st class from 1 May 1786 at 12*l*. 10.
*Pierre BARBIER. Fusilier at 10*l*. 5. Supplement 4*l*. 15. Gunner 1st class at 12*l*. 10. Drowned at Baie des Français, Northwest Coast of America, on 13 July 1786.

SUPERNUMERARIES

*Jean-Marie KERMEL, from Crozon. Ship's Steward at 21*l*. Died on 7 September 1787 from a gunshot wound.
Pierre CANEVET, from Crozon. Cooper at 21*l*. 5.
René RICHARD, from Brest. Butcher at 21*l*. 5.
Nicolas BOUCHER, from Orleans. Baker at 21*l*. 5.
Jacques LERAND, from Brest. Master Armourer at 40*l*. + 5. Increased to 42*l*. from 1 May 1786. To be paid to his wife: 20*l*. per month.
François-Marie OMNES, from Brest. Blacksmith at 40*l*. Increased to 50*l*. from 1 May 1786. Paid to his mother, Marie Guillemette Kermenguy, from 1786 to 1791 four instalments of 135*l*. and one of 220*l*. on 21 June 1791.
F. MORDELLE, from Tréguier. Ship's Boy at 7*l*. 10. Increased to 18*l*. in August 1788.

APPENDIX II

SERVANTS

For Mr de Langle, captain:
†Yves RIOU, from St Brieuc. Discharged at Tenerife 30 August 1785.
Simon-Georges DEVEAU, from Etampes. Cook.
For Mr de Monty:
★Jean GERAUD, from Avranches, at 15*l*. Killed by the natives at Tutuila Island, Samoa, on 11 December 1787.
For Mr de Vaujuas:
★Jean LE FOL, from Lameur, at 15*l*. Died on board 11 August 1786.
For Mr de La Borde:
†Jean-Louis DROUX, from Brienne, at 15*l*. Discharged at Macao on 1 January 1787 and joined the *Marquis de Castries*.
For Mr Daigremont:
François POTORELLE, at 15*l*.
For Mr Blondela:
Joseph HEREAU, from Tours, at 15*l*.
Total at time of departure from Brest: 113; at time of departure from Botany Bay: 91.
Losses: drowned at Port des Français on 13 July 1786, 11 including 3 officers; killed at Tutuila (Samoan Islands) on 11 December 1787, 9 including the commanding officer and Fr Receveur who died from his wounds; on board ship: 1 from a longstanding illness, 1 officer from dysentery, 1 from an accident with a firearm, 1 fallen overboard; deserters: 2 at Concepción, Chile.

APPENDIX III

The Death of Father Receveur

It has been widely assumed that Father Receveur, the *Astrolabe*'s chaplain and a respected naturalist, died in Botany Bay of wounds received in Samoa less than two months earlier. Watkin Tench, in his *A Narrative of the Expedition to Botany Bay, with an Account of New South Wales*, published in London as early as 1789, gives some details on the French stay in the bay and refers to Receveur's death, adding that Captain Phillip had told him that this was the result of wounds received at Tutuila.[1] On the face of it, this seemed a reasonable conclusion. La Pérouse, as his letters indicate, was still deeply affected by the attack on De Langle and his men, and undoubtedly would have spoken about it in conversations with English officers and mentioned by name those who died or had been hurt in the affray.

Yet, Vaujuas, in the report he made to La Pérouse about the Tutuila tragedy and which is reproduced in the journals, did not include Receveur among those who were seriously wounded. The worst affected from the *Astrolabe*'s landing party were, according to him, the surgeon Lavaux, the master-at-arms and an unnamed soldier. All three recovered, including Lavaux who had to be trepanned. In fact, the only mention of Fr Receveur's wound appears elsewhere in La Pérouse's journal, in which the expedition's commander speaks of a badly bruised eye, *une forte contusion*. 'This does not give the impression of anything particularly serious.'[2]

It could be that internal bleeding occurred and that eventually a clot formed which reached his brain. Other complications, not

---

[1] See above, 'La Pérouse at Botany Bay'. The comment is found in Watkin Tench's *Narrative*, p. 98.

[2] Jean Royer, 'Mystères autour de la vie et de la mort du Père Receveur, aumonier de l'*Astrolabe*', in *Colloque Lapérouse Albi 1985*, p. 120.

APPENDIX III

outwardly visible, may have developed. However, one would have expected these developments to cause some discomfort – headaches, nausea, or some neurological imbalance. Another possibility is a minor fracture of the upper cheekbone, leading to osteitis (inflammation of the bone) and other complications, but once again one would have expected an increasing physical discomfort to affect the patient.[1] Yet, neither La Pérouse nor Receveur himself mentioned in their letters that there was anything untoward when the expedition put into Botany Bay.

On 5 February 1788, La Pérouse wrote to the Minister of Marine: 'We reached Botany Bay without a single case of sickness in either vessel, eighteen of the twenty wounded we had when we left Mahouna had completely recovered and Mr Lavau, the *Astrolabe*'s senior surgeon who had been trepanned and another sailor from that frigate leave us with no anxiety about their state of health.'[2]

Two days later, Fr Receveur himself wrote to his brother Claude-Ignace Receveur in France: 'I would not have mentioned this misfortune if I did not worry that you might learn through the papers that I was present at this scene of horror [the attack on the landing party at Tutuila] and that I was wounded there, but very slightly. So, whatever you may be told about me, you can rest easy. My wounds, which were nothing serious, healed within the space of seven or eight days. We shall be back in France in the spring of 1789 and possibly earlier. So write to me at Brest or Rochefort.'[3]

Fr Receveur died ten days after writing this letter. Stressing that his family were not to worry about him and that he was in good health could be interpreted as a charitable man's endeavour to reassure others, and to be as self-effacing as his Christian background required him to be, and indeed all reports indicate that he was a kindly person more concerned with his duties than about his own welfare; but we still have La Pérouse's own assurances that he had no case of sickness on board and that all those who had been wounded, with only two exceptions, neither of them Fr Receveur, had fully recovered. If the chaplain was suffering from any after-

---

[1] Ibid. But note Vaujuas's comment on his prior state of health at Tutuila, *supra*, p. 406.
[2] See 'Correspondence'.
[3] Correspondence reprinted in A. Gautier, 'Le Père Receveur, aumonier de l'expédition La Pérouse', *Courrier des Messageries Maritimes*, 140 (May-June 1974), pp. 24–34.

effects of the injury he had sustained, he was able to conceal it without difficulty from all his friends, while he was living in cramped quarters with them in the *Astrolabe* and sharing in expeditions ashore at Botany Bay.

He had had one further mishap, shortly after leaving Norfolk Island, when a crate of chemicals caught fire, endangering the ship and sending out a thick cloud of acrid and probably toxic fumes, but there is no evidence that he or anyone else suffered in any way from this accident. This occurred three weeks before the above-quoted letter to his brother and gives us no additional reason to believe that his state of health was in any way impaired at the time.

One is left with the possibility of some event taking place in Botany Bay which had fatal consequences for the chaplain. We unfortunately have no letter from La Pérouse or any of his officers dated later than 7 February, a full ten days before Fr Receveur's death. All we have is Watkin Tench's book and the chaplain's tomb at Botany Bay, with the inscription '*Hic jacet L. Receveur E.F. Minoribus Galliae sacerdos, physicus in circumnavigatione mundi, duce de La Pérouse, obiit 17 februarii 1788.*' If his health was restored when the expedition reached Botany Bay, then we must assume that something happened on board, while the ships were at anchor in the bay, or on land. The latter is much more likely, with a keen botanist like Fr Receveur who seized every opportunity to add to his collections when the expedition put into a new place. As De Langle stated in letters to the Minister of Marine, Receveur worked on meteorological and astronomical observations while at sea and on natural history when ashore.[1] When the expedition reached Tenerife, he joined the others on their five-day excursion up the Peak; at Trinidade he briefly went ashore, and had 'just time enough to bring back a few stones and shells'; he went inland in Chile, Easter Island, the Northwest Coast, the Tartary Coast, Kamchatka (going with others up the icy slopes of Avatscha volcano) and, of course, at Tutuila. There is every reason to suppose that he did the same at Botany Bay, whose very name had been given by James Cook on account of 'the great quantity of New

---

[1] Langle to the Minister, 14 March 1786, written on board, sent from Concepción; 18 January 1787 from Macao; 25 September 1787 from Avatscha.

Plants &c' collected there by the eminent botanists Sir Joseph Banks and Daniel Solander.[1]

The French spent more than six weeks in the bay, building two longboats to make up for the losses incurred at Tutuila, erecting a wooden palisade to protect them from the roaming aborigines, and coping with convicts seeking to obtain a passage to some other part of the Pacific or to France.

The aborigines must have viewed these activities with growing concern. Firstly, a fleet of English vessels had arrived, landing sailors, soldiers and convicts. After a few days, these had left, but merely to enter a new harbour just a few miles to the north where they had begun to settle ashore, putting up tents and starting to erect dwellings. The appearance of a permanent settlement – which indeed it was – on their own land cannot have failed to arouse their concern and their anger. But even before the English abandoned Botany Bay, a new fleet, smaller admittedly, consisting of merely two frigates, had dropped anchor; the new arrivals had given every sign that they too intended to stay, setting up an observatory ashore and constructing a strong wooden fence behind which they began building something which the aborigines might have seen as some kind of dwelling. Both the English in Port Jackson and the French in Botany Bay stationed sentries ashore and went fishing daily, while small parties of naturalists collected plants and flowers, a matter of additional concern to the indigenous residents who eked out a precarious existence on what the shore and the rivers could offer. There was a steady traffic along the track between the two settlements, suggesting that some concerted invasion was being organised.[2] Tribal lands were defended as a matter of course from serious or threatening encroachments by roaming neighbours, and the European idea that Australia was a kind of unclaimed territory with no boundaries of any kind, with a few nomadic inhabitants, was quite wrong. It would have been a natural reaction to the happenings at Botany Bay and Port Jackson for the aborigines to

---

[1] It was originally named Stringay Habour, then Botanists Harbour and Botanists Bay. See on these various names J.C. Beaglehole (ed.) *The* Endeavour *Journal of Joseph Banks 1768–1771*, Sydney, 1962, II, p. 61n, and Cook, *Journals*, I, pp. ccix and 310n.

[2] On 'Frenchman's Road', see Alec Protos, *The Road to Botany Bay: The Story of Frenchman's Road, Randwick, through the Journals of Lapérouse and the First Fleet Writers*, Randwick, 1988.

consider the Europeans as dangerous enemies to be repelled whenever the opportunity presented itself.

From the European point of view, the aborigines were poor, 'weak and not numerous' as La Pérouse put it[1]; they needed to be kept at a distance, and once this was done the superiority of the European weapons, the imposing size of the ships, the very presence of the soldiers, sailors and convicts who greatly outnumbered the locals, would be enough for work to proceed in relative safety. La Pérouse had never been a great believer in the innate goodness of the Noble Savage, and the events in Samoa had destroyed any shred of illusion he might have had about the theory.

Not every member of his expedition, however, shared his opinion. Some still felt that what had happened at Easter Island, where there had been numerous thefts, but above all at Tutuila, had been due to temptation and to circumstances. It was important not to act in a thoughtless manner, placing temptation in the way of natives who, after all, were rather like children, innocent and mischievous, often greedy and quick-tempered, but not necessarily evil. Fr Receveur came from a devout religious background: not only was he a priest himself, but so was his brother Claude-Ignace, his cousin Joseph, while his brother Antoine was to marry the sister of a priest, Claude-Ignace Tournier. During the Terror, the family was to oppose the secularisation of French life imposed by the Republican government and at least one was executed for his actions in 1793. Fr Receveur is recorded by all those who knew him on the La Pérouse expedition as an amiable, kindly man, a keen botanist who may have been too trustful, too generously-minded towards the aborigines he could see watching the French from a distance.

His keenness may have led him to go botanising alone too far from the French camp. His eagerness and his tendency to go on long excursions was well-known.[2] He died on 17 February, three weeks after the two frigates entered Botany Bay. Familiarity with the place may have encouraged him to wander further afield than was wise, unaccompanied, in that botanist's paradise. The aborigines

---

[1] Letter to Fleurieu, 7 February 1788, AN. M. 3JJ 389, 22.
[2] 'Robust, used to navigations and to an explorer and botanist's long and strenuous marches, Receveur was the archetypal scientist Lapérouse was looking for.' F. Bellec, *La Généreuse et tragique expédition Lapérouse*, Rennes, 1985, p. 60.

may well have seen their chance to rid themselves of at least one of the alien invaders. If so, Fr Receveur did not die of some unexplained complications from the wound he had suffered at Tutuila: he was killed while collecting natural history specimens.

How this happened will never be known. Cook and his men had lances and stones thrown at them by groups of aborigines who disputed their landing. One episode related by Banks gives a not improbable impression of what may have taken place in 1788:

> Our surgeon, who had strayd a long way from the people with one man in his company, in coming out of a thicket observed 6 Indians standing about 50 yards from him; one of these gave a signal by a word pronounced loud, on which a lance was thrown out of the wood at him which however came not very near him. The 6 Indians on seeing that it had not taken effect ran away in an instant, but on turning about towards the place from whence the lance came he saw a young lad, who undoubtedly had thrown it, come down from a tree where he had been Station'd probably for that purpose; he descended however and ran away so quick that it was impossible even to attempt to pursue him.[1]

---

[1] Beaglehole, *Banks*, 1962, II, p. 60.

APPENDIX IV

Principal Monuments Erected to La Pérouse

– *Albi*. Statue on stone plinth to 'Jean François Galaup de Lapérouse... Victim of his courage and his devotion to science' erected in 1848 with at its foot the first anchors brought back from Vanikoro. Albi now has its 'Association Lapérouse Albi-France' with its archives and displays. The house at Le Gô, outside Albi, remains the property of the family.
– *Botany Bay*. Commemorative column erected in 1825 by Hyacinthe de Bougainville and Du Camper, commanding the *Thétis* and the *Espérance*, bearing the inscription 'To Lapérouse and his companions. This land where he called in 1825 is the last place from which he sent out news of his expedition.' Close by is Father Receveur's tomb, where an annual commemorative service is held by the 'Père Receveur Commemoration Committee'. This is also the site of the Lapérouse Museum and the centre for the 'Friends of the Lapérouse Museum'.
– *Hudson's Bay*. A plaque set up at Port Prince of Wales by the Historical Sites and Monuments Board of Canada in 1951 records La Pérouse's capture of this outpost in 1782.
– *Monterey*. A stele placed at the entrance to Carmel Bay church in 1848 records La Pérouse's visit.
– *Tutuila*. A memorial was erected in 1887 by the French Navy to the memory of De Langle and his companions.
– *Vanikoro*. The first monument commemorating the loss of the frigates was erected in 1828 in Manevai Bay by Dumont d'Urville; in 1923 Captain Gaspard found that it had collapsed and he re-erected it. It had practically disappeared by the 1950s. In 1959 a new monument was erected at Paiou through the good offices of Reece Discombe and Haroun Tazieff and formally inaugurated in June by men of the naval vessel *Tiaré*.

SOURCES FOR THE JOURNAL

For close on two centuries, the voyage of La Pérouse was known only from the four-volume edition published in 1797 by order of the French Government and edited by Louis-Marie-Antoine Destouff, Baron de Milet-Mureau (1751–1825), a senior officer in the French army who had lost his command after the Italian campaign in 1792 on account of his moderate views (he had been sent by the town of Toulouse to attend the States-General in 1789) and been forced to return to Paris. He rose later to the position of Minister for War and Imperial Prefect.

Although the book was not a financial success at the time, it was translated into a number of languages and reprinted in full or abridged form over the years. Milet-Mureau had had full access to La Pérouse's papers and to reports and letters written by other members of the expedition. La Pérouse sent sections of his journal back from various ports of call, but not a copy of his shipboard log which he kept with him and which was certainly lost in the shipwreck of 1788. His journal, however, is a narrative which contains numerous observations on the places and people he came across as well as some navigational data. It is clear from his letters to Fleurieu and the Minister of Marine that he was confident his journal would eventually be published, as well as seen by Louis XVI, and he wrote it with this in mind. It caused him to express concern on a number of occasions that, as written, it was not in a fit state to be published or even read by the king and he repeatedly urged Fleurieu to ensure that some editing work was done on it.

This insistence is partly a form of courteous modesty and partly an awareness that he was a man of action, not a man of letters. It was customary for accounts of voyages to be rewritten in a form acceptable to the substantial reading public made up of educated people which expected such works to meet the literary standards of the time. La Pérouse was conscious that his most famous French predecessor, Louis de Bougainville, had written an elegant work which the French salons had widely discussed and praised. Bougainville's account of his voyage was, like Milet-Mureau's work, translated and reprinted over the years; it was not until 1977 that the original text was made available in a scholarly edition by Etienne Taillemitte which gave readers the opportunity of comparing the two versions.

A comparable edition of La Pérouse's journal appeared unlikely because the originals had not been found. It was assumed that, as Milet-Mureau had worked during the troubled years of the French Revolution, they had been lost, or that he had kept them until such time as things quietened down and that they had been discarded with other personal papers. Other possibilities were that they had been handed to the Government printers, the *Imprimerie de la République*, itself going through difficult times, and that they were still somewhere in the vaults of its successor, the *Imprimerie Nationale*; or that they had been returned to the wrong administrative body, such as the War Ministry on which Milet-Mureau depended.

One had therefore to take Milet-Mureau's version on trust, knowing all the time that La Pérouse had asked for his work to be modified so that the style and presentation were acceptable to the king and other eventual readers. There was always the danger that his writings had suffered the same fate as James Cook's journal of the *Endeavour* voyage, and been recast by a French equivalent of the unfortunate John Hawkesworth. At the very least, it was realised that La Pérouse and Milet-Mureau had each written in widely different circumstances and under different forms of pressure. La Pérouse had been working on board ship at a time when Louis XVI was on the throne and his overthrow and that of so many ancient institutions was still unthinkable in the minds of most people, especially officers in the royal navy. Milet-Mureau, on the other hand, was an army man, with a scant knowledge of ships and the technique of navigation, and forced to tread warily at a time of great political unrest – his version, for his own sake, and if it was to be published at all, needed to be politically sensitive.

Milet-Mureau had not been selected because of any expertise or special interest in the history of navigation, but because he was available. The preferred choice had been Claret de Fleurieu, La Pérouse's friend, the director of Ports and Arsenals, influential at court and with the French naval administration and already the author of a major work on Pacific exploration, *Découvertes des Francois en 1768 et 1769 dans le Sud-Est de la Nouvelle-Guinée et reconnaissance postérieure des mêmes terres par des navigateurs anglois qui leur ont imposé de nouveaux noms*, published in Paris in 1790 and translated into English and published in 1791. But Fleurieu became Minister of Marine in October 1790, a post he retained until May 1791; his main preoccupations were to promote and organise a search for La Pérouse's lost ships, and look after the interests of the various families affected by the expedition's disappearance. An alternative was a senior naval officer, François-Etienne de Rosily-Mesros (1748–1832) who had sailed to the South Indian Ocean with Yves de Kergelen in 1772. Rosily, however, was not interested, and the choice fell again on Fleurieu who by now was becoming anxious about the way political events were developing in France. Milet-Mureau was in Paris at the time, free to start work on what at first seemed to be a straightforward task, and he accepted the commission.

He was held up by the need to be circumspect and, as he himself put it, the problem of 'reconciling memoirs written before the Revolution and under the influence of all forms of prejudices, with the austere principles of republicanism'. Frequent changes in the administration and the reluctance of officials to take the personal risk of giving him advice or directives delayed publication until 1797. It was obvious that there would have been some alterations made to La Pérouse's original text, for political as well as for stylistic reasons. Milet-Mureau comments on this in the Preface to the *Voyage*. To his credit, it must be said that Milet-Mureau did everything he could to make up for his lack of naval training, and that he must have spent a great deal of the time reading up accounts of navigators and land travellers; this enabled him to place the expedition in its historical context and, on occasions, to add his own comments to La Pérouse's narrative.

The originals came to light in 1977. They form part of a collection of miscellaneous papers entitled 'Documents Scientifiques' in the series 3JJ of the Archives Nationales (Dépôt du Service Central Hydrographique). The documents cover a variety of places and topics, and in some cases their scientific value is minimal. There is no summary of their contents in the lists or catalogues. Towards the end of the series a number of papers relating to the La Pérouse expedition are to be found in folders 386 to 389. Inside No. 387 and bound with other sundry papers and files concerning the expedition are five *dossiers* or files which can be identified as parts of the journals despatched by La Pérouse from his various ports of call:

Dossier 7, pp. 121–30
This contains La Pérouse's first chapter, copied by someone on board but amended by La Pérouse himself.

Dossier 8, pp. 131–7
This is an extract from La Pérouse's account in his journal of the loss of the boats in Lituya Bay, together with Boutin's report on the tragedy and a list of the dead.

Dossier 9, pp. 138–283
This large file contains the entire first part of the journal, from the preface to the period spent in Monterey. The last five pages, however, have been cut out and the original of the next chapter (La Pérouse's Chapter XIII) has not been found, although the text appears in the Milet-Mureau edition. It must be assumed therefore that he had it in his possession when he was preparing the narrative for publication, but one cannot be sure that the five folios that have been quite plainly cut out represented the beginning of this chapter: the possibility must remain that some report or some comments were included at the end of Chapter XII which it was not considered wise to send as part of the journal.

This file contains La Pérouse's insistent request that a fair copy of the journal should be made to be submitted to the Minister of Marine.

Dossier 10, pp. 284–378; dossier 11, pp. 379–464
These two files complete the journal, from Chapter XIV to Chapter XXI when the expedition arrived at Botany Bay.

Other manuscript sources relating to the expedition are held in various sections of the Archives Nationales. Personal files and muster rolls are kept in the Archives des Colonies and the section of the Archives de la Marine which covers the period prior to the Revolution. De Lesseps's personal file is held in that part of the Archives de la Marine which covers the period following the Revolution and in the Archives des Affaires Etrangères.

The Département des Cartes et Plans of the Bibliothèque Nationale and the Archives Nationales hold charts drawn by members of the expedition as well as the engravings made for the Milet-Mureau edition. The library of the Service Historique de la Marine holds a number of original drawings sent back to France during the voyage, most of which were subsequently engraved.

It is now possible to compare the original with the published version. Milet-Mureau was as respectful of La Pérouse's work as one could have expected him to be. Only occasionally does he intrude and work in a personal view derived, one suspects, from what he had read or discussed with friends: when he does, he frequently expresses his opinion and his reservations in a footnote, sometimes a lengthy one. Most of his amendments were stylistic. In a number of cases, they were straightforward corrections of the spelling or syntax, but even here one must issue a warning. La Pérouse had been educated by Jesuits in a provincial college, and his spellings often reflects the trends and fashions of his young days. Accepted forms were undergoing a number of changes in the late eighteenth century, and although there are numerous errors of language and lapses in the originals there are also a number of variants that would have been perfectly acceptable in his younger days. Milet-Mureau, on the other hand, was writing in Paris at the end of the century and he uses the new forms which had become the norm in the literary world.

Stylistic changes needed to present the narrative in a more literary form were made and these were what La Pérouse would have expected. They do not cause distortions of the original except when Milet-Mureau, as he himself admits, uses some idea to link two passages or give added emphasis to some point he considers important. His lack of expertise in navigational matters shows at times, especially when he has to deal with astronomical observations at sea when he finds it hard to cope with La Pérouse's reflexions. He occasionally inserts whole passages by way of explanation or expansion, and omits others. A significant cut is the entire description of the expedition's stay in Madeira and the hospitality of the British consul. Whether he considered this to be an unnecessary *longueur* or whether he believed it to be too generous towards the French republic's enemy we shall never know.

Political considerations certainly made him amend the text. References to Louis XVI whose interest in the expedition was very considerable and to his ministers were cut out, as was practically every mention of the King of Spain. Titles were omitted and comments made by La Pérouse which could be construed as favourable towards the Old Regime or the aristocracy were recast to present a more neutral or even a critical point of view. Fortunately, La Pérouse was a moderate, neither a blind supporter of the *philosophes* nor a defender of the old traditions; he was first and foremost a seaman and an explorer, eager to succeed in his mission and bring fame to his country. And to this extent, Milet-Mureau's task was made easier and his editorial function less intrusive than it might have been.

SELECT BIBLIOGRAPHY

Note: A detailed bibliography of books and articles relating to the La Pérouse expedition and its aftermath has been compiled by Ian F. McLaren, and published by the University of Melbourne in 1993, *Lapérouse in the Pacific, including Searches by d'Entrecasteaux, Dillon, Dumont d'Urville: An Annotated Bibliography*, p. 285.

1. MILET-MUREAU EDITION AND TRANSLATIONS

MILET-MUREAU, M.L.A., *Voyage de la Pérouse autour du monde, publié conformément au décret du 22 avril 1791*, 4 vols and atlas, Paris, 1797.

[ANON] *Podroz na Odkrycia Nowych Krajow W Latach 1785–1788 Wydana Przez M.L.A. Milet-Mureau.* [A Voyage for the Discovery of New Lands in the Years 1785–88 published by M.L.A. Milet-Mureau]. 3 vols, Krakow, 1801–3.

BUMSTEAD, J., *A Voyage Round the World, performed in the Years 1785, 1786, 1787, 1788 by M. de la Peyrouse, abridged from the Original French Journal of M. de la Peyrouse, which was Lately Published by M. Milet-Mureau in Obedience to an Order from the French Government, to which are Added A Voyage from Manilla to California by Don Antonio Maurelle, and an Abstract of the Voyage and Discoveries of the Late Capt. G. Vancouver.* Printed for Joseph Bumstead, Boston, 1801.

FORSTER J.R. and SPRENGEL C.L., *Lapeyrouses Entdeckungsreisen in den Jahren 1785, 1786, 1787 und 1788. Heraugegeben von M.L.A. Milet-Mureau.* 2 vols, Berlin, 1799.

GOLENITSCHEFF and KUTUOFF, *Puteshestvie Laperuza v Juzhnom i Severnom Tikhom Okeane v Prodolzhennie 1785–1788.* St Petersburg, 1800.

HAMILTON, *A Voyage Round the World, performed in the Years*

1785, 1786, 1787 and 1788 by the Boussole and Astrolabe under the Command of J.F.G. de la Pérouse, published by Order of the National Assembly under the Superintendence of L.A. Milet-Mureau, 2 vols and atlas. Printed by A. Hamilton for G.G. and J. Robinson and others, London, 1799. [This edition was reprinted in facsimile by N. Israel, Amsterdam, 1968 as No. 27 in the *Bibliotheca Australiana*.]

HAMILTON, *Voyage de la Pérouse autour du monde, publié conformément au décret du 22 avril 1791 et rédigé par M.L.A. Milet-Mureau*, 2 vols and atlas. Printed by A. Hamilton for G.G. and J. Robinson and others, London, 1799.

[HORREBOW, O. and BRUN J.] *La Pérouses Reise Omkring Verden Oversat Efter Milet-Mureaus Franske Original*. Copenhagen, 1799.

JOHNSON, *A Voyage Round the World in the Years 1785, 1786, 1787 and 1788, by J.F.G. de la Pérouse: published conformably to the Decree of the National Assembly of the 22nd of April 1791, and edited by M.L.A. Milet-Mureau*, 3 vols. Printed for J. Johnson, London, 1798.

LEMOINE, *A Voyage Round the World Performed by Captain de la Pérouse in the Years 1785, 1786, 1787 and 1788, written by himself and sent by Several Expresses to Europe, the Last of Which is Dated from the English Settlement at Botany Bay Where he Arrived January 23, 1788*. Printed for Ann Lemoine, London, 1798.

MOIR, *A Voyage Round the World which was Peformed* [sic] *in the Years 1785, 1786, 1787 and 1788 by Mr de la Péyrouse, Abridged from the Original French Journal of M. de la Péyrouse which was Lately Published by M. Milet-Mureau...to which are added A Voyage from Manilla to California by Don Antonio Maurelle and an Abstract of the Voyage and Discoveries of the Late Captain G. Vancouver*. Printed by J. Moir for T. Brown, Edinburgh, 1798. [This work was reprinted in Boston in 1801].

[Odmann, S.] *Resa Omkring Jorden af Herr de la Pérouse. Aren 1785 och Följande*. Stockholm, 1799.

PETRACCHI, A., *Viaggi di La Perouse Intorno al Mondo*, Milan, 1815.

REINICKE & HINRICHS, *La Perousens Entdeckungsreise in den Jahren 1785, 1786, 1787 und 1788*, 2 vols and atlas. Leipzig, 1799.

STOCKDALE, *The Voyage of La Perouse Round the World in the Years 1785, 1786, 1787 and 1788, with the Nautical Tables, arranged by M.L.A. Milet-Mureau. To which is Prefixed, Narrative of an Interesting

Voyage from Manila to St. Blaise. And Annexed, Travels over the Continent, with Dispatches of La Pérouse in 1787 and 1788 by M. de Lesseps. 2 vols. Printed for John Stockdale, London, 1798.

VAN DER LINDEN, *Reize van de la Perouse in de Jaaren 1785, 1786, 1787 en 1788…naar het Fransch door Mr Johannes van der Linden.* 3 vols, Amsterdam, 1801.

II. OTHER PRINTED ACCOUNTS

[Anon.] *Sketch of a Voyage of Discovery undertaken by Monsieur de la Pérouse under the Auspices of the French Government*, London, 1798.

Fairburn's Edition of the Voyages and Adventures of La Pérouse. London, 1800.

[Barbou & Cie] *Voyage de La Pérouse autour du monde 1785–1788*, Limoges, 1885.

Jean-François de la Pérouse: Voyage autour du monde sur l'Astrolabe et la Boussole (1787–1788). Introduction by Hélène Minguet, Paris, 1980.

Voyage de La Pérouse (1785–1788). Preface by C. Farrère, Paris, 1930.

Voyage de La Pérouse autour du monde, publié d'après tous les manuscrits de l'auteur… Edited by B. Guégan, Paris, 1930.

Voyage de la Pérouse autour du monde 1785–1788. Preface by P. Deslandres, Paris, 1933.

Voyage de Lapérouse autour du monde pendant les années 1785, 1786, 1787 et 1788. Preface by M.R. de Brossard, Paris, 1965.

Voyage de Lapérouse autour du monde pendant les années 1785, 1786, 1787 et 1788. Preface by P. Sabbagh, introduction and postface by M.R. de Brossard, Geneva, 1970.

Voyage autour du monde 1785–1788 par La Pérouse. Edited by A. Duponchel for the *Nouvelle Bibliothèque des voyages anciens et modernes*, Paris, 1841.

ARVENGAS, H., *L'Exploration et le mystérieux naufrage de Lapérouse*, Albi, 1941.

BANCAREL, F., *Voyage de Lapérouse pendant les années 1785,…1788..* 2 vols, Paris, 1809.

BELLEC, F.M., *La Généreuse et tragique expédition La Pérouse*, Rennes, 1985.

BENOIT-GUYOT, G., *Au temps de la marine en bois: sur les traces de Lapérouse*, 2 vols, Paris, 1942–4.
BLANCHARD, V., *Voyages de La Pérouse autour du monde*, Limoges. 1848.
BROSSARD, M.R. de and DUNMORE, J., *Le Voyage de Lapérouse 1785–1788*, 2 vols. Paris, 1985.
CHATENET, E. du., *Voyage de La Pérouse autour du monde 1785 à 1788*, Limoges, n.d.
FLEURIOT DE LANGLE, I., *Le Voyage extraordinaire de La Pérouse*, 3 vols, Nice, 1971–2.
FLEURIOT DE LANGLE, P., *La Tragique Expédition de La Pérouse et Langle*, Paris, 1954.
GASSNER, J.S., *Voyages and Adventures of La Pérouse, from the Fourteenth edition of the F. Valentin Abridgment*, Honolulu, 1969.
GIRAULT DE COURSAC, P. AND P., *Le Voyage de Louis XVI autour du monde: l'expédition La Pérouse*, Paris, 1985.
HAPDE, J.B.A., *Expédition et naufrage de Lapérouse: recueil historique*, Paris, 1829.
HYENNE, R., *La Pérouse: aventures et naufrage*, Paris, 1859.
LAHARPE, J.F. de, *Histoire abrégée du voyage de La Pérouse pendant les années 1785, 1786, 1787 et 1788*, Leipzig, 1799.
LESSEPS, J.B.B. de, *Journal historique de M. de Lesseps, Consul de France, employé dans l'expédition de M. le Comte de la Pérouse en qualité d'interprète du Roi*. 2 vols, Paris, 1790.

Travels in Kamtschatka during the Years 1787 and 1788, translated from the French of M. de Lesseps, Consul of France and interpreter to the Count de la Perouse, 2 vols, London, 1790. [Other translations of Lesseps's narrative appeared in German (in 1791 in Berlin, and Leipzig, in 1792 in Vienna), Dutch (in 1791 in Utrecht), Swedish (in 1793 in Upsala), Italian (in 1794 in Naples) and in Russian (in 1801–2 in Moscow); there were new editions of the French text in Leipzig in 1799 and in Paris in 1831 and 1880, and of the English translation in London in 1798.]

LOCATELLI, A., *La Spedizione di La Pérouse nel Grande Oceano*, Turin, 1929.
MANTOUX, G., *Voyage de La Pérouse, capitaine de vaisseau, autour du monde (années 1785, 1786, 1787 et 1788), raconté par lui-même*, Paris, 1882.
MARCEL, G.A., *La Pérouse: récit de son voyage; expédition envoyée à sa recherche*, Paris, 1888.

MONTEMONT, A.E. de, *Voyage de La Pérouse autour du monde*, 2 vols, Paris, 1885.
SAUVAN, J.B.B. de, *Voyage de Lapérouse rédigé d'après ses manuscrits originaux, suivi d'un appendice renfermant tout ce que l'on a découvert depuis le naufrage jusqu'à nos jours*, Paris, 1831.
SHELTON, R.C., *From Hudson Bay to Botany Bay: the Lost Frigates of Lapérouse*, Toronto, 1987.
VALENTIN, R.F., *Voyages et aventures de La Pérouse*, Tours, 1839.
VATTEMARE, H., *Vie et voyages de La Pérouse*. Paris, 1887.

III. STUDIES AND RELATED WORKS

'The Alaskan Adventures of Jean François Galoup [sic] de la Pérouse', *Alaska Magazine* (1927), pp. 109–43.
Historical Records of Australia, I:2, Sydney, 1892.
ALLEN, E.W., *Jean François Galaup de Lapérouse: a check list*, San Francisco, 1941.
The Vanishing Frenchman: The Mysterious Disappearance of Lapérouse, Rutland, Vt, 1959.
ARMSTRONG, T., *Russian Settlement in the North*, Cambridge, 1965.
ASSOCIATION LAPEROUSE ALBI-FRANCE, *Bicentenaire du voyage de Lapérouse: actes du colloque d'Albi Mars 1985*, Albi, 1988.
BANCROFT, H.H., *Alaska 1730–1885*, San Francisco, 1886.
History of California, San Francisco, 1884.
BARRATT, G., *Russia and the South Pacific 1696–1840*, 4 vols, Vancouver, 1988–92.
BARRES, A., 'L'hypothétique bateau de secours de Lapérouse', *Bulletin de la Société d'études historiques de la Nouvelle-Calédonie*, 57 (Oct. 1983).
BARRINGTON, DAINES, *Miscellanies*, London, 1781.
BARTHES DE LAPEROUSE, N. de, 'La vie privée de Lapérouse', *Bulletin de la Société de géographie*, II (1888), p. 28.
BEAGLEHOLE, J.C. (ed.), *The Journals of Captain James Cook on his Voyages of Discovery*, 3 vols in 4 and portfolio, Cambridge, 1955–69.
The Exploration of the Pacific, 3rd ed., London, 1966.
The Life of Captain James Cook, London, 1974.
BEERMAN E. AND C., 'The Hospitality of the Spanish Governor

of Monterey, Pedro de Fages, to the Ill-fated French Expedition of the Conte [sic] de la Pérouse', *Noticias del Puerto de Monterey*, XX:2 (June 1976), pp. 9–13.

BELLEC, J.F., 'Le Naufrage de l'expédition La Pérouse: une nouvelle analyse d'un dossier mal connu', *Neptunia*, 149 (1983), 1–11 and 150 (1983), pp. 1–14.

BELLESORT, A., *La Pérouse*, Paris, 1926.

BERIOT, A. *Grands voiliers autour du monde: les voyages scientifiques 1760–1850*, Paris, 1962.

BROC, N., *La Géographie des philosophes, géographes et voyageurs français au XVIIIe siècle*, Lille, 1972.

BROSSARD, M.R. de, *Lapérouse: des combats à la découverte*, Paris, 1978.

Rendez-vous avez Lapérouse à Vanikoro, Paris, 1964.

BROUGHTON, W.R., *A Voyage of Discovery to the North Pacific Ocean, in which the Coast of Asia,... the Island of Insu (Commonly Known under the Name of the Land of Jesso), the North, South and East Coasts of Japan, the Lieuchieux and the Adjacent Isles, as well as the Coast of Corea, have been Examined and Surveyed, Performed in His Majesty's Sloop Providence and her Tender in the years 1795...1798*, London, 1804.

BROWNING, O., *Despatches from Paris 1784–1790*, 2 vols, London, 1909.

BUACHE, J.N., 'Mémoire sur les terres découvertes par La Pérouse à la côte de Tartarie et au nord du Japon', *Mémoires de l'Académie des sciences morales*, V:1 (1803), pp. 1–42.

CARLETON, F.R.L., 'Père Receveur Bicentenary Commemoration 1788–1988', in *An Australian Mosaic*, ed. Leo J. Ansell, Toowoomba, N.S.W., 1988, pp. 71–9.

CHAPIN, S.L., 'Scientific Profit from the Profit Motive: the Case of the La Pérouse Expedition', *Actes du 12e congrès international d'histoire des sciences, 1968*, XI (1971), pp. 45–9.

CHAPMAN, C.E., *A History of California: the Spanish period*, New York, 1921.

CHINARD, G., *Le Voyage de Lapérouse sur les côtes de l'Alaska et de la Californie (1786)*, Baltimore, 1937.

COLLINS, D., *An Account of the English Colony in New South Wales*, London, 1798.

COOK, W.L., *Flood Tide of Empire: Spain and the Pacific Northwest 1543–1819*, New Haven, 1973.

SELECT BIBLIOGRAPHY

CORDIER, H., 'Deux compagnons de Lapérouse [Lamartinière and Clonard]', *Bulletin de la Société de géographie*, XXXI (1916), pp. 54–82.
— *Le Consulat de France à Canton au XVIIIe siècle*, Leiden, 1908.
COXE, W., *An Account of the Russian Discoveries between Asia and America*, London, 1780.
DELIGNIERES, E., 'Note sur Gaspard Duché de Vancy', *Réunion des sociétés des beaux-arts des départements* (1910), 72–90.
DERMIGNY, L., *La Chine et l'Occident: le commerce à Canton au XVIIIe siècle 1719–1833*, 3 vols, Paris, 1964.
DIXON, G., *A Voyage Round the World; but more particularly to the North-West Coast of America, Performed in 1785...1788 in the King George and Queen Charlotte, Captains Portlock and Dixon*, London, 1789.
DODGE, E. S., *Beyond the Capes: Pacific Exploration from Captain Cook to the Challenger 1776–1877*, Boston, London, 1971.
DONDO, M.M., *La Pérouse in Maui*, Maui, Hawaii, 1959.
DUNMORE, J., *French Explorers in the Pacific*, 2 vols, Oxford, 1965–9.
— 'Le Vrai Journal de Lapérouse', *Revue du Tarn*, 38 (1980), pp. 175–80.
— 'The Louis-Napoleon La Pérouse', *Turnbull Library Record*, 17:2 (1984), pp. 106–7.
— *Pacific Explorer: The Life of Jean-François de la Pérouse*. Palmerston North, New Zealand, Annapolis, Md., 1985.
— *Who's Who in Pacific Navigation*, Honolulu, Melbourne, 1992.
EMMONS, G.T., 'Native Account of a Meeting between La Pérouse and the Tlingit', *American Anthropologist*, 13 (April-June 1911), pp. 294–8.
ESTAMPES, J. D'., *Catalogue de l'exposition du centenaire*, Paris, 1888.
FISHER, R.H., *The Russian Fur Trade 1550–1770*, Los Angeles, 1937.
FITZHARDINGE, L.F. (ed.), *Sydney's First Four Years, being a Reprint of A Narrative of the Expedition to Botany Bay...by Captain Watkin Tench of the Marines, with an introduction and annotations*, Sydney, 1961.
GARNIER, J., 'Traces du passage de La Pérouse à la Nouvelle-Calédonie', *Bulletin de la Société de Géographie* (Nov. 1869), pp. 407–13.

GAUTIER, A., 'Le Père Receveur, aumonier de l'expédition La Pérouse', *Courrier des Messageries maritimes* (May-June 1974), pp. 24–34.

GAZIELLO, C., *L'Expédition de Lapérouse 1785–1788: réplique française aux voyages de Cook*, Paris, 1984.

GOEP, E., 'Lapérouse', in *Les Grands Hommes de la France* Paris, 1873, pp. 85–212.

GOLDER, F.A., *Russian Expansion on the Pacific 1641–1850*, Gloucester, Mass, 1960.

GORDON, M., *The Mystery of La Pérouse*, Christchurch, New Zealand, 1961.

GUNTHER, E., *Indian Life on the Northwest Coast of North America as seen by the Early Explorers and Fur Traders during the Last Decades of the Eighteenth Century*, Chicago, London, 1972.

HALLWARD, N.L., *William Bolts: A Dutch Adventurer under John Company*, Cambridge, 1920.

HARDY, J.P. AND FROST A.(ed.), *European Voyaging towards Australia,* Canberra, 1990.

HUARD P. AND ZOBEL, M., 'La Société Royale de Médecine et le voyage de La Pérouse', *Actes du 87e congrès national des sociétés savantes,* 1962 ,(1963), pp. 83–91.

HUNTER, J., *An Historical Journal of the Transactions at Port Jackson and Norfolk Island, with the Discoveries which have been made in New South Wales and in the Southern Ocean since the Publication of Phillip's voyage.* London, 1793.

INGLIS, R., *The Lost Voyage of Lapérouse,* Vancouver, 1986.

'The Effect of Lapérouse on Spanish Thinking about the Northwest Coast', in Cook, W. L. (ed.), *Spain and the North Pacific Coast*, Vancouver, 1992, pp. 46–52.

JACKSON, D., 'Ledyard and Lapérouse: A Contrast in North Western Exploration', *Western Historical Quarterly*, 9 (1978), pp. 495–508.

KERALLAIN, R. de, 'La Pérouse à Botany-Bay', *La Géographie*, 33 (1920), pp. 41–8.

KERNEIS, A.A., 'Le Chevalier de Langle, ses compagnons de l'*Astrolabe* et de la *Boussole*; expédition envoyée à la recherche des bâtiments', *Bulletin de la Société académique de Brest*, 2–XV (1900), pp. 221–88.

KING, P.G., *The Journal of Philip Gidley King R.N.* Edited by Paul G. Fidlon and R.J. Ryan, Sydney, 1980.

LAMB, W. KAYE (ed.), *The Voyage of George Vancouver 1791–1795*, 4 vols, London, 1984.
LE PAUTE, G.J., *Notice sur la famille Le Paute*, Paris, 1869.
LEPAUTE, G., 'Notice biographique sur Lepaute-Dagelet', *Bulletin de la Société de géographie* (1888), p. 149.
LINNEKIN, J., 'Ignoble Savages and Other European Visions: The La Pérouse Affair in Samoan History', *Journal of Pacific History*, XVI:1 (1991), pp. 3–26.
LOIR, M., *La Marine Royale en 1789*, Paris, 1892.
MACKANESS, G., *Admiral Arthur Phillip, Founder of New South Wales*, Sydney, 1937.
MAINE, R., *Lapérouse*, Paris, 1946.
MARCHANT, L.R., *France Australe*, Perth, 1982.
MARGOLIN, M., *Monterey in 1786: The Journals of Jean François de La Pérouse*. Berkeley, 1989.
MARTIN-ALLANIC, J.E., *Bougainville navigateur et les découvertes de son temps*, 2 vols, Paris, 1964.
MCKENNA, J.F., 'The Noble Savage in the voyage of La Pérouse', *Kentucky Foreign Language Quarterly*, XII (1965), pp. 33–48.
MEARES, J., *Voyages Made in the Years 1788 and 1789 from China to the North West Coast of America*, London, 1790.
MEISSNER, H.O., *La Pérouse, le gentilhomme des mers*, Paris, 1988. Translation by Raymond Albeck of *Die Verschollenen Schiffe des Lapérouse*, Munich, 1984.
NISBET, A.M. AND BLACKMAN, M., *French Navigators and the Discovery of Australia*. Sydney, n.d.
ORD, A. de la G., *Occurences in Hispanic California*, transl. and edited by Price, F. and Ellison, W.H., Washington, D.C., 1956.
PETIT, G., 'Le Chevalier Paul de Lamanon 1752–1787', *Actes du 90e congrès des sociétés savantes*, (1965), pp. 47–58.
PHILLIP, A., *The Voyage of Captain Phillip to Botany Bay, with an Account of the Establishment of the Colonies of Port Jackson and Norfolk Island*, London, 1789.
PISIER, G., 'Lapérouse en Australie', *Bulletin de la Société d'études historiques de la Nouvelle-Calédonie*, 23 (April 1975).
POMEAU, R., 'Lapérouse philosophe', in *Approches des Lumières: mélanges offerts à Jean Fabre* (Paris, 1974), pp. 357–70.
PROTOS, A., *The Road to Botany Bay: the Story of Frenchmans Road*.

RICH, E.E., *The Fur Trade and the Northwest to 1857*, Toronto, 1967.
ROY, B., *Dans le sillage de La Pérouse*, Paris, 1946.
RUDKIN, C.N., *The First French Expedition to California: Lapérouse in 1786*, Los Angeles, 1959.
SCOTT, E., *Lapérouse*, Sydney, 1912.
Terre Napoléon: A History of French Explorations and Projects in Australia, London, 1910.
SHARP, C.A., *The Discovery of the Pacific Islands*, Oxford, 1960.
SOCIETE DE GEOGRAPHIE DE PARIS, *Centenaire de la mort de Lapérouse*, Paris, 1888. *Bulletin de la Société de Géographie* (Nov. 1869), pp. 407–13.
SPATE, O.H.K., *The Pacific since Magellan*, 3 vols, Canberra, 1979–88.
STEPHAN, J.J., *Sakhalin: A History*, Oxford, 1971
The Kuril Islands: Russo-Japanese Frontier in the Pacific, Oxford, 1974. Sydney, 1988.
TAILLEMITE, E., *Bougainville et ses compagnons autour du monde 1766–1769*, 2 vols, Paris, 1977.
TENCH, W., *A Narrative of the Expedition to Botany Bay, with an Account of New South Wales*, London, 1789.
THOMAS, J., 'La Pérouse ou Lapérouse: légitimité d'une orthographe', *Bulletin de la Societé d'études historiques de la Nouvelle-Calédonie*, 71 (April 1987).
VARSHAVSKII, A.S., *Laperuz*, Moscow, 1957.
VIEULES, P.M., *Centenaire de Lapérouse: notice sur la famille et la vie privée du célèbre marin*, Albi, 1888. A *Supplément* was published at Albi in 1892.
VINATY, J.A., *Eloge de Lapérouse*, Paris, 1823.
WAGNER, H.R., *The Cartography of the Northwest Coast of America to the Year 1800*, 2 vols. Berkeley, Calif., 1937.
WHITE, J., *Journal of a Voyage to New South Wales*, London, 1790.

IV. THE SEARCH FOR LA PEROUSE

BAYLY, G., *Sea-life Sixty Years Ago: A Record of Adventures which led up to the discovery of the Relics of the Long-Missing Expedition commanded by the Comte de la Pérouse*, London, 1885.
BERGASSE DU PETIT-THOUARS, *Aristide-Aubert du Petit-Thouars, héros d'Aboukir*, Paris, 1937.

SELECT BIBLIOGRAPHY

BERTHELOT, P., 'Antoine-Raymond-Joseph Bruny, chevalier d'Entrecasteaux, contre-amiral (1737–1793)', *Bulletin de la Société d'études scientifiques et archéologiques de Draguignan*, XIII (1968), 75–107.

BROSSARD, M.R. de, '179 ans après son départ de Brest, la frégate de Lapérouse est retrouvée', *Cahier des explorateurs* (June 1964), pp. 8–9, 12–13.

BURNEY, J., *A Memoir on the Voyage of d'Entrecasteaux in search of La Pérouse*, London, 1820.

CHEVALIER, A., 'L'Expédition d'Entrecasteaux à la recherche de La Pérouse', *Pays Bas-Normand*, 46 (1953) pp. 5–6.

DAVIDSON, J.W., *Peter Dillon of Vanikoro* [edited by O.H.K. Spate]. Melbourne, 1975.

DELATTRE, F.P., *Rapport sur la recherche à faire de M. de la Pérouse, fait à l'Assemblée nationale*, Paris, 1791.

DILLON, P., *Narrative and Successful Result of a Voyage in the South Seas, performed by Order of the Government of British India to Ascertain the Actual Fate of La Pérouse's Expedition*, 2 vols, London, 1829.

DUMONT D'URVILLE, J.S.C., *Voyage de découvertes autour du monde et à la recherche de La Pérouse... sur la corvette l'Astrolabe pendant les années 1826, 1827, 1828 et 1829*, 5 vols and atlas, Paris, 1832–3.

Voyage de la corvette l'Astrolabe exécuté par ordre du Roi pendant les années 1826, 1827, 1828, 1829 sous le commandement de M. J. Dumont d'Urville, capitaine de vaisseau, 5 vols and atlas, Paris, 1830–33.

DUPETIT-THOUARS, A.A., *Prospectus d'un armement particulier pour la recherche de M. de la Pérouse*, Paris, 1790.

DUPONT, M., *D'Entrecasteaux: rien que la mer, un peu de gloire*, Paris, 1983.

FREMINVILLE, C.P., *Nouvelle Relation du voyage à la recherche de Lapérouse, exécuté par ordre du Roi pendant les années 1790...1794 par M. Dentrecasteaux*. Brest, 1838.

HULOT, E.G.J., *Le Contre-amiral Dumont d'Urville (1790–1842)*, Paris, 1892.

LABILLARDIERE, J.J.H. de, *Relation du voyage à la recherche de La Pérouse fait par ordre de l'Assemblé constituante pendant les années 1791, 1792 et pendant la 1ère et la 2de année de la République française*, 2 vols and atlas, Paris, 1800.

MARCEL, G.A., *Une expédition oubliée à la recherche de Lapérouse*, Paris, 1888.

OVIGNY, P.L., 'A la recherche de La Pérouse: relation sur l'expédition de l'amiral d'Entrecasteaux de 1791 à 1795', *Revue économique française*, 101 (1979), pp. 19–25.

PARIS, F.E., 'Comment on a retrouvé les restes de l'expédition de Lapérouse à Vanikoro', in Société de géographie, *Centenaire de la mort de Lapérouse* Paris, 1888, pp. 191–207.

RICHARD, H., *Une Grande Expédition scientifique au temps de la Révolution française: le voyage de d'Entrecasteaux à la recherche de la Pérouse*, Paris, 1986.

ROSSEL, E.P.E. de, *Voyage de Dentrecasteaux envoyé à la recherche de La Pérouse*. 2 vols and atlas, Paris, 1807–8.

INDEX

Abbreviations used: LaP for La Pérouse; DeL for Fleuriot de Langle

Aborigines in New South Wales, LaP's views on, ccv, 539–40; and LaP, ccvi, 567–9
Acapulco, 171, 201, 377n
Acqueis, Akkeshi, Ternay Bay (?), 277, 323, 514
Adams, George, supplies compasses, cviii
Admiralty Bay, Alaska, see Yakutat Bay
Admiralty Islands, rumours of LaP at, ccxv
Africaine, LaP sails in, to Madagascar, xliv–xlv
Aigle, Bouvet's ship, 5n
Aguilar, Martin de, navigator, cxvii
Ainu, early inhabitants of Japan and the northern islands, 284n, 290n, 292, 320–2, 338; vocabulary, 339–43
Alaska, North-West Passage through, xxxvi; Russian exploration and settlement, xxxv, 337n, 372; British intentions, xciv; LaP in, cvi, clix–clxv, 96–160; other references, 336, 366n, 381n
Albatross, encountered, 32, 65, 281, 471; on menu, cliv, 34
Alberoni, Giulio, Spanish politician, 242
Albi, France, xi, xxxviii-xl, lii, lvii-lviii, lxxv, ccxxiv, 570
Aleutian Islands, in instructions, cxxxi, cxxxviii; mentioned, clx, 146, 158, 188, 337n, 347n, 360n, 366n, 372, 378n, 475, 481, 484n, 492
Alexander Archipelago, Alaska, clxiv
Amazone, frigate, LaP commands, li, lii, lxv, 275n
American Indians, California, LaP reports on, cciv, 169, 170–1, 175–87; at Carmel mission, 178–87, 198; language, 197–200; cannibalism, 186; disappearance, ccxxxv, 180n, 188n; (see also: Araucanians, Tlingit)

Amur River (Segalien River), 282n, 284n, 290–1, 294, 296, 306–7, 312, 321, 332–3, 337n, 340, 375, 510
Anderson, William, surgeon on Cook's *Resolution*, 88, 525
Anian, Strait of, xxxi-xxiii, xxxviii, 331
Aniva Bay, Sakhalin, 323, 514
Anson, George, cxvi, 23, 35, 38, 77, 202, 204, 258, 466
Anthonioz, Pierre, finds an anchor, ccxxiv-ccxxv
Apolima, Samoan islands, discovered, ccxxx, ccxxxviii, 413–14
Araucanian Indians, Chile, LaP describes, 49–50
Ariel, in American war, 247n
Arsacides, Terre des, see Solomon Islands
Ascençaon, rumoured island, Atlantic, cxiv, 23–4, 29, 466
Astrée, frigate, LaP commands during American war, liii, lxv, lxxxii; DeL takes over for Hudson's Bay raid, liv-lvi, lxi, 10
Astrolabe (formerly *Coquille*), Dumont d'Urville's ship, ccxxii-ccxxiv
Astrolabe (formerly *Autruche*), accommodation in, clii; boats, xcix-xc, 11; appointments to, lxvi-lxvii, lxxi-lxxiv, lxxx, lxxxiv, lxxxvi, lxxxix, xc, xcii-xciii, civ, 8, 247, 488; better sailer than the *Boussole*, 468; departure, cliii; sends boat ashore at Trinidade, 22; receives new signals, 30; anchors first at Easter Island, 55; thefts sustained at Easter Island, 66; off Maui, 82–6, 202; sends boat ashore at Maui, 85; swamped by natives at Maui, 82–3; sends boat ashore at Yakutat Bay, 98; at Lituya Bay, 102; losses sustained at Lituya Bay, 122;

589

reaches Monterey, clxvi, 175; calls at Asunción, 210–12; at Macao, clxxi, clxxiii; encounters Japanese vessels, 268–9; Russian governor dines on board, 355; damaged, 385; losses at Tutuila, 401, 403, 410; fire off Norfolk Island, 444, 565; in Australia, cxcii–cxcviii; DeL reports on condition of, 454, 526; death of sailors, clxxxviii, 246, 378, 526; of D'Aigremont, 246, 509, 519; changes of commander, lxix, cxc; final itinerary, ccxxvi; due at the Ile de France, ccvii; search for, ccxi; rumoured to be near Vavau, ccxvii; wreck of, ccxxii, ccxxiv

Asunción Island, Marianas, call at, clxx, 208–12

Atrevida, Bustamente's ship, ccviii

Australia, St Allouarn's landfall, xlii; in LaP's instructions, cxxix, cxxxvii, cl, cliv, 530–1; first sighting of, cxcii, 445–6; convicts, cxciii–cxcvi; French names, ccxviii–ccxix; proposed plan of exploration by LaP, 467, 476, 492, 517, 521, 542; other mentions, 158, 535, 541; (see also: Botany Bay, New South Wales, Tasmania)

Autruche (later *Astrolabe*), fitting out, xcvi–xcvii; name change, xcix

Avacha, Avatsha, see Petropavlovsk

Avacha Bay, 329, 348, 361, 365, 368, 374, 376, 510, 513, 532

Babuyan Islands, Philippines, conquered by the Spanish, 522

Bachaumont, Louis de, gossip writer, lxi

Baffin Island, Hudson Strait, lv

Baffin Bay, 515

Bailly, Jean-Sylvain, astronomer, helps with planning, xxviii–xxix

Ball, Henry Lightbird, naval lieutenant, 446n, 447n

Banks, Sir Joseph, on Cook's first voyage, ccxxx, 34, 567, 569; at Tierra del Fuego, 36; assists with preparations of LaP expedition, xci, cix, 9; concerned about LaP's fate, ccviii

Baranof Islands, Northwest Coast, Chirikov at, 101n; reported Chinese landings on, 148n; LaP sights, clxiv

Barkley, Charles William, captain, ccviii

Baros, Francisco de, governor of Santa Catarina, 28–30

Barreto, Lorenzo, probable discoverer of Vanikoro, ccxvii

Barrington, Daines, amateur geographer, xxxvi, 515

Bass, George, English navigator, asked to look for LaP, ccxix

Batan Islands (Bashee Islands), sighted by LaP, 213, 257; on English ships' route, 217; Spanish possession, 234n

Batavia, possible rendezvous, cxix, cxx, ccviii; D'Entrecasteaux expedition survivors go to, ccxxviii; mentioned, 263n, 273, 331

Baudin, Nicolas, French navigator, asked to look for LaP, ccxix; problems with scientists, ccxxxi

Bay of Islands, Alaska, 150

Bayonnaise, Tromelin's ship, ccxxiv

Beaglehole, J.C., on Cook, ccxxix, 404n, 440n

Bear Islands, Siberia, 376

Beautemps-Beaupré, Charles-François, assists with charts, xxx; and D'Entrecasteaux expedition, ccxviii, ccxxxix; islands named after him, ccxxxix

Behm, Magnus von, Russian official, 329, 346, 355n, 358n, 372

Bellec, François, ccxxxvi, 568n

Bellegarde, Gabriel-Jean Dupac de, biography, lxxi; transfers to the expedition, clxxvi, 498, 504; in Castries Bay, 304; goes ashore at Tutuila, 392; praised by DeL, 527; promotion mooted, 524, 528

Belle-Poule, LaP sails in to Mauritius, xli, returns, l; Clonard sails in, lxviii

Bellin, Jacques-Nicolas, hydrographer, 433, 519

Benier, captain, obtains remains of French victims, lxv

Benyowski, Maurice de, xlix

Bering Bay (Behring Bay) cxxxvi; named by Cook, 98n, 99

Bering Island, 374, 531

Bering Strait, 374

Bering, Vitus, explorations in northern Pacific, xxxv, clx, clxxxiv, 96n, 151n, 155, 360, 371n, 372; LaP comments on, 101

Bermudès, commander of Philippines naval forces, 235, 251–2

INDEX

Bernarda, Pedro de, Spanish navigator, 148n

Bernizet, Gérault-Sébastien, surveyor, lxxxiii, 488; appointed, 10; charts of Trinidade, 23; of Chile, 42; of Yakutat Bay, 98; of the Batans, 213; of Quelpaert, 263; at Easter Island, ccxxxiii, lands at Maui, 85; in Lituya Bay, clxii; work on the Northwest Coast, 100–1, 109, 110, 111–12, 197; lands at Suffren Bay, 282; in Castries Bay, 304; ascends volcano at Petropavlovsk, 349–52; presents drawing to Kozlov, 361; observations at Vavau, cxci, 434; praised by LaP, 4, 487, 523

Berthoud, Ferdinand, watchmaker, timepiece tested by Fleurieu, cviii, 15n; gives temperature table to Dagelet, 53, 458; chronometers praised by LaP, cix, 76, 150, 161–2, 197; by DeL, 484

Bichy (Bitchy), name given by LaP to Castries Bay residents, 306, 312–13, 321, 332–5

Billings, Joseph, 360–1, 534n

Bligh, William, captain, at Tonga, ccxvii

Blondela, biography, lxvi–lxvii; accommodation, clii; health, 463; work at Lituya Bay, 111; in California, 197n; sketches in Sakhalin, 296; in Castries Bay, 304; at Petropavlovsk, 516; in Samoa, 416; praised by LaP, 505, 511, 523; by DeL, 469, 483, 496–7, 526–7; promotion, 479, 483, 499, 524, 528

Bodega y Quadra, Juan Francisco, Spanish navigator, clx, 103n, 146n, 147n, 151n, 152n, 155n, 165n, 518n

Bolts, William, works for French government, xxi–xxii

bonitos, 382, 424; provide a meal, clxxxviii, 383; follow the frigates, 75, 91

booby, sighted, 212; indication of land, 204, 384

Borda, Jean-Charles de, astronomer, helps with planning, xxviii–xxix; praised, 15–16, 484; criticised, 458–9

Bordeaux, France, xcix, 256, 453, 460

Boro, Francisco, governor of Santa Catarina, 466

Boscawen, see Cocos Island

Botany Bay, New South Wales, new aim of voyage, clxxxv, cxci, ccxxxviii, 405, 427, 435, 441, 445, 530; LaP arrives at, lxix, cxcii, ccxxviii, 446–7, 538, 564–9; stay at, cvi, cxcii–cxcviii, ccvi, 446–8; drinking water, lxiii; discomfort, cxcviii; French stockade, cxcvii–cxcviii, ccvi, 539, 567; boats built at, 535–6, 539, 567; English fleet at, 446–7, 536; English leave, 447–8, 536, 539, 567; LaP leaves, ccxxvi, 544, 546; lost report on, 531n; monuments, 570

Boudeuse, Bougainville's ship, cii, 6

Bougainville, Hyacinthe de, 570

Bougainville, Louis-Antoine de, route, xxv, cliv, 69n, 380, 413n, 416, 418, 427–8, 433, 520–1, 541; at Tahiti, clxx, 51n; at Samoa, clxxxviii, 386–7, 389n, 390, 404, 411, 418, 423; health of crews, cii; opinion on geographers, cxcix; best-selling works, xxi, 571–2; social theories, ccii–cciii; belief in Polynesian travels, 87n; influence, xliv, 6, 133n; proposed for an expedition to search for LaP, ccix; mentioned vii, xvii, xliv, cxv, cxxi, cxxviii–cxxix, 6, 39n, 40n, 93n, 474n

Bougainville Strait, 517

Bouman (Bauman) Islands, see Manua group

Bourbon (now Réunion), island in Indian Ocean, xlv

Boussole (formerly *Portefaix*), boats, xcix–xvx; appointments to, lxvii, lxix, lxx, lxxi, lxxii, lxxv, lxxvi, lxxvii, lxxxi, lxxxiii, lxxxiv, lxxxv, lxxxvi, lxxxviii, lxxxix, xc, xci, xcii, 8, 247; sails from Brest, cliii; boat sent ashore at Trinidade, 22; at Hawaiian Islands, 82–5, 202; at Lituya Bay, 102–26; loses sight of the *Astrolabe* on Northwest Coast, 154, 158; puts in at Monterey, clxvi, 175; at Macao, clxxi, clxxiii; sustains damage, 385; loses longboat at Tutuila, 403; losses sustained, 122, 410; Monty transfers to, lxxvii; health on board, 246, 431, 525; in Australia, cxcii–cxcviii; English officers visit, cxcvi–cxcvii; final itinerary, ccxxvi; due at the Ile de France, ccvii; search for, ccxi; rumoured to be near Vavau, ccxvii; wreck, ccxxii, ccxxiv

591

Boussole Strait, Kuril Islands, lcxxxiii, 326–7, 516, 520, 523

Boutin, Charles-Marie Fantin de, biography, lxvii; at Madeira, 16; goes ashore at Trinidad, 22, 465; at Santa Catarina, 28; lands at Maui, 85; at Yakutat Bay, 98; in Lituya Bay, 104–5, 110, 112, 114–15; his account of Lituya Bay tragedy, 116–20; lands on Asunción, 210–12; sent ashore in China, 215, 217; goes to Manila, 234; takes soundings at Dagelet Island, clxxviii, 267; lands on Sakhalin, 287; takes soundings in Tartary Strait, 301–2; Point Boutin named, 336; left in charge, 396; sent ashore at Tutuila, 397–401, 406, 411, 426; wounded 400, 402, 408–10; decorated, 499; recommended for a pension, 524

Bouvet de Lozier, Jean-Baptiste-Charles, navigator, cxiv, cxxxiii, 5, 476

Branciforte, Miguel de la Grue Talamanca de, governor at Tenerife, 18

Brazil, cxxvi, 19, 26, 463, 465

breadfruit, in Samoa, 411, 425; in Tonga, 436, 440; references omitted in printed account, 436n

Breskens, Dutch ship, 273n

Brest, France, naval port, lix, lxi, lxv, lxvii, lxx–lxxiii, lxxxix, cxxiv, 376; LaP studies at, xxxix–xli; medical school, lxxx–lxxxi; epidemics, lxxviii, lxxxii; LaP sails for America from, li; Eléonore Broudou at, l; preparations for LaP expedition at, lxxvii, xcvii–xcx, 346, 453, 455, 485n; expedition leaves from, cli, clxxxv, 11–12, 18, 383, 431, 457–8, 467, 468–9, 474, 479, 485, 506, 519, 543, 544–6; expected date of return to, cxlv, ccvii, 517, 540; trouble during the Revolution, ccix, ccxv

Briggs, Henry, mapmaker, xxxiii

Brisson, H.J., ornithologist, helps with planning, xxviii–xxix

Brossard, Maurice de, viii–ix, xliii, lxvi, cxci, ccvii, 404n

Brossard, Pierre de, biography, lxvii–lxviii; praised by DeL, 484, 497; promotion, 497, 528

Broudou, Abraham, LaP's father-in-law, xlviii

Broudou, Eléonore, see Eléonore de la Pérouse

Broudou, Elisabeth, LaP's sister-in-law, xlviii, lxvii

Broudou family, xlvii, lxvii

Broudou, Frédéric, LaP's brother-in-law, biography, lxvii, 114n; praised by LaP, 500, 525; promoted, clxiii, 146, 479, 481–2

Bruat, at Vanikoro, ccxxiv, ccxxvii

Buache, Philippe, geographer, xxiii, xxxiv, 95n

Buache de Neuville, Jean-Nicolas, hydrographer, assists with planning, xxii, xxix–xxx, xxxvii, 156n

Buckhardt (Bushart), Martin, lands on Tikopia, ccxix–ccxx

Buffon, Georges-Louis Leclerc de, naturalist and writer, xviii, ccii, 30; gives his advice on the expedition, xxviii, xxix–xxx, xc; advises Thouin and Collignon, lxxxiv; recommends Lamartinière and Du Bosc, lxxxix; on the natives of Tartary, clxxix; on the otter, 107; and dogs, 135; and Kamchatka puffin, 155

Burriel, Andrés Marcos, historian of California, 170n; refutes Fonte theories, xxxiv–xxxv, 95n

Bustamente, José, ccviii

Byron, John, voyage of 1764, xxxv, cxxii, 213, 384n, 385

Cabo Virgenes, Strait of Magellan, 35, 471

Cahors, France, wine, 455, 459–60, 520

Calcutta, English company at, 507; authorities organise search voyage, ccxx

California, shown as an island, xxxiii, LaP in, ccxxxiv–ccxxxv, 167–74; description, 169, 172; LaP's opinions on clviii, clxix, cciv; reports on, ccxx, ccxxxv, 491; (see also: American Indians, missionaries, Monterey)

Callander, John, urges Pacific exploration, xix

Callao, Peru, xxxiii, cxvi, 94n

Canton, China, Marchand at, ccxxxiv; in LaP's instructions, cxxxi, cxxxix, cxl; LaP comments on trade at, 219, 220, 224–5, 363, 502–4; mentioned,

INDEX

cxviii, cxix, clxxii, 190n, 214n, 224–5, 244, 501n, 503–4, 508, 515
Cape Bojeador, Luzon, 252
Cape Buache, Northwest Coast, ccxxxiv, 156
Cape Choiseul, Solomon Islands, 520
Cape Circumcision, discovered by Bouvet, cxiv, cxxxiii, 5; proposal by LaP to investigate, 467, 476, 517, 521
Cape Crillon (Mys Kriljon), Sakhalin, clxxxiii, 318–19, 322–4
Cape Edgecumbe (Cape Engaño), Alaska, 147, 150–1
Cape Engaño, Alaska, 147, 150–1
Cape Fairweather, Alaska, 101
Cape Flattery, Washington, 160, 161n
Cape Fleurieu, Vancouver Island, clxv, ccxxxiv, 157–8, 165
Cape Hector (Cape St James), Queen Charlotte Islands, British Columbia, named, clxv, ccxxxiv, 155; LaP sails past, 152, 156–7, 165, 494, 515
Cape Horn, rounded, cliv, 36, 38, 42, 367, 473, 480; LaP's opinion on, cliv, 38, 473; effect of temperature on timekeepers, 53; mentioned, cxv, cxxvi, cliv, ccxxxii, 31, 33, 54, 84, 93, 462, 467, 470
Cape Lamanon, originally Pic Lamanon, Sakhalin, 285
Cape Mendocino, California, 163n, 171n
Cape Nabo, Japan, 266, 324, 514
Cape Noto, Japan, clxxviii, 269–71
Cape of Good Hope, cxxxiii, cxli, cxlv, cl, ccvi, 477, 517, 521
Cape Rollin, Marikan Island, 326–7
Cape Verde Islands, LaP cancels plan to call at, 18–19; Dupetit-Thouars at, ccxiii
Carmel, Franciscan mission at Monterey Bay, clxvi, 174, 177; LaP visits, 178–88; destroyed, ccxxxv
Caroline Islands, cxxxii, cxl, 146, 383n, 421, 475, 492, 516–17, 520–2
Carpentaria, Gulf of, Australia, in LaP's later plans, 476, 517, 521, 542
Carteret, Philip, navigator, cxxiii, cxxvii, ccxvii, 6, 433, 518
Carvajal, Ciriaco Gonzalez, intendant of Manila, 245, 248, 506
Castricum (*Kastricum*), Vries's ship, clxxxiii, 273, 276, 314, 323, 513; Cape Castricum, 325

Castries, Anne-Jean-Jacques-Scipion de La Croix de Vagnas, lxxii, 246–7, 507, 508–9, 525
Castries Bay (De Kastri), Tartary coast, clxxxii; French reach, 302, stay in, 303–17, 319, 512; description, 302, 314–16; natives, 306–12, 332–3, 375; fauna, 314, 333, 335
Castries Canal, Castries Strait, see La Pérouse Strait
Castries, Charles-Eugène-Gabriel de, and planning of expedition, xxi, xxvii, xxix, lvii-lviii, xciv, xcv, xcvi, xcvii, 8–9, 305; LaP meets, lii; approves LaP's marriage, lvii-lviii; proposes Lesseps as interpreter, xciii; farewells the expedition, cli-clii; commends LaP, liii, 10; letters to, 453–4, 455, 457, 460, 463, 468, 470, 476, 477, 479, 480, 482, 485, 487, 489, 490, 496, 498, 500, 501, 503, 505, 508, 510, 520, 522, 523, 524, 526, 529, 531, 532, 533, 534, 535, 541; mentioned, lxxix, lxxxv, clxxxv
Cavite, Manila Bay, LaP at, clxxv, 231–48, 506, 515; details on, 234–5
Chabert-Cogolin, Joseph de, hydrographer, xxii
Chaigneau, Eugène, French consul, ccxx
Chandernagore, French India, lxxiv; LaP at, xlvi
Cheju-do, see Quelpaert Island
Chichagof Island, Northwest Coast, Chirikov at, 101n; Spanish at, 103n
Chile, call at, cxv, cliv-clvi, clviii, 40–52, 59, 111, 474; report on, ccxxx
Chiloe Island, Chile, cxvi, 44
China, LaP reaches, 202, 213–14, 261–2, 367, 464, 509; LaP's opinion on, cciv, 215, 219–21, 227, 241, 507; closed world, clxxvi, ccxxxvii, 95n, 218, 222n, 227, 254, 331, 376n; trade possibilities, cxxxix, clxxii, clxxv, ccxxxvi, 220n, 242, 504; fur trade in, cxii, cxxxviii, cl, ccxxxiii, ccxxxiv, ccxxxvi, 146–7, 172, 189, 293n, 373, 376n, 533; conquest of Tartary, 278, 280n, 294n
Chinese, reported landings on Northwest Coast, 147–8; sailors recruited by LaP in Macao, clxxiii, 227, 293, 296; in Manila, 236, 241, 249; garrison on the Pescadores, 257; sailor killed in Samoa, 410

593

Chirikov (Tchirikov), Aleksei Ilich, explorations in northern Pacific, xxxv, 96n, 101n, 151–2, 359n, 360, 372; — Bay (Chatham Strait), 151; Cape — (Cape Ommaney), 151–2
Chotka (Chiocka), name used for Sakhalin, 338, 340, 343, 513–14
Christian Sound, Alaska, clxiv, 151n
Christmas Sound, Tierra del Fuego, cxv–cxvi, cxxvi, 467
chronometers, instructions, cxlii–cxliii; tested at Brest, 11–12; at Madeira, 16; at Talcahuano, 47, 53, 227; on Northwest Coast, 99, 150; at Monterey, 195–7; at Macao, 227; at Cavite, 248, 376; at Castries Bay, 304–5; at Langle Bay, 317; at Petropavlovsk, 376; at Tonga, 440–1; at sea, 20, 76, 93, 385; praised, 270, 484
Clerke, Charles, captain of the *Resolution*, 203n; LaP cuts track of, 203, 207, 382; in Kamchatka, 351; LaP visits tomb of, clxxxiv, 359–60
Clonard, Robert Sutton de, biography, lxviii–lxix; first meets LaP, lxviii; meets Lauriston, lxxiv; sails in *Guyane*, lxxiv; wounded at Mahé, xlvii, lxviii; oversees the preparations at Brest, lxii, civ, 11; reports on supplies, 474; at Easter Island, 65–6; buys a helmet at Maui, 90; work on the Northwest Coast, 93, 100, 123; attempts rescue in Lituya Bay, 115–16; lands at Langle Bay, 287; lands on Sakhalin, 298; goes ashore at Tutuila, 392–3; attempts to land on Norfolk Island, cxcii, 443–4; at Botany Bay, cxcvii; takes over the *Astrolabe*, lxix, 443, 536, 541, 547; praised by LaP, 499, 524, 541; promoted, clxxxv, 499; mentioned, clxxxi, 502, 505
coconut, at Asunción, 210–12; in Samoa, 387, 391, 393, 394, 403, 411, 417, 425, 429, 431; in Tonga, 430, 438, 440
Cocos Island (Tafahi), cxc, 416, 427–9, 440, 540
cod, found in Ternay Bay, 277, 279, 281; along the Tartary coast, 282, 283; in the Strait of Tartary, 297, 299; in Kamchatka, 353
Collignon, Jean-Nicolas, gardener, works with Thouin, lxxxiv, ccxxxv; sows plants at Easter Island, clvi, 61, 70; climbs at Lituya Bay, 132; identifies plants at Monterey, 193; sows plants at Monterey, clxix; 192; goes ashore at Suffren Bay, 282; at Langle Bay, 287; captures a seal at Castries Bay, 316; wounded, lxxxiii–lxxxiv, clxxxiii, 316; goes ashore at Tutuila, 426; relations with Lamartinière, lxxxiv, lxxxix, 8–9
Collinet (Colinet), biography, lxix–lxx; in Samoa, 392, 399; sick, 397, 406; wounded at Tutuila, clxxxix, 401, 408, 410–11; praised by LaP, 500; promotion, 479, 499, 524
Commerson, Philibert, naturalist with Bougainville, views on Tahiti, cciii, 51n
Concepción, Chile, call at, cliv–clv, clxvii, ccxxii, 40–52, 474–6, 477–9; new site, 41–4; festivities, 48, 50–1, 478–9; LaP comments on missionaries, 47, 51, 175; Dufresne refused passport, 469; letters from, 468, 470, 476, 477
Condorcet, Marie-Jean-Antoine-Nicolas Caritat de, mathematician, approched re scientific aims, xxvii–xxix; invites Lamanon to join, lxxxviii, 456n; letter from Lamanon, 461–3; mentioned, clxxxii, cxcviii, 487
Cook Islands, 385n
Cook, James, his ships, xcv; contacted by Latouche-Tréville, lxxix, 155n; similarity in career with LaP, xli, ccxxix, ccxl; LaP to continue his work, xxiii, xxv–xxvi, c, cix, cxv, cxvii–cxviii, cxx, clxv, cvxxx, 452, 491, 511, 515; and Joseph Banks, ccxxx, 9; and the Forsters, ccxxx; and scurvy, c–ciii, cv; and South Georgia, 33n; at Tierra del Fuego, 31n, 36; and Davis Land, 55; at Easter Island, cxvi, cxxxvi, ccxxxii–ccxxxiii, 55, 59, 62–4; at Hawaiian Islands, 80, 81n, 84, 86–9, 91, 404; on Northwest Coast, clx, 95–6, 101, 137n, 144n, 147, 150–1, 155n, 160–1, 491; at Tonga, ccxvii, 436–41, 538; in New Zealand, ccxxxviii; at Norfolk Island, 443–5; at Botany Bay, 446, 566, 569; at Cape of Good Hope, 477; killed, xx, clvii, clviii–clix, ccvi, ccxxviii, 85, 400n, 404; French transla-

INDEX

tions of his Voyages, xliii, 34n; high regard in France for, 7-8, 78; compared with Kerguelen, 7; with Maurelle, 517; LaP praises, 78, 189, 440, 511, 515, 520; LaP follows his watch system, 25; his pig salting methods, 92, 246; his spruce beer recipe, 431n; other references to his voyages, xix, xxxvi-xxxvii, xlii, xliv, xci, cviii, cxi, cxii, cxvii, cxix, cxx, cxxi, cxxiii, cxxv, clxix, clxxxiv, cxci, ccxxxvi, 3, 6, 31n, 34, 35, 54, 76, 80, 84, 140, 167, 189, 228, 276, 308, 329, 346n, 347, 357-8, 364, 372, 380, 388, 396, 401n, 413, 418n, 430, 432-3, 435, 465, 467n, 472n, 491, 502, 510, 512, 539, 572
Cook's River, Northwest Coast, cxxxi, 374, 502, 515
Cook Strait, New Zealand, cxxx, 492n
Copper Island, Siberia, 364, 531, 533
copper, on Northwest Coast, 105-6, 132, 139-40; inland, 148n
cormorants, as indication of land, clxxxvii, 378; in Lituya Bay, 130, 131n; in California, 173; in Siberia, 315
Cossigny, David Charpentier de, army officer, 216, 504
Cowley, William Ambrose, buccaneer, cxiv, 33n
Coxe, William, author, 190n, 364, 373
Crillon, Félix-François Berton de, 109n, 149, 318
Crillon Bay, Sakhalin, 320, 335
Cross Sound, Alaska, sighted, clxiv, 149-52, 157, 165, 166, 514-15

D'Aigremont Pépinvast, Prosper-Philippe, biography, lxv-lxvi, meets LaP during American war, lxv; accommodation, clii; on Northwest Coast, 100; moderately praised by DeL, 468; re-appraised, 483, 496-7; death, lxvi, clxxv, 246, 509, 525
Dagelet, Joseph Lepaute, astronomer, sails with Kerguelen, lxxxiv-lxxxv, recommended by Borda, xxix; by Lalande, lxxxiv; appointed 8; Guéry as his assistant, lxxxvi; Roux as his assistant, 525; helped by officers, 34, 469; role on land, 458; observations at sea, 76; compared with Lamanon, lxxxviii; with Monge, xc; in Paris, 53; at Brest, 11; at Madeira, 16; at Tenerife, 458; at Talcahuano, 42, 52; at Easter Island, 61, 70; takes bearings at Maui, 91; calculates height of Mt St Elias, 99; calculates position of Lituya Bay, 127; explores Lituya bay, 110-11; Mount Dagelet named, 109n; latitude of Cape Fleurieu, 158; at Monterey, 195-6; determines position of Necker Island, 205; of French Frigates Shoal, 207; observations at Macao, 217; at Cavite, 248; off the coast of Korea, 264; discovers Ullung-do, 267; named, clxxviii; sets up observatory in Castries Bay, 304; checks positions on Tartary coast, 317; draws up inscription for Delisle memorial, 359; observations at Petropavlovsk, 376; at Traitors Island, 430, at Tonga, 434, 436, 440-1, at Vavau, cxci; at Pylstart, 441, 442n; in Botany Bay, cxcvii; health, 525; LaP praises him, lxxxv, clxxi, 4, 487, 525, 539
Dagelet Island (Ullung-do), clxxviii, ccxxvii, 267-8
Dalmas, François-Marie-Léon de Lapérouse, xii-xiii
Dalrymple, Alexander, believer in southern continent, xix; and Easter Island, 54; chart, 214n, 228
Dampier, William, English navigator, 55, 213, 257n
Danaïde, hears rumours of a French boat at Ponape, ccxxviii
D'Angelis, Giacomo, (Fr des Anges), 276, 336
Dangerous Islands (Pukapuka), 385
D'Après de Mannevillette, Jean-Baptiste-Nicolas-Denis, cartographer, 23, 29, 214n, 228, 229, 232, 256, 258, 466
Daubenton, Louis, naturalist, helps with planning, xxviii
D'Auribeau, Alexandre d'Hesmivy, D'Entrecasteaux's officer, ccxv, ccxviii
David, Edward, English buccaneer, 54-5
Davis Land, mythical island, clvi, ccxxxii; LaP's views on, 54-5
De Brosses, Charles, historian, xix, 31n, 79n, 202n, 382n, 417
De Langle, see Fleuriot de Langle

Delattre, François-Pascal, politician, ccx
Delisle, Joseph-Nicolas, geographer, 259; and Fonte's voyage, xxxiv-xxxv, 95n
Delisle de la Croyère, Louis, death, 357n; LaP visits tomb of, clxxxiv, 359–60, 530; LaP names islands after, 152
Delphin, Thomas, French merchant in Concepción, 476
D'Entrecasteaux, Joseph-Antoine Bruny, French navigator, biography, ccxiii, 216; helps Fleurieu with expedition plans, ccxiv; expedition to Macao and the Philippines, lxxii, lxxxii, clxxii, ccxiv, 247, 504, 507; at Canton, clxxvi, 246–7, 507–8; searches for LaP, xciv, ccix, ccxii-cxxix, 543; problems with scientists, ccxxxi, at Tonga, ccxxiii; passes near Vanikoro, ccxviii, ccxxviii; completes LaP's work, ccxix; death, ccxviii
Dentrecasteaux Islands, ccxxxix
Descartes, René, philosopher, cxcix
Descubierta, Malaspina's ship, ccviii
D'Escures, Charles-Gabriel Morel, biography, lxx-lxxi; first meets LaP, lxx; supervises fitting out, xcvi, 11; at Madeira, 16; at Easter Island, 60; explores Lituya Bay, 104–5, 112–14; death, lxxi, clxii-clxiii, 114–15, 117, 497, 499, 524, 528; LaP criticises, 112, 116, 119n, 481; sister, 481
Des Anges, Fr, see D'Angelis
Des Mazis, Alexandre-Jean, lxx
Des Moulins (Dumoulin), Fougoux, French official at Canton, 503–4, 508
D'Estaing, Henri-Charles-Hector-Théodat, French admiral, 294
D'Estaing Bay, Sakhalin, clxxxi, 298, 322; named, 295; French land at, 296–7
D'Hector, Charles-Jean, Comte, commands at Brest, lxi, 10, 543; chairs Kerguelen court-martial, lxi; during American war, lii, lxxviii; DeL weds his niece, lxi; and LaP expedition preparations, lxii, xcvi-xcvii, xcix, civ-cv, clxv, 10, 455; and departure of expedition, clii, 11; LaP names a cape after him, clxv, 155
Diderot, Denis, philosopher, xviii
Diego Ramirez Island, 39, 473

Diligent, Dupetit-Thouars sails on search expedition in, ccxiii
Dillon, Peter, discovery of wrecks, ccxix-ccxxiii, ccxxiv, ccxxxix; brings LaP relics to Paris, xciv, ccxxii; decorated, ccxxii
Discombe, Reece, searches at Vanikoro, ccxxv-ccxxvi, 570
Discovery, Cook's ship, 7, 202, 266n, 346n, 355n, 357n, 359n, 363n
disease, veneral, detected, 12, 467; other, 470, 481, 485, 490; (see also: scurvy, venereal disease)
Disgraciada, rumoured island, 79
Dixon Entrance, Northwest Coast, 153–4
Dixon, George, navigator, clxix
Dodds, Arthur, amateur geographer, and North-West Passage, xxxiv-xxxv, 95n; his New Albion, xxxv
Dolphin, Byron's ship, xxxv, 359n, 385n, 440
Dorset, John Frederick Lord, British ambassador, xxvi, cli, clxiii
Drage, Theodore Swain, mapmaker, xxxiv
Drake, Francis, xxxi, cxvi, 39–40, 473n, 473–4
Drake's Island, cxvi, cxxvi, 40, 473–4
Drake's Land, xxv, cxxvi, ccxxxii, 38–9, 473
Du Camper, Nourquet, French captain, 570
Du Chayla, Claude-Joseph, commands the *Africaine*, xliv; at the Ile de France, xlvii
Duché de Vancy, Gaspard, artist, lxxxvi, clxxx, 9, 110; at Easter Island, 61, 70; lands on Sakhalin, 287; drawings, 23, 26, 48, 62, 124, 137, 178n, 355, 361, 488, 511, 516; praised by LaP, 4, 505, 515, 523; possible survival, lxxxvi
Dufourny, Léon, architect, 9
Dufresne, supernumerary naturalist, lxxxvi-lxxxvii, cli, 9, 469, 477, 489, 500–1; accommodation on board, clii; at Concepción, clv, 469; lands at Easter Island, 61, 70; sells the furs, 146–7, 495, 497, 501–2; left at Macao, lxxxvii, ccxxxi; takes despatches, 505–6, 512, 522, 523; report on fur trade, clx, ccxxxiii, 147n; praised by DeL, 483

INDEX

Du Halde, Jean-Baptiste, missionary and writer on China, 282n, 283n
Dumont d'Urville, Jules-Sébastien-César, expedition, ccxxi-ccxxv; searches for LaP, ccxxi-ccxxiiv; in Tonga, ccxxiii; at Vanikoro, ccxxii-ccxxiv, ccxxxix, 570; opinion on wreck, ccxxvi
Dunkerquoise, at Vanikoro, ccxxv
Duperrey, Louis, French navigator, ccxix, ccxxii
Dupetit-Thouars, Abel, French navigator, 178n
Dupetit-Thouars, Aristide-Aubert, search for LaP expedition, ccxi-ccxii
Dutch, in Formosa, 253-4, 256; prisoners in Korea, 263; explorations in Northwest Pacific, 273, 277, 285, 314, 323-5, 337, 377, 510, 513

Easter Island, call suggested by Louis XVI, cxci; in instructions, cxxvi, cxxxvi; LaP calls at, clvi-clvix, cciii, ccxxxii, 53-75; 383, 473, 491, 509; thefts, 60, 65, 66, 73, 568; statues, 57, 61-3, 71, 488; islanders compared with Hawaiians, 85; with Samoans, 390; LaP reports on, ccxxx; DeL reports on 70-4; other, 417n
Edwards, Edward, captain of the *Pandora*, sights Vanikoro, ccxvii, ccxxviii
Eendracht, Schouten's ship, 429n
Ellis, Henry, English explorer, 5
Endeavour, Cook's ship, xliii, 36n; journal of, adapted, 572
Endeavour Strait, Australia, 476, 492, 541
Engageante, La Jaille's ship on Hudson's bay raid, liv-lvi
Eskimos, LaP distinguishes from Tlingit, 138-40; 143; distinguishes them from Kamchatdales, 365; compared with Californian indians, 189
Espérance, Kermadec's ship, ccxiv-ccxviii
Espérance, Du Camper's ship, 570
Esperance Rock, ccxxxix
Etoile, Bougainville's ship, cii, 6

Fabert, at Vanikoro, ccvi
Fagés, Pedro de, governor of California, clxvii-clxix, 176-7, 180n, 181, 189, 193

falcon, in California, 164, 173
Falkland Islands, and Bougainville, xvii, 93n; on LaP's itinerary, cxv, cxxxvi; mentioned, cliv, 39, 474
Farallon de Medinilla Island (Abreojos), Marianas, 382
Favorita, Spanish ship at Monterey, 176
Fawcett, John, dramatist, ccxix, ccxl
Feuillet, Louis, scientist, 52
Finau Ulukalala Feletoa, Tongan chief, 439-40
Flassan, Joseph-Ignace Raxi de, biography, lxxi-lxxii; accommodation on board, clii; report on Lituya Bay, 102-4; praised by DeL, 468-9; death, clxii, 481-2, 499, 524
Flavie, proposed for a search expedition, ccxi
Fleurieu, Charles Claret de, biography, 10n, 511; work at Madeira, 15; LaP first meets, l; friendship with LaP, 452; plans Hudson's Bay raid with LaP, liii, liv; meets William Bolts, xxi, xxii; draws up plans for an expedition, xxii, lvii-lviii, ccxxxiii, 10; plans LaP voyage, xxiv-xxx, xcv, 382; appoints Lamatinière, lxxxix; negotiates with Lamanon, lxxxviii; LaP complains to, clxxi; attitude towards natives, clxxxviii; LaP names cape after him, clxvi, 157; acts to organise search for LaP, ccviii-ccix, ccxiii-ccxiv, 573; appointed Minister of Marine, ccvii, ccxiii, 543-4, 573; action over master rolls, 543, 545; asked to edit LaP's journal, 572; letters to, 455, 458-60, 464, 493, 509, 512, 530, 536, 571; mentioned, xci, xcii, cvi, cviii, 156n, 196
Fleuriot de Langle, Georgette-Marie-Françoise, née De Kérouartz, wife of Paul-Antoine, lxi, 155
Fleuriot de Langle, Paul-Antoine-Marie, biography, lix-lxv; protégé de Duc de Chartres, lx-lxi; meets LaP, lix; service during American war, lx-lxi, 10;commands the *Astrée* on Hudson's Bay raid, liv-lvi, lxi, 10; marriage, lxi, lxii; appointed to LaP expedition, lxi, 10; his distilling stove, xcvii-xcviii; plans for a mill, 94; on Lavaux and Guillou, lxxx; care over recruitment, civ; accepts sick servant, 246; final

arrangements in Brest, clii, 11; urges Monge to leave, xc, 458; at Easter Island, ccxxxiii, 58–62, 66, 70–4; forgets to warn off islanders at Maui, 82–3; goes ashore at Maui, 85; favours anchoring in Lituya Bay, 103; attempts rescue at Lituya Bay, 115–16; distress at losses incurred at Lituya Bay, 115; gives a mill to Carmel mission, 182, 485; lands at Asunción, 210–12; goes ashore at Macao, 217; visits Manila, 236, 248; attempts to hunt in Ternay Bay, 279; lands on Sakhalin, 287–93; lands at D'Estaing Bay, 296–7; in Castries Bay, 302–17; in Avatcha Bay, 368; lands on Tutuila, 391; discovers a watering cove, 392–3, 396, 406; death, lxiii-liv, clxxxix, cciii, 398–400, 408, 410, 535; burial, lxv;work done on astronomy , 197n, 487, 539; opinion on missionaries, clxvii-clxviii; theories on water, xcviii, clxxxix, 396–7, 536, 539; character, lxii-lxiii, 403; bay named after him, clxxx, 512; peak named after him, 337; praised by LaP, 4, 403, 500, 522–3, 525, 541; fondness and praise for LaP, 458, 485, 528; letters from, 454, 457, 468, 482, 496, 526, 530–2; mentioned, lxvii, lxviii, lxxiii, 505

Flinders, Matthew, English navigator, asked to look for LaP, ccxix

Fonte (Fuentes), Bartolomeo de, alleged voyage by, xxxiii-xxxvi, xxxviii, 95n, 148, 165, 331n

Fonualei Island, Tonga, Maurelle's Amargura, 433–4

Formosa (Taiwan), in instructions, cxxxi, 474, 506; LaP sails close to, clxxvi-clxxvii, ccxxxvii, 252–61; anti-Chinese rebellion, clxxvi, ccxxxvii, 253–5, 507; natives, 421; LaP reports on, clxxvi-clxxvii, ccxxx, ccxxxvii, 511, 534

Forster, George and Johann Reinhold, naturalists with Cook, ccxxx, 34, 59, 61–2, 440n, 515n

Fortune, Kerguelen's ship, xlii

Fort York, see York Factory

Fouasse, François-Maximilien, swordmaker, ccxx

Franciscans, see missionaries

French Frigates Shoals, discovered, lcxx, ccxxx, ccxxxvi, 206–7, 382, 490, 492

Freycinet, Louis de, French navigator, 389n; and scientists on ships, ccxxxi; asked to look for LaP, ccxix

Frézier, Amédée-François, army engineer, 22, 26, 41

Friendly Islands, see Tongan Archipelago

frigate birds, sighted, 19, 78, 204, 212, 384; mentioned 445

Frobisher, Martin, xxxii

Fuca, Juan de, claims to have sailed through Anian, xxxii, xxxiv, 95n; — Strait, xxxv-xxxvi, xxxviii, cxvii, clxv-clxvi, 161n, 164n

Fuentes, B. de, see Fonte

fur trade, as early motivation to LaP expedition, xxi, cxii, cxvii-cxix, ccxxxiii; in instructions, cxxxvii-cxxxix, cl; Dufresne's role in, clv, clxxii, 501–2; in Alaska, clx, clxi, 107, 128, 140, 495, 502, 533; in California, clxviii, 172, 189, 503; in China, cxii, cxxxviii, cl, clxviii, clxxii-clxiii, 225, 244–5, 501–2, 533; reports on, ccxxx, 147n, 490; various references to, ccxi-ccxii, ccxxxiii, 188, 337n, 346n, 347n, 362n, 363n, 367, 507

Galapagos Islands, cxvi

Galaup, LaP's family, xii-xiii, xxxix, xlviii, lvii

Galaup, Victor-Joseph, LaP's father, xxxix, xlviii, l

Galvez, Bernardo de, Viceroy of Mexico, 177n, 190, 434n

Gama, Antonio de, Portuguese officer, 29

Gama, João de, Portuguese navigator, 377n

Gaubil, Antoine de, Jesuit missionary and writer, 259–61, 336

Gaulin, Jean, master gunner, rewarded, 484, 558

Gaziello, Catherine (Mme Hustache), and LaP journal, viii; and purpose of the voyage, xx; and Bolts plan, xxii, and Monneron's role, xci; and Lituya Bay wreck, xcix

Gente Hermosa (Rakahanga), Cook Islands, 385, 440, 540

Gentille, Mengault's ship, li

INDEX

Gilbert, Humphrey, xxxi–xxxxii
Gironde Estuary, France, xcix, 256
Goa, India, clxxii, 217–18, 222–4, 238
Gonzalez y Haedo, Felipe, Spanish navigator, 60n, 63n
Good Success Bay, Tierra del Fuego, agreed as rendezvous, 31, 472; avoided, 36–7
Goos, Abraham, mapmaker, xxxii–xxxiii
Gore, John, captain, 225, 472
Gough Island (Tristan da Cunha), cxiv, cxxxiii
grapefruit (rima, shaddock), traded in Tongas, 425, 429
Grenier, Jacques-Raymond de Géron de, xlv, l
Grosset, Jean, sailmaker, praised, 484, 559
Gros-Ventre, St Allouarn's ship, xlii–xliii
Guam, Marion du Fresne's expedition at, xliv; LaP planned to call at, clxxxv, 516–17; mentioned, 209n, 518n, 520
Guéry, Pierre, watchmaker, lxxxv, lxxxvii
Guignes, Joseph de, historian, 147–8, 508
Guillou, Jean, assistant-surgeon, lxxviii, lxxx, ccxxvii; praised by DeL, 497, 527
Guimard, Madeleine, dancer, 356
gulls, sighted, 53, 130, 131n, 212, 281, 315; as indication of land, 53

Hainel, Anne-Frédérique, dancer, 356
Halley, Edmund, astronomer, 23, 41n, 208
Hamon, sailors, brothers, 545, 560
Haro, Gonzalez Lopez de, Spanish captain, clxix
Harrison, John, watchmaker, cviii
Huaping Hsu (Hoapinsu), 261
Haüy, crystallographer, helps with planning, xxviii
Hawaiian Islands (Sandwich Islands), and Cook, clvii, ccxxxiii, 76n, 491; and Spanish, 83–4, 87; LaP at, ccxxxiii, 80–91, 480, 509; supplies obtained, 82, 92; LaP comments on islanders, 81, 83, 390, 421; compared with French West Indies, 91; mentioned, cxvi–cxviii, cxxx, clviii, clxix, 76–7, 202, 383, 475, 493

Hawkesworth, John, 3n, 34n, 572
Hearne, Samuel, explorations by, xxxvi, 148, 166n; taken prisoner by LaP, lv–lvi; humane treatment by LaP, cix
Hermione, sails to America under Latouche-Tréville, liii, xcv
Hezeta, Bruno de, Spanish navigator, 103n
Hodges, William, painter on Cook's second voyage, 56, 61, 488
Howe, Richard Lord, learns of proposed voyage, xxv
Hudson's Bay, possible way through to the Pacific, cxvii, cxxx, cxxxviii, 472; raid by LaP during American war, lii–liii, liv–lvi, xci, cv, 9; whale oil, 65; size of trees, 128; seals, 189; plaque, 570
Hungfou Shu (Koto Sho), LaP's Bottol Tabagoxima, 257–8
Hunga Haapai and Hunga Tonga islands, LaP at, 437–8, 538
Hunter, Dillon on, ccxix, ccxx
Hunter, John, English captain, cxcvi, ccxv–ccxvi, 447
Huygens, Christiaan, astronomer, cviii

Iphigénie, coastal vessel, LaP commands, to Madagascar, xlix
Ile (Isle) de France (now Mauritius), Bougainville at, cii–ciii; LaP at, xli–xlix; compared with Easter Island, 58–9; intended final place of call for LaP expedition, cxx, cxxxiii, ccvii, ccx, 464, 476, 492, 497, 499, 503, 516–17, 522, 527, 540, 542–3
Isle Grande, island believed to have been discovered by La Roche, search for, xxiv, cxiv, cxxv, cxxxv, cxxxii, 31–3, 467, 470–1
Iturup, Kuril Islands, 273n, 276n, 324n, 366n
Ivashkin, Pieter Matteios, 357–9

Jaolsissima (Nanatsu-shima) island, 269–70
Japan, exploration of coast and islands, cxviii–cxix, cxxxi–cxxxii, cxxxviii–cxxxix, cxl, cl, clxxvi–clxxix, 146, 264, 284n, 331, 492, 506, 510, 523; piracy near, xcl, 220n; LaP off the coast, 268–71, 276, 319; closed to foreigners, clxxvii–clxxix, ccxxxvii,

275, 331, 337n, 376n, 512–13; trade with Sakhalin, 289, 295, 321–2, 335
Java, cxxix, 241, 423
Jesuits, in America, 170, 187–8; in the East, 209, 221n, 235, 259n, 264, 282, 331–2, 336n; laP educated by, 575
Johnston, William, English merchant at Madeira, 13–15
Juan Fernandez Islands, LaP decides to make for, 36, 38, 472–4; changes his mind, 40, 52, 474; mentioned, 54
Juan y Santacilla, Jorge, scientist, 42
Jussieu, Antoine de, botanist, lxxxix, 8

Kaborov, commander of Petropavlovsk, 347–9, 352–3, 372, 374, 376
Kaempfer, Engelbrecht, doctor and naturalist, 336
Kaloohawe, Hawaiian island, ccxxxii, 90
Kamchatka, LaP makes for, 327, 492; exploration of the coast, lxii, lcxxiv-lcxxxviii, 146, 158, 286, 303–18, 345, 476, 510; French call, xciii, cv, clviii, 328–9, 443, 502; LaP comments on inhabitants, 308, 346, 375; on the administration, 346–8, 354, 362, 372; and the economy, 352–4, 361, 362–3; English attempts at trade, 363–4, 533–4; letters awaited, 348, 464, 467
Kao, Tongan Islands, LaP at, 436–7
Kastricum, see *Castricum*
Kauai, Hawaiian Islands, clix, 81
Keppel, see Traitors Island
Kerguelen, Yves-Joseph de, voyages, cv, cxxi, 573; ships, xcv, believes himself the discoverer of the southern continent, xix-xx, xlii, 6–7; at Ile de France, xli-xliii, xlv; court-martial, xliii, lxi, 10; Dagelet sails with, lxxxiv
Kermadec, François Huon de, Villeneuve's uncle, lxxii
Kermadec Islands, discovered, ccxxxix
Kermadec, Jean-Michel Huon de, sails on search expedition, ccxiv; in *Résolution*, ccxv; death, ccxvii-ccxviii
Kermel, Jean-Marie, accidental death, 526–7, 562
Kerouart Islands, Northwest Coast, ccxxxi, 155–6
Kiakhta, Mongolia, 167, 190, 337n, 363–4, 373, 502, 533
King, James, captain, clxxxiii, 77, 88, 225, 228, 266, 269, 271, 329n, 357n, 512
Kodiak Island, Alaska, clx, 140n, 372, 373n
Korea, mentioned in instructions, cxxxii, cxxxix; LaP along the coast of, 264–8, 276, 492, 510; description, 265; closed to foreigners, clxxvii, ccxxxvii, 263, 337n; piracy near, xcl
Kotzebue, August von, dramatist, ccxix, ccxl
Kozlov-Ugenin, Grigor, governor of Okhotsk, 347–8, 352–7, 359, 360–1, 363–5, 367–8, 375, 534
Krashenninikov, Stepan Petrovich, geographer, 349n, 364n, 376
Kruzof Island, Northwest Coast, Spanish at, 103n, 147n
Kume Shima, 260
Kunashir, Kuril Islands, 366n
Kuril Islands, mentioned in instructions, cxxxi, cxxxviii, cxxxix, cxl; Dutch exploration of, 273n, 324–5; LaP sails towards, clxxxiii, ccxxxvii, 284, 326, 492, 513, 516; LaP sails through, 326–9; French outpost suggested, cxviii-cxix, cxxxix; origin of inhabitants, 338; native name, 290n; mentioned, 319, 332, 337–8, 366–7, 373, 377, 378n, 513, 520, 523

La Billardière, Jacques-Julien Houtou de, on D'Entrecasteaux expedition, ccxv, ccxviii, ccxxxix
La Borde, Jean-Joseph de, lxxii, lxxiii, clxiii, ccix, ccxi
La Borde de Boutervilliers, Ange-Auguste-Joseph de, biography, lxxii-lxxiii; accommodation on board, clii; reports on Lituya Bay, 102–4; explores the bay, 110; death, clxii-clxiii, 115, 481–2, 499, 524; praised by DeL, 469, 482
La Borde Marchainville, Edouard-Jean-Joseph de, biography, lxxiii; accommodation on board, clii; goes ashore at Santa Catarina, 28; on the Northwest Coast, 100, 110, 112; death, clxii-clxiii, 114–15, 116, 119–20, 481–2, 499; praised by DeL, 468, 482
La Croyère Islands, Northwest Coast, ccxxxiv, 152

INDEX

Ladrones (Thieves) Islands, see Marianas
La Jaille, Marquis de, commands the *Engageante*, on Hudson's Bay raid, liv-lvi, lxi
La Jonquière, Clément Taffanel de, commands the *Célèbre*, xl, 299n; LaP's protector, xlvi, l, 299n; influence on LaP, xxxix
La Jonquière Bay, clxxxi-clxxxii, 299
La Jonquière, Pierre-Charles Taffanel de, xxxix
Lamanon, Jean-Honoré-Robert de Paul, biography, lxxxvii-lxxxix; follower of Rousseau, cciii, 540; character, ccxxxi; appointed to expedition, 8; clashes with LaP, lxxxviii, cliii, clxxi, ccxxxi, 452, 457, 459; teased by sailors, ccxxxi, 459; work at Brest, 12; at sea, 13; at Tenerife, 17, 456-7; lands at Trinidade, 22; lands at Santa Catarina, 28; at Easter Island, ccxxxiii, 61, 70; work in Lituya Bay, 124, 130, 132; describes the otter, 107; writes text for Lituya Bay memorial, clxiii, 121; report on Tlingit language, 141-4; report on Californian Indian language, 197-200; ill-health at Cavite, clxxv, 246, 314; other references to his health, clxxxii, 349, 397; describes land crab, 211; work in Castries Bay, 314; at Avacha Bay, 368; projected book, 4; memoir on temperature and other reports, ccxxxviii, 462-3; letter to Condorcet, 461-3; letters to Castries, 455-7, 463, 456-7, 485; death, lxiv, lxxxviii, clviii, clxxxix, 401, 403, 406, 410, 535, 540
Lamare, François, boatswain, rewarded, 470, 483, 557
Lamartinière, Joseph-Hughes de Boissieu de, biography, lxxxix; recommended by Jussieu, 8; accommodation on board, clii; on leader's appointment, lxi; and Prévost, xcii, 469; clashes with Collignon, lxxxiv; at Tenerife, cliii, 17; at Trinidade, 21; at Easter Island, ccxxxiii, 61, 70; work in Lituya Bay, 129, 132; lands at Asunción, 210-11; at Macao, clxxi; at Suffren Bay, 282, 315; at Castries Bay, 315; goes ashore at Tutuila, 398, 401, 406, 426; reports by, ccxxxii, ccxxxv; peak named after him, 298; praised by LaP, 4; praised by DeL, 469, 483, 497, 527; mentioned, 488
La Mesa, rumoured island, 77-8, 209n
La Mira, rumoured island, 209, 492
land crabs, at Asunción, 211
Langle, see Fleuriot de Langle
Langle Bay (Tomari), Sakhalin, clxxx-clxxxi, clxxxiii, 286-95, 297, 298-9, 317-18, 322, 328, 512; meeting with islanders, 287-95, 296, 319; vocabulary, 339
Langle Peak, 318
La Pérouse, Eléonore, née Broudou, LaP's wife, meets LaP in Mauritius, xlviii; in Brittany, l-li; marriage, lvii; in Albi, lviii; separation from LaP, clii, clxxii; financial situation, ccxi, ccxix; pension, lviii; death; lix; in fiction, ccxl
La Pérouse, Jean-François de Galaup de, biography, xxxviii-lix; family name, xi-xiv; second name, xxii-xiii, xxxix-xl, 544; family, lxxiv; 544; accent, 339n; poor horseman, 15, 50; at the Ile de France, xli-l; influence of his stay at the Ile de France, xliv; sails to Madagascar, xliv, xlix; expeditions to India, xlv-xlvii; in American war, li-lvi, lxv-lxvi, lxxxii, 364n, 402n; meets DeL, lix; Hudson's Bay raid, lii-liii, liv-lvi, cv, cix, clvii, 93, 128; in Paris, l, lviii; marriage, xlviii, lvii; early planning of expedition with Fleurieu, xxii; aims, ccxxx; seeks Russian interpreter, xciii; instruments required, cviii; goes to Rochefort, xcv; Spanish co-operation, 177, 475; and the *philosophes*, clvi, clviii, cxc, cxcix, ccii, ccxxxvi, 133, 175n, 395, 430, 452, 507n, 540, 576; and Noble Savage theories, cxcix-ccvi, 67n, 133, 395, 429, 568; anti-clericalism, 459n; opinion on theoretical geographers, 5, 38, 133, 166, 473; views on trade, clxviii, 27, 46, 172, 190, 223, 237, 241-3; and fur trade, ccxxxiii, 146-7, 501-2; theories on water, xcviii, clxxxix, 396-7, 539; and scurvy remedies, ci-cvi, 93-4, 317, 431-2, 520, 537-9; his health, 452, 464; attitude and comments on native people, clvi-clviii, clxii, clxviii, clxxxviii, cxc, cxcviii, ccv, 8, 56, 57, 59, 67, 69, 85, 88, 111,

132–4, 137, 291–3, 306–7, 364–5, 394–5, 405, 413, 418–19, 427, 429, 430, 440, 486, 537, 539–40, 568; views on missionaries, 170, 174–5, 187–8, 238; high regard for James Cook, 78, 246, 510, 518, 521; compared with James Cook, ccxxix, ccxl, 93n, 390, 510, 515, 519, 521; LaP's opinion of DeL, 403, 498, 500, 522–3, 525, 536–7, 539, 541; LaP in literature, ccxix

La Pérouse, voyage, farewelled by Louis XVI, cli, 506, 510; departure from Brest, cli, clxxv, 11–12, 18; LaP urges Monge to leave, xc, 458; at Tenerife, cliii, 15–18, 20; clashes with Lamanon, cliii, clxxi, ccxxxi, 452; LaP goes ashore at Santa Catarina, 28; in Chile, 40–52, 477–9; at Easter Island, 55–69, 491; bay named after LaP, 55n; LaP at Hawaiian Islands, ccxxxiii, 80–91; goes ashore on Maui, 85–90; bay named after LaP, ccxxxiii, 83; exploration of Northwest Coast, clx–clxvi, ccxxxiii, 96–161, 367, 472, 475; LaP's disbelief in a North-West Passage, xxxvii, clxiv, 110, 148, 165; comment on Bering Bay, 101; discovery of Lituya Bay, 103; LaP meets Tlingit, 106; comments on Tlingit society, 132–3; sadness over losses sustained at Lituya Bay, 112, 481, 541, 564; LaP's instructions to D'Escures, 112–13; memorial erected, 121–2; mountain named after LaP, 109n; LaP in California, clxvii–clxix, 167–94, 491; arrives at Macao, 214; lands in China, 217; stay at Macao, 214–27; reaches the Philippines, 227; stay at Cavite, 230, 232–48; goes to Manila, 236–44, 248; report on Manila, 485n, 506, 534; report on fur trade, 502; LaP off Korean coast, 264–8, 276, 492, 510; off coast of Tartary, 274–85, 299, 302, 314, 317, 326–7, 331, 353, 371, 384; meets islanders on Sakhalin, 288–94; compares them with Tlingit, 292; LaP in Castries Bay, 304–17; reaches Petropavlovsk, 329; stay in Kamchatka, 346–69, 532; promoted, clxxxv, 357, 532, 544, 546; changes to his programme, 475–6, 491–3, 520, 541–2; in Samoa, 386–427; birds named after him, 389n; LaP's reaction to Tutuila massacre, 403–4, 427, 430, 537, 568; stay in Botany Bay, cvi, cxcii–cxcviii, ccvi, 446–8; district named after him, ccxxxix; proposed route from Botany Bay, cxcviii, 540–2; planned date of return, 476, 521, 540; possible survival, ccxxvii; letters from, 453, 455, 458, 460, 464, 470, 476, 479, 487, 489, 490, 493, 498, 500, 501, 503, 505, 508, 509, 510, 512, 520, 522, 523, 524, 529, 530, 531, 533, 534, 535, 536, 541, 571; places named after LaP, clxxxiii, ccxxxiii, ccxxxix, 55n, 83n, 109n, 318n

La Pérouse Strait (Castries Strait, Strait of Sangar), clxxxiii, 318n, 322–3, 335, 337, 510, 512

Laplace, Pierre-Simon de, astronomer, 462

Laprise-Mouton, Jérôme, biography, lxxiii–lxxiv; at Lituya Bay, 114, 120; in Castries Bay, 304; goes ashore at Tutuila, 397–401, 406, 408–9, 411; praised by LaP, 500, 525

La Roche, Antoine de, merchant and navigator, xxiv, cxiv, cxxv, cxxxv, ccxxii, 31, 33, 467

La Rochefoucault-Liancourt, François-Alexandre, Duc de, intervenes for Lamanon, lxxxiii

Las Mojas, see Los Mojos

Lasúen, Fermin Francisco, missionary in California, clxvii–clxix, 178, 182n, 188n, 192n

Late (also Lape), Tongan Islands, 435, 541

Latouche-Tréville, Louis de, proposes voyage of exploration, lxxix, 155n; in charge at Rochefort, xcv; bay named after him, 155–6

Lavaux (Lavau), Simon-Pierre, surgeon, lxxviii–lxxx, 313, 485; works on vocabulary, lxxx, 14n, 311–12, 319, 527; goes ashore at Tutuila, 398, 406; wounded, lxxxiii, 401, 410, 535, 564–5; praised by LaP, 525; praised by DeL, 470, 497, 527

La Villeneuve, Pierre-Louis Guyet de, biography, lxxii; joins the Boussole, clxxvi, 247, 509; praised by LaP, 525

Lavoisier, Antoine-Laurent, chemist, helps with planning, xxviii–xxix; executed, clxiii

INDEX

Law de Lauriston, Jean-Guillaume, biography, lxxiv; accommodation on board, clii; library, lxxiv; helps, then takes over astronomical work, lxii, lxxiv, cxc, 197n, 484, 488, 496, 498–9, 525, 539; in Lituya Bay, clxii, 109; in Castries Bay, 304–5; praised by LaP, 539; by DeL, 469, 483, 497, 527–8; promoted, 479, 483; death, xciv

Le Cor (Le Corre), Jacques-Joseph, second surgeon, lxxx-lxxxi

Ledru, Nicolas-Philippe, physicist, 9

Ledyard, John, urges French expedition, xxi

Le Fol, servant, dies, 481, 485, 526, 563

Le Gal, Robert-Marie, master carpenter, praised, 484, 558

Le Gobien, Pierre, biography, lxxv, joins the *Astrolabe*, clxxvi, 247, 509; goes ashore at Tutuila, 397, 399, 406, 411; wounded, 401, praised by DeL, 527–8; promotion mooted, 524

Le Maire, Jacob, Dutch navigator, 39, 418n, 473–4

Le Maire Strait, cxxvi, cliv, 31–2, 36–7, 53, 335, 471–2

Lemos, Bernardo Alexis, Portuguese governor of Macao, clxxii, 216–18, 226; his wife's friendliness, 217–18

Le Noir, Jean, police chief, xciii

Léon, Mathurin, senior pilot, praised, 484, 557

Le Roy, Pierre, watchmaker, cviii, 15n

Lesseps, Jean-Baptiste Barthélémy de, biography, xciii-xciv; personal file, 574; accommodation on board, clii; fluency in Russian, 347, 349, 357n, 366, 470, 529; attacked in Lituya Bay, 134; works on Tlingit language, 141; supplies information on fur trade, ccxxxiii; in Petropavlovsk, 347–69, 529; takes despatches to Europe, 511–12, 522, 529, 531, 535, 546; travels across Siberia and Russia, xciii-xciv, clxxxvi-clxxxvii, ccxxxviii, 367–9, 529; praised by LaP, 529; praised by DeL, 483, 497, 527; helps with plans for D'entrecasteaux expedition, xciv; identifies relics of LaP expedition, ccxxii

Lima, Peru, trade with Chile, 45; earthquake, 54

Lind, James, treatise on scurvy, ci, cii

Lisianski Inlet, Alaska, Chirikov in, 101n

Lituya Bay, Alaska, discovered, clxi, 102–3; French stay at, clxi-clxiii, ccxxxiv, 103–26; description, 109–110, 127–144; vegetation, 128–9; fauna, 130–1, 312; trade carried on at, 105, 107, 111; observatory set up on island, 107–8; thefts, 108–9, 133–4, 195; charts, 490; losses sustained, clxxiii, clxxx, ccviii, ccxxxvi, 111–22, 491, 493, 505; text of memorial, 121–2; island purchased, ccxxxiv, 110–11; suitable as an outpost, 128; mentioned, 101n, 166, 302, 500, 539, (for native people, see: Tlingit)

Locu Yesso (L'Oku Yesso), see Oku Yesso

Lopatka Point, Kamchatka, 327, 513

Loreto, Baja California, 170

Lord Howe Island, LaP suspects existence of, 446

Los Jardines, rumoured islands, 209, 492

Los Matelotes (Fais), Caroline Islands, 382n

Los Mojos (Los Majos, Las Mojas), rumoured islands, 77n, 80, 202n

Los Remedios, Puerto de, cxxxvii, 103, 146, 150, 165

Louis XVI, interest in geography, xx, xxiii-xxiv; assists American colonists, lx; pleased with LaP's actions in Hudson's Bay, lvi; share in planning LaP's voyage, xxii-xxiv, cx-cxx, 8–9, 510, 575; his instructions concerning native people, lxiv, clxxxix; approves final instructions cx; last meeting with LaP, cli, 10; generosity, clvi; concern about the fate of the expedition, ccvii, ccviii, ccxi-ccxxii; death, ccxiii; court, cc; references to him deleted in Milet-Mureau edition, viii, 575

Louisiades Archipelago, cxxix, cxxxvi, ccxxxix, 380, 517, 521, 541

Loyalty Islands, ccxxvi-ccxxvii, ccxxxix

Luzon, Philippines, sighed, 228; LaP in, clxxii-clxxvi, 227, 228, 231–49; inhabitants, 421; raids by 'Moors', 230; LaP on trade, 243; mentioned, 522

Macao, in instructions, cxxxi, cxxxix; voyage to, 201, 212–14; LaP at, clxxi-clxxv, clxxxv, clxxxvi, ccxxxvi, 215–

603

27, 508; observatory at, 217, 227; LaP's description and comments, 218–26, 241; letters sent from, 487, 489, 496, 498, 500, 501, 503; mentioned, 244, 246, 294, 363n, 377n, 506

Madagascar, LaP at, xlv-xlvi, xlix; mentioned, xliii, 287

Madeira, in instructions, cxxv, cxxxiv; crossing to, cliii, 11, 453, 455, 457; LaP at, 13–15, 453–5; price of wine, 15, 453, 455, 461; letters sent from, 453–4, 461; description omitted by Milet-Mureau, 575

Magellan, Strait of, sighted, cliv, 38, 471; mentioned, clx, 39, 95n, 472–4

Magon de Villaumon, French captain proposed to command a search expedition, ccxi

Mahé, French India, LaP at, xlvi-xlvii

Makatea, Tuamotu Archipelago, 429n

Malaspina, Alessandro, Spanish navigator, ccviii, 192n

Malays, ancestors of Polynesian people, 422–3

Maldonado, Lorenzo Ferrer, Spanish navigator, and Strait of Anian, xxxii

Malouines Islands (Malvinas), see Falkland Islands

Manchu Tartars, under Chinese rule, 280; trade with Sakhalin, 293, 296–8; burial site, 277–8; compared with North American Indians, 277, 280; mentioned, 272, 276, 293–4, 306, 321–2, 332, 338

Mang (Maug) Islands, 77n, 209, 212

Manila, Philippines, Marion du Fresne expedition at, xliv, 23; in instructions, cxxxi; LaP's intended timetable, 145; LaP sails to, 225, 228, 506; LaP at, clxxiv-clxxvi, clxxxvi, 236–45, 515–16; LaP prefers Cavite for his ships, 233; defences, 249; description, 236–40, 248–9; trade at, 189, 243, 252; French reports on, ccxxx, ccxxxvi, 485n, 506, 511, 534; letters sent from, 505, 508, 509; mentioned, 176, 201, 228–9, 330, 355, 421, 432, 446, 464, 475, 505–6, 510, 518n, 525

Manila Fair, 243

Manono, Samoan island, discovered, ccxxx, ccxxxviii, 413–14

Manua Islands, Samoa, 416–17

Marchand, Etienne, French captain, circumnavigation, cviii; and fur trade, ccxxxiv

Mariana Islands, in instructions, cxxxii, cxl; on LaP's route, clxx, 146, 207, 516, 520–1; LaP reaches, 208–12, 492; mentioned, 227n, 249, 382n, 389n, 421, 518

Marie, Bouvet's ship, 5

Marie-Antoinette, Queen, lxxxv, lxxxvii, clii, clxxii-clxxiii; cc, ccxiii, 11, 390, 500

Marikan, Kuril Islands, 325–8, 339, 366, 377, 513

Marion du Fresne, Marc-Joseph, French navigator, xli, xliii-xliv, cxxi, clvii, ccxxxviii, 428n, 432; death, ccvi

Mariveles Island, Manila Bay, LaP anchors at, 228–31

Marquesas Islands, cxvi, cxxx

Marquis-de-Castries, Marion du Fresne's ship, xliii, 428n; Richery's ship at Macao, clxxii, 504; Bellegarde transferred from, clxxvi, 498

Martinez, Fernandez Esteban José, Spanish captain, clxix, 164, 176, 194, 484

Martin Vas Islands, cxiv, 20, 24, 461, 465

Mascarin, Marion du Fresne's expedition, xliii, 428n

Maskelyne, Nevil, astronomer, cvii

Maui, Hawaiian Islands, LaP calls at, clix, ccxxxiii, 79–86, 480, 491; French officers go ahore, 85; mentioned, 189n

Maurelle, Francisco Antonio, Spanish navigator, clxiv, cxci, 146n, 151n, 152, 154, 165, 432–5, 515, 517, 538

Mauritius, see Ile de France

Mayer, Johann Tobias, astronomer, cvii

Meares, John, English navigator, at Nootka Sound, 144n

Melanesians, identified by LaP, 421–2

Mendaña, Alvaro de, Spanish navigator, cxxxvii, cxxx, cxxxvii, ccxvii, 76, 379, 385n, 517, 521, 541

Mengault de la Hague, French officer, buys a property with LaP at the Ile de France, xlvii-xlviii; plans campaign in the East, l; raids shipping, li

Middleton, Christopher, navigator, xxxiv

Midway Island, 379n, 381n

Milet-Mureau, Louis-Marie-Antoine Destouff de, army officer, edits, LaP journal, vii-viii, xi, xcix, ccxxx, 215n,

INDEX

451, 571–6; proceeds of sale, lviii; adds his comments, clviii, 87n, 124n, 166n, 240n, 377n, 573, 575; effects changes to the text, 69n, 131n, 132n, 136n, 155n, 159n, 177n, 200n, 204n, 305n, 318n, 354n, 357n, 358n, 395n, 420n, 423n, 425n, 429n, 431n, 434n, 436n, 440n, 447n, 480, 496, 505, 510, 521n, 524, 573, 575; reprints letters, 479, 480, 482, 485, 487, 490, 493, 496, 498, 503, 505, 508, 509, 510, 512, 520, 524, 526, 529, 530, 535, 536, 541; reprints Maurelle's journal, 434, 461, 515n

Mindanao, 243–4, 249, 383n; in instructions, cxxxii-cxxxiii, cxxxix, cxl

missionaries, in California, Franciscans, clxvi, clxix, ccxxxv, ccxxxvi, 169, 170–1, 174n; LaP visits, 178–86; mill given to, 182, 485; language problem, 198; LaP comments on, cciv, ccxxxv, 174–5, 182, 187–8, 191, 480; compared with missionaries in Chile, 175n; Dominicans, 170; Jesuits, 170, 187–8, 209, 221n

Mocha, island off Chile, 40

Moerenhout, Jacques, French consul, his opinion on Tutuila attack, lxiv

Molokini, Hawaiian Islands, ccxxxiii, 80–1, 90

Moluccas, cxxxiii, cxxxix, cxli, ccxvi, 158, 241, 492

Monge, Louis, astronomer, biography, lxxxix-xc; recommended by Borda, xxxix; at Brest, 11; in Tenerife, 458–9; leaves the expedition, ccxxxi, 197n, 457–8, 498

Mongez (Mongès), Jean-André, chaplain and physicist, biography, xc; edits *Journal de Physique*, lxxxviii, xc, 8; writes to LaP, clxxi; appointed, 8; cooperates with Lamanon, 4, 12, 457, 463; stock of chemicals, 444; lands at Santa Catarina, 28; lands at Easter Island, 61, 70; climbs at Lituya Bay, 132; works on Tlingit vocabulary, 141n; lands at Suffren Bay, 282; at Langle Bay, 287; at Castries Bay, 314; in Kamchatka, 354; ascent of Mt Avachinskaya, 349–52; gift to Kozlov, 361

Monneron Island, named, clxxxiii, 318

Monneron, Paul-Antoine de, engineer, biography, xc-xci; with LaP on Hudson's Bay raid, xci, 9; goes to London, xci, c-cx, cli, clxii, 9; at Tenerife, 17–18, 459–60; lands at Trinidade, 22; lands on Santa Catarina, 28; report on Chile, clv, ccxxxii; on Northwest Coast, 100, 109, 110–12, 123, 141n; report on California, ccxxxv; comments on the Philippines, 529; goes ashore at Suffren Bay, 282; island named after him, clxxxiii, 318; sword nearly stolen, 395; reports on military installations, 489; letters from 529; praised by LaP, 4, 489

Monsiau, Nicolas, painting of LaP and Louis XVI, cli

Montarnal, Pierre-Armand-Léopold Guirald de, biography, lxxv-lxxvi; uncle, 481; death, clxii, 114, 481, 499, 524

Monterey, California, in instructions, cxxx, cxxxvii; no early call at, clvi; proposed date of arrival, 146; LaP approaches, 164–5; LaP calls at, clxvi-clxix, ccxxxiv, ccxxxv-ccxxxvi, 167–94, 491, 505, 570; presidio, 167, 170, 172, 175, 192n; French donate a flour mill, xcviii, 182, 485; other gifts, 192, 479; gifts received, 193–4, 479; plan of, 493; LaP leaves, 202–3; political importance of LaP's call, 178n; LaP sends letters from, clxxxvi, 479, 480, 491; DeL sends letter from, 482; steele, 570

Monterey Bay, California, LaP arrives in, 164, 485; description, 167–75; leaves, 194

Montero (Monteiro), Pedro Jorge, Funchal, 13–15

Montesquieu, Charles-Louis de Secondat de, philosopher, ccxxxvi

Monty (Monti), Anne-Georges-Augustin de, biography, lxxvi-lxxvii; accommodation on board, clii; explores Lituya Bay, 110, 123; goes ashore in Tartary, 274–5; in Samoa, 392–3; at Norfolk Island, 444; transferred, cxc, 535–6, 547; praised by LaP, 499, 524, 536, 541; praised by DeL, 468, 483, 496–7, 526, 527; bay named after him, clxi, 98; mentioned, 505

Moor, William, navigator, xxxiv

Moresby Island, British Columbia, sighted, clxv, 154n, 155n, 156n

605

Morlaix, Brittany, home town of a number of sailors, 545
Mount Avachinskaya, Kamchatka, 328; French ascent of, 349–52; believed by natives to have poisonous atmosphere, 350
Mount Crillon, Alaska, 109n, 149, 151–2, 157
Mount Edgecumbe, Alaska (Mount St Hyacynth), 151
Mount Fairweather, Alaska, clxi, 101, 110, 149, 156
Mount St Elias, Alaska, in instructions, cxxxi; Cook sights, 95–6; LaP reaches, clx, 96, 159, 480, 491, 493, 505, 509; Dagelet calculates height of, 99; mentioned, 188, 488
Mount St Elias, Alaska, 151
Murray, Charles, British consul, Madeira, 13–15

Nagasaki, Dutch trading outpost, cxix, cxxxix, clxxvii–clxxix, 263n, 331
Nairne and Blunt, instrument makers, provide barometers, cix
Nairne, Edward, instruments lent to LaP, cix; comment on 272
Napoleon Bonaparte, possible appointee to LaP expedition, lxx; grants a pension to Eléonore LaP, lviii
Navigators Islands, see Samoan Islands
Necker Island, Pacific Ocean, discovered, clxix–clxx, ccxx, ccxxxvi, 204–5, 207, 382, 490, 492
Necker Islands (Orford reef), Oregon, discovered, ccxxxiv, 162, 163n
Necker, Jacques, economist, clxix, 46n, 162n, 204
Nepean Island, LaP sights, 442
Nerchinsk, Treaty of, 337n, 375n
Neve, Felipe de, governor of California, 181
Nevelskoi, Gennadii, Russian navigator, 313n
New Albion, America, xxxv, 473
New Britain, south-western Pacific, 421
New Caledonia, Louis XVI suggests surveying, cxx, 380; in instructions, cxxviii, cxxxvi, clxxxv, 541; on new route, 476, 492, 517, 521; likely call by LaP at, ccxxv–ccxxvi, ccxxxix; D'Entrecasteaux at, ccxvi–ccxvii; mentioned, 444–5

New Guinea, to be surveyed, cxx, cxxix; possible destination of LaP, ccxvi, 158, 517, 521, 541; Dumont d'Urville makes for, ccxxiii; mentioned, 421
New Hebrides, cliv, clxxxv, 380, 421
New Holland, see Australia
New Ireland, south-western Pacific, Maurelle at, 433; D'Entrecasteaux at, ccxvi
New South Wales, xciii, clxxxv, cxcii, ccv, ccxxxviii; (see also: Botany Bay)
New Zealand, and Surville, cii; Marion du Fresne killed in, xliv; and Cook, cxv; included in instructions, xxv-xxvi, cxxix–cxxx, cxxxvii, clxxxv, ccxxxviii, 476, 492, 512, 517, 520, 522, 531; rumours of a settlement by LaP, cli; Dillon in, ccxxi–ccxxii; mentioned, 444–5, 518
Niuatoputapu, Tongan archipelago, cxc
Nomuka, Tongan archipelago, LaP at, ccxxiii, ccxxvi, 430
Nootka Sound, Vancouver Island, clxv, 95, 103, 143, 144n, 146–7, 159–60, 164n, 491
Nootka Sound Controversy, clxix, 103n, 164n, 346n, 484n
Norfolk Island, cxcii, 446, 538; French attempt a landing, 442–3; departure from, 445, 565; pine trees, 444–5
Northwest Coast of America; French outpost envisaged, cxvii–cxviii; exploration by LaP, clxi–clxvi, ccxx, ccxxxiii, ccxxxiv, 96–161, 367, 472, 475, 480, 490–1, 493; programme affected by Lituya Bay losses, 145; and by the season, 157–8, 494
Nuestra Señora de la Gorta, rumoured island, clxix, 202–4, 492

Oahu, Hawaiian Islands, clix, 81, 90
Oberea, Tahitian queen, 521
Ofu, Samoan Islands, 386n, 387n
O'Higgins, Ambrosio, 477; LaP meets, clv, 48–51
Oiseau, on Kerguelen's expedition, xliii
Okhotsk, Siberia, clxxxiv–clxxxviii, 337, 346, 355, 356, 358, 360, 361n, 367n, 368, 371n, 372–5, 516, 530–1, 534; Sea of —, clxxxi, clxxxvi, 273, 284, 319, 322n, 355, 374, 376n
Oku Yesso (Sakhalin), 273–4, 284, 319,

323, 326, 332–6, 337–8, 353, 510, 513, 523
Olosega, Samoan Islands, 386n, 416, 426
Omai, Cook's Tahitian passenger, xliv, cxxxvi
Orford Reef, Oregon, 162, 163n
Orotchy, name given to Castries Bay natives, 306–16, 320, 321, 332–5; houses, 308–9; tombs, 308–9; women, 310–11; dogs, 312

Palau Islands, 389n
Pandora, Edwards's ship, passes near Vanikoro, ccxvii, ccxxviii
Paratunka, Kamchatka, 359–60, priest of, 329, 354, 357, 365–6; — River, 353, 365, 374
Paramusir, Kuril Islands, 513
Paris, Peter, convict, cxcv, 448n
Patagonia, coast of, 32, 34–5, 471
Paul, Jean-François, master caulker, praised, 484, 559
Paulaho, Fatafehi, Tongan chief, 433, 439–40
pelicans, at Monterey, 168, 173
Pennedo (Penedo) de San Pedro, Atlantic Ocean, 19
Pennevert, engineer at Rochefort, xcvi-xcvii
Pepys Island (Falkland Islands), cxiv, 33
Perez, Juan, Spanish navigator, clx, 160n, 164n, 171n
Pescadores Islands (Penghu Liehtao), Formosa Strait, cxix, clxxvii, 254–6
Peters, William, English trader, 363–4, 533
petrels, encountered, 32, 78, 471; on menu, cliv, 34
Petropavlovsk, Kamchatka, in instructions, cxxxi-cxxxii, cxxxviii, cxxxix; LaP makes for, 328; stay at, lxii, cxvii, clxxxiv-clxxxvi, ccxxxvii, 329, 345–69, 516; observatory set up, 349; balls at, 355–6, 366, 368; saunas, 375; description, 365, 371n, 374n; mail received, 356–7, 530–2; Lesseps disembarks at, xciii, 511, 535; reports and letters sent from, 485n, 510, 512, 520, 522, 523, 524, 526, 529, 530, 531, 532, 533, 540
Philippines, LaP reaches, 212, 367; stay and description, 233–49; tobacco tax, 240–1; New Company of the, 235, 241–3, 244; languages and vernacular, 421, 425; LaP's comments on, clxxiv-clxxv, cciv, 237–44, 522
Phillip, Arthur, captain commanding the First Fleet to Australia, cxcii-cxcviii, ccvi, ccviii, 442n, 447–8, 536, 564
Phoenix Islands, 384n
Physiocrats, influence on LaP, xviii, cciv, ccxxxvi, 45n, 243n; views on China, 220n
Pierrevert, Ferdinand-Marc-Antoine Bernier de, biography, lxxvii; sent ashore at Santa Catarina, 28; lands on Maui, 85, 88; on the northwest Coast, 101; death, clxii, 114, 117, 481, 499, 524
Pingré, Alexandre-Guy, astronomer, 15n, 55
pirates, in eastern seas, cxl, 103, 220; in the Pacific, 84; in the Philippines, 244
Pitcairn Island, ccxxvii
Pitt Island, see Vanikoro
Pitt, William, learns of proposed voyage, xxv
Point Banks, Botany Bay, 446n, 448
Poissonnier, Pierre-Isaac, develops a distilling machine, xcvii-xcviii, ciii, cv
polygamy, in California, 184
Polynesians, 58n, at Easter Island, 64, 67–8, 73n; origin of, 380, 421
Ponape, Caroline Islands, possible final end of LaP's expedition, ccxxviii
Poncel de la Haye, French navigator, 467
Pondicherry, French India, xlvi, xlviii, li, lxxvi, lxxix
Port Bucareli, Northwest Coast, cxxxi, cxxxvii, clxiv, 152, 165, 515
Port des Français, Alaska, see Lituya Bay
Portefaix (later *Boussole*), selected, xcv; fitting out, xcvi-xcix; sails to Brest, xcvi; name change, xcviii
Port Jackson, New South Wales, English arrive in, cxcii-cxcviii, 448, 531n, 536, 539, 567
Portlock, Nathaniel, English navigator, names Admiralty Bay, 98n; meets Etienne Marchand, ccxxxiv; describes Northwest Indian women, 137n
Portuguese, in China, 217–24, 226
Postigo de, Spanish captain, 477
potato, sweet, 63–4, 67, 70, 72, 386
Pote, Iachan Daniel, Russian official, 358

Pratas Shoal, China Sea, 228, 232
Prévost, Guillaume ('The Elder'), artist, biography, xcii; appointed, 9; accommodation on board, clii, ccxxxi; lethargy, lxxxix, xcii, ccxxxiii, 452, 469, 488, 527; at Lituya Bay, 110; at Asunción, 210–11; at Langle Bay, 287; thanked by LaP, 4; criticised by DeL, 483, 497
Prévost, Jean-Louis-Robert, artist, xcii, ccxxxiii, 4, 9; in Castries Bay, 304; praised by LaP, 488, 505; praised by DeL, 497
Prince of Wales Fort, Hudson's Bay, captured by LaP, lv–lvi, xci, 128, 148n, 570
Princesa, Spanish ship, 164n, 176, 484n, 485, 515n, 518n
Prince William Sound, Alaska, in instructions, cxxxi; sighted, clxiv; and James Cook, 137n
puffin, in Kamchatka, 155, 281
Pylstart (Pijlstaert) Island (Ata), LaP at, 441–2, 538

Queen Charlotte Islands, British Columbia, cxix, clx, clxiv-clxv, 154n, 156n
Queen Charlotte Islands, Carteret's islands of that name, cxxxvi, ccxvii
Queen Charlotte Sound, British Columbia, 157n
Queen Charlotte Sound, New Zealand, in instructions, cxxxvii, clxxxv, 476, 492n, 517, 520
Quelpaert Island (Cheju-do), clxxvii, 262–3, 265, 268n, 276, 330
Quexada, Chilean officer, 47, 477
quinine, as anti-scorbutic, 93–4, 317
Quiros, Pedro Fernandez de, Spanish navigator, cxix-cxx, ccxvii-cxxviii, 76, 379, 385, 440, 540

Ramsden, James, supplies instruments, cviii; LaP comments on, 196
Raoul Island, Kermadec group, ccxxix
Receveur, Claude-François-Joseph, chaplain and naturalist, biography, xcii-xciii, 568; co-operates with Lamanon, 4; on Peak of Tenerife expedition, 566; goes ashore at Trinidade, 21–2, 566; lands at Santa Catarina, 28; on Easter Island, ccxxxiii, 61, 70, 72, 566; explores Lituya Bay, 110, 132, 566; works on Tlingit language, 141n; lands on Asunción, 210; at Macao, clxxi; explores Castries Bay, clxxxii, 314, 566; in Suffren Bay, 282, 566; in Kamchatka, 349–52; gift to Kozlov, 361; goes ashore at Tutuila, 398, 410n, 566; wounded at Tutuila, clxxxix, 401, 410n; health, 406, 410n, 565; chemicals on fire, 444; death, cxcviii-cxix, ccxxxviii, 564–9; tomb, 446n, 570; praised by DeL, 469, 483–4, 497, 527
Recherche, D'Entrecasteaux's ship, ccxiv-ccxviii
Recherche Island, see Vanikoro
Refreshment Island (Makatea), Tuamotu archipelago, 429
Research, Dillon sails to Vanikoro, ccxx
Resolution, Cook's ship, clxxxii, 7, 36, 202, 203n, 266n, 329n, 346n, 355n, 357n, 441, 488n
Résolution, commanded by D'Entrecasteaux, lxxii, lxxv, lxxxii, ccxiv, ccxv; at Macao, 246–7
Rességuier, Marguerite de, LaP's mother, xxxix, lxxv
Réunion Island, see Bourbon
Revilla Gigedos Islands, Mexico, 79n, 382n
Rica de Oro, mythical island, ccxxxviii, 377, 523
Rica de Plata, mythical island, ccxxxviii, 377, 523
Richery, Joseph de, captain of the *Marquis-de-Castries*, at Macao, clxxii, 216, 498, 504
Robin, Laurent, help with wheat mill, 94
Rochambeau, Jean-Baptiste de, lii-liii, lxv
Rochefort, French port, lxviii, lxix, lxxiii, lxxvi, lxxviii; early preparations for the expedition at, xciv-xcvi
Roggeveen, Jacob, Dutch navigator, 54, 60n, 63n, 69n, 386n, 413n, 417–18, 429
Rolland, Kerguelen's ship, xliii, xcv
roller, bird, shot, 30
Rollin, Claude-Nicolas, senior surgeon, lxxx-lxxxii; opinion on venereal disease at Maui, 86; proposes anti-scorbutic measures, 93; works with

INDEX

Lamanon, 107; exhausted after Pacific crossing, 217; at Suffren Bay, 282; and natives of Sakhalin, 292; tends Collignon, clxxxiii, 316; attacked at Tutuila, 395; reports by, lxxxiii, ccxxxii, ccxxxv, ccxxxviii, 4n, 530; cape named after him, 326–7; praised by LaP, 4, 107, 525, 530

Romulus (later the *Résolution*), lxxxii

Rosily-Mesros, François-Etienne de, French captain, 573

Rossel, Elisabeth-Paul-Edouard de, D'Entrecasteaux's officer, ccxv, ccxviii, ccxxxix

Rousseau, Jean-Jacques, philosopher, xviii; theories, lxxxviii, clvi-clviii, clxxix, clxxxviii, cc-cciii, 67n, 133n, 220n, 395n; romanticism, 124n

Roux d'Arbaud, biography, lxx; at Tenerife, 458; at Lituya Bay, clxxii, 109; goes ashore at Suffren Bay, 282; in Castries Bay, 304; praised by LaP, 500, 525; promoted, clxiii, 146, 479, 482

Russia, LaP to ascertain extent of penetration into America, cxxxviii, ccxxxv; advance and settlements in Siberia, 284n, 293n, 371–2, 376n; in Kamchatka, clxxxiv, 107n, 347n, 362, 371; in Alaska, clx, ccxxxiv, 106, 107n, 110n, 140, 293n, 372–3, 516; fur trade with China, 190, 293n, 502

Ryukyu Islands (Likeu), 259–61

Sabatero, Chilean major, 47

Saint-Céran, Henri-Marie-Anne-Jean-Baptiste Mel de, biography, lxxv; explores Lituya Bay, 110; ill-health, lxxv, clxxvi, 247, 509, 525; sails for India, lxxv; praised by LaP, 500

Sakhalin, LaP reaches, 284; LaP surveys coast of, clxxx-clxxxiii, ccxxxviii, 286–300, 317–18; lands on 286, 288, 296; flora, 292, 298–9; fauna, 339; LaP and islanders, cciv, 289, 319–22, 337–8; considered to be an island, clxxxii, ccxxxvii, 284n, 289, 296, 300, 313, 319, 321, 340; Strait of —, 274, 303; mentioned, 307, 310

Sakishima Islands, LaP passes through, clxxvii

salmon, caught in traps on Northwest Coast, 124; purchased, 129; caught on Sakhalin, 287, 289, 298; traded by Sakhalin islanders, 288, 297, 320, 322–3; in Siberia, 304, 306–11, 335; in Kamchatka, 346, 353

Salon-de-Provence, Lamanon, mayor of, 455–6

Samoan Islands, Bougainville reaches, 386; LaP at, 386–427, 537–8; LaP meets islanders, 386–97, 411–13, 419; barter trade, 387, 389–93, 398–9, 403, 412, 426, 538; thefts, 387, canoes, 390, 422–4, 439; houses, 394, 425; language, 421; diseases, 388–9; attacks by islanders, 393–403, 405–11; mentioned, lxiii, lxxxviii, cxxviii, clviii, clxxxv, clxxxviii, cciii, cciv, ccxxxviii, 416, 417n, 418, 428, 440–1, 540–1

Samwell, David, surgeon on Cook's *Resolution*, on language of Nootka, 144n; on Chinese boat, 269n; on Kamchatka, 357n, 359n, 361n

San Ambrosio and San Felix, cxvi; LaP identifies as Davis Land, 55

San Carlos, Spanish ship, clxix, 484n

San Carlos Islands (Forrester Islands), Alaska, 153–4

San Carlos mission, California, see Carmel

San Diego, California, 171, 188n, 189n

Sandwich Islands, see Hawaiian Islands

Sandwich, John Montagu, Earl of, 467, 491

Sandwich Land (South Sandwich Islands), cxiv, 31, 467

San Francisco Bay, California, not sighted by Spanish explorers, ccxxxiv; mentioned, 180n, 188–9, 488, 514

Sangar, reputed strait north of Japan, 303, 319–20, 335

Santa Catarina Island, Brazil, call at, cliv, 22–4, 26–30, 36, 470, 472; description, 26–8; supplies taken on, 29–30, 466; letters sent from, 30, 461, 463–4

Santa Cruz Islands, cxxviii, ccxvii, ccxx, ccxxii, ccxxvii, 517, 521, 541

Santa Maria, island off Chile, 40–2

Santiago, Spanish ship, 103n, 164n

Sanvitores, Diego Luis de, missionary and explorer, 209–10

Sarmiento y Gamba, Pedro, Spanish historian and navigator, cxvi

Sartines, Antoine-Raymond-Jean de, Minister of Marine, 7, 159n

609

Sartines Islands (Scott Islands), British Columbia, ccxxxiv, 158
Sarychev, Gavriil, Russian navigator, 534n
Savai'i ('Pola'), Samoan island, LaP discovers, cxc, ccxxx, ccxxxviii, 413–14, 416; LaP sails along coast of, 426–7
Sceptre, LaP's ship in Hudson's Bay raid, liv–lvii, xci; Montarnal in, lxxvi; Monneron in, xci
Schmalev, Vasilii Ivanovich, governor of Kamchatka, 346, 353, 354, 357
Schouten, Willem Corneliszoon, Dutch navigator, cxc, 39, 418n, 427–8, 429, 473–4
scurvy, c–cii, clxxxiii, cxci, ccxii, ccxviii, 4, 93, 111, 138, 203, 217, 279, 316–17, 396–7, 431–2, 490, 519, 520, 525, 537, 538
sea-larks, caught by *Astrolabe*, 203; indication of land, 203–4
sea otter, caught and described, 107; in California, 173, 189–90; in Northwest Pacific, 287, 293, 311, 335–6, 366–7, 374; trade in their furs, 128, 140, 167, 189, 363, 371, 373, 495; LaP reports on, 147n, 490, 502; fur sent to France, 500–1
sea swallows, see terns
seaweed, as indication of land, 32, 96, 378–9, 471; in Castries Bay, 306; in Strait of Tartary, 336
Sébir, French merchant in Manila, 236, 244–5, 248
Segalien (Sakhalin) River, see Amur river
Seine, LaP commands, to India, xlv–xlvii, clxxii, 217
Selvagens Islands, Atlantic, 15–16
Serin, corvette, LaP commands, li
Serra, Junípero, missionary in California, 171n, 176n, 182n, 191n
Seychelles Islands, Indian Ocean, La P at, xlv
sharks, as food, 381, 384
Shelikov, Gregorii, Russian trader, clx, 140n, 373n
Shumagin Islands, Alaska, cxxxi, 140
Siberia, xciii–xciv, cv, clxxxiv, clxxxvi–clxxxvii, 272n, 346n, 362n, 381n; La P off the coast of, 274–99, (see also: Tartary Coast)
Simusir Island, Kurils, lcxxxiii

Sirius, Hunter's ship, 446
Sitka Sound, Baranof Island, Alaska, clx, clxiv, 151n
Smith, Francis, navigator, xxxiv
Society Islands, cxxvii–cxxviii, clxxxv, 39, 380, 389, 405, 418, 420–1, 476, 517, 521, 537, (see also: Tahiti)
Solander, Daniel, Swedish botanist, sails with Cook, 34, 446n, 567
Solomon Islands, Surville's Arsacides, cxix, cxxviii–cxxix, cxxxvi–cxxxvii, cln, clxxxv, ccix, 76n, 380, 433, 476, 492, 517, 520–1, 541
Sonora, Spanish ship, 103n, 146n, 165n
South Georgia, Antarctic, cxiv, cxxv, cxxxv, 467; LaP's opinion on, 33n
South Sandwich Islands, see Sandwich Land
Spanberg, Morten Petrovich, Russian navigator, 325n
Sparwer, Dutch ship in Korean waters, 263
St Allouarn, François de, French captain, discoveries in South Indian Ocean, xlii–xliii
Staten Eylandt (Iturup), Kuril Islands, 324
Steinheil, Ivan, former governor of Kamchatka, 348
St Jean-Baptiste, Surville ship, 433
St Lazarus, Strait of, shown on Buache's map, xxxiv, xxxvii; and Chirikov, xxxv; LaP comments on, 165, 490, 493; and rumoured islands, 331
Staten Land (Island), Southeast Pacific, cxxvi, 37, 40, 514
Steller, Georg Wilhem, naturalist, 155n, 364, 376
Stockenstrom, Swedish agent in Canton, aids LaP, 245, 503, 508
St Patrick, Dillon's ship, ccxx
St Paul (St Pavel), Chirikov's ship, 96n, 152n, 359n
Strait of Tartary, 274, 289, 290–1, 300–2, 318–19
Strange, James, captain and fur trader, 533
Subtile, French ship commanded by Vagnas de Castries, sails for the Philippines, lxxii, lxxv, clxxvi, ccxiv, 246–8, 507; Villeneuve transfers from, lxxii, clxxvi, 248; Le Gobien transfers from, lxxv, clxxvi, 247

INDEX

Suffren Bay, Tartary Coast, clxxx, 282–3, 333
Suffren Saint-Tropez, Pierre-André de, lxxvii, lxxix, ccxiv, 94, 485; — Bay, clxxx, 283–3, 333
Sumusir, Kuril Islands, see Marikan
Supply, Ball's ship, 446n, 447n
Surville, Jean-François-Marie de, expedition of lxxiv, xci, cxix, cxxi, cxxviii, cxxxvi, ccix, ccxxxviii, 433, 476, 517, 541; explores new seas, xix, 380; scurvy on his ship, cii
Swedish traders in China, 225, 245, 503, 508

Tafahi Island, see Cocos Island
Tahiti, Bougainville at, clxx, ccii–cciii, 51n, 389n; in Instructions, cxxvii–cxxviii, clxxxv, 467, 517; to be rendezvous point, 31; mentioned, xliii, cliv, 69n, 411, 521, 537
Talcahuano, Chile, LaP's anchorage, 41–52, 195, 478
Talin, Pierre, master-at-arms, killed, 401, 408, 410
Tamar, Byron's expedition, xxxv
Tartary Coast, Siberia, sighted, 271–2; mirage, 273–4; exploration, lxii, cxxxii, cxxxix, clxxix–clxxxi, ccxxx, ccxxxvii, 146, 274–85, 299, 302, 314, 317, 326–7, 331, 353, 371, 384, 476, 492, 506, 510–12, 516, 523, 531; natives met, cciv, 332; piracy danger, xcl; reports on, ccxxx; Strait of — , 274, 289, 290–1, 300–2, 318–19
Tasman, Abel, Dutch navigator, cxci, 377n, 379, 418n, 430n, 435n, 437n, 440n, 441n
Tasmania (Van Diemen's Land), xliv, cxxix, ccxvi, ccxx, 517, 542
Tau ('Opouna'), Samoan Island, 386, 416, 426
Tench, Watkin, English officer, book on the French in Botany Bay, cxciii–cxciv, cxcvi–cxcviii, 564, 566
Tenerife, call at, cliii, 15–18, 20, 351, 454, 455, 456–7, 458–60, 467; problems at, ccxxxi, 452, 464; wine, 453–5, 520; letters sent from, 455–60
tern (sea swallows), at Easter Island, 72; in the Pacific, 212, 445; indication of land, 204, 384, 446
Ternate, Moluccas, cxxxii

Ternay Bay, Tartary Coast, clxxx, 275–81, 289, 319, 333, 512
Ternay, Charles-Henri-Louis d'Arsac de, LaP's protector, xl, xlvi, 275; influence on LaP, xxxix, l, 275; LaP sails with, xl–xli, lii; at the Ile de France, xlii–l, 275n; appointed LaP's guardian, xlviii, lvii; abortive plan for an eastern seas squadron, li, lxxvi; and American war, lii, 275n; death, liii, lxv, 275; bay named by LaP in his honour, clxxx, 275–81
Tessier, Alexander-Henri, scientist, 9
Tessoy, rumoured strait in northwestern Pacific, 271, 273–4, 276, 300, 336, 513
Thévenard, Antoine, Minister of Marine, ccxiii–ccxiv
Thienhoven, Cornelis Bouman's ship, 417–18
Thérien, François, French official at Canton, 503–4
Thétis, Hyacinthe de Bougainville's ship, 570
Thomas, Jacques, ix, xi, xiii
Thouin, André, botanist, helps with planning, xxix, 30; Lamartinière's teacher, lxxxix; selects Collignon, lxxxiv, 9; Lamartinière corresponds with, cliii; Collignon reports to, ccxxxv
Tierra del Fuego, cxv, cxxvi, ccxxxii, 35–7, 39–40, 471, 473–4
Tikopia Island, Santa Cruz group, Dillon at, ccxx–ccxxi
Timor, cxxix, cxxxiii, clii
Tinian, cxxxii, cxl, 209, 475
Tlingit people, Lituya Bay, Alaska, meet the French, clxi, 105; trade, 105; LaP describes, 105, 132–40; fishing methods, 124; thefts, 108–9, 133–4; dogs, 135–6; women's lip saucers, 136–7; language, 141–4, 200; cannibalism, 144; gambling, 133, 141; funeral practices, 124–6; sell the French an island, clxii, 110–11; fight the Russians, 111n; LaP's opinion of, cciv, 123, 132, 134, 137; reports on, ccxxx
Tobias, former governor of the Marianas, 239, 522
Tofua, Tongan islands, LaP at, 436–7
Tokelau Islands, 384n
Toku, Vavau group, Tongan Islands, discovered by LaP, 434–5

Tongan Islands (Friendly Islands), in instructions, cxxviii–cxxix, cxci, 380, 476; LaP approaches, 384; La P among, 420–1, 427–40; canoes, 439–40; mentioned, xliv, cxcviii, ccxvi, ccxxi, ccxxiii, ccxxvi, 384, 389, 418, 476, 517, 518, 521, 537, 541

Tongatapu Island, Tongan Islands, 432, 434, 436; LaP reaches, 437, description, 438–9; inhabitants, 438–40

Toulon, Mediterranean port, lxxi, lxxiv, lxxv, lxxvii, ccxiii, 102

Traitors Island (Niuatoputapu), 416, 427–30, 440, 540

Trinidade, island, Atlantic ocean, in instructions, cxiv, cxxv, cxxxv; LaP at, cliii–cliv, 19–21, 465; described, 21, 23, 29, 461; garrison at, 21–2; supplies hoped for, 18, 21

Tripods Island, Pacific Ocean, cxvi

Tristan da Cunha, South Atlantic, cxiv, cxxxiii, 33

Trobriand, Jean-François-Silvestre-Denis de, officer with D'Entrecasteaux, ccxv

Trobriand Islands, Southwest Pacific, ccxxxix

Tromelin, Legoarant de, French navigator, ccxxiv

tropic bird, sighted, 78, 212, 384, 445; Dutch name, 441n

Troughtons, suppliers of instruments, cviii, cix

trumpeter, fish caught at Norfolk Island, 445

Tssissa Island, name for Hokkaido, 318–19, 322–3, 328, 338, 342

tunny fish, 424; encountered, 19

Turgot, Anne-Robert-Jacques de L'Aulne, economist, xviii

turtles, 538; sighted, 378; caught, 78; hoped for in vain at Asunción, 210

turtle doves, in Luzon, 231; in Tartary, 281; in Samoa, 389–90, 394

Tutuila ('Mahouna'), Samoan Islands, expedition reaches, clxxxviii, 390–1, 416, 538; description, 391, 394, 396, 425–6; pigs bought at, cxci, 391–4, 431–2, 540–1; diseases, 388–9; massacre at, clxxxix, ccxxxviii, 133n, 398–411, 430, 535, 539; Receveur wounded, xciii, 401, 564; Lamanon killed, lxxxviii, 401, 410, 535; DeL killed, xcviii, ccv, 399–400, 410, 535; theories on massacre, lxiv–lxv, ccv–ccvi, 395n, 403, 407, 568; LaP's opinion on natives, ccv, 67n, 394–5, 418–19, 536; memorial, 570

Ullung-do (Dagelet Island), named, clxxviii; French sail past, ccxxii, 267–8

Unalaska, Aleutian Islands, 372, 378n

Unamuno, Pedro de, Spanish navigator, 377n

Upolu ('Oyolava'), Samoan Islands, cxc, ccvi, 404n, 411–13, 416, 426, 440

Uracas, Marianas Islands, 209, 212

Urup, Kuril Islands, clxxxiii, 273n, 276n, 366n

Utile, fitting out begun for expedition, xcv; abandoned, xcvi

Valerianos, Apostolos, xxxii, (see also: Juan de Fuca)

Valparaiso, Chile, avoided, clvi

Vancouver, George, English navigator, xxxvii–xxxviii, clxv, 136n, 166n

Vancouver Island, clxv, 161n, (see also: Nootka Sound)

Van Diemen's Land, see Tasmania

Vanikoro, Santa Cruz group, site of wreck, 448n; wreck found at xciv, ccxx–ccxxi, 570; likely events at, ccxxi–ccxxii, ccxxiv, ccxxv; LaP's likely attitude towards islanders, ccv; likely cause of massacre, ccvi; items lost at, cxcix, 448n, 571; island sighted by D'Entrecasteaux, ccxvii; Dillon at, ccxix–ccxxiii, ccxxiv, ccxxxix; Dumont d'Urville at, ccxxii–ccxxiv; erects monument, ccxxiii, 570; Tromelin at, ccxxiv; Discombe at, ccxxv–ccxxvi, 570

Vargas, Basco y, governor of Luzon, clxxv, 234

Vassadre y Vega, Vicente, Spanish official in California, 189

Vaudreuil, Comte de, commander in Caribbean, liv, lxi, lxvi

Vaugondy, Robert de, cartographer, 202n, 290n, 382n

Vaujuas, Jean-François Tréton de, accommodation on board, clii; works on astronomy, 197n, 484, 488; lands at Trinidade, 21–2, 465–6; lands on Maui, 85; at Yakutat Bay, 98; at

Asunción, 210; in Manila, 245; lands at Suffren Bay, 282–3; takes soundings in Strait of Tartary, clxxxii, 301–2; climbs Cape Crillon, 322; goes ashore on Tutuila, 399–401; account of and opinion on attack in Samoa, lxiv, 388n, 405–11, 464;counsels against reprisals, 402; health, 397, 405, 406, 411; poverty, lxxi; praised by LaP, 498; praised by DeL, 468, 483, 496–7, 526; decorated, 496, 499; recommended for a pension, 497, 524, 528

Vavau group, Tonga Islands, cxci, 428n, 430n; LaP at, 432–5, 441, 537, 541

Vela Rete Rocks, Shichesi Seki, 257

Venegas, Fr Miguel, historian, 170

veneral disease, 12, 45, 86–7, 467

Verdun de la Crenne, astronomer, 15

Vereshagin, Romaan Feodorovich, priest in Kamchatka, 325, 354, 357

Vésian, Mademoiselle de, proposed bride for LaP, l, lvii

Vieillard, Philippe, French consul at Canton, cxxxix, clxxiii, 225–6, 501–2, 505, 508

Villalobos, Ruy Lopez de, 382

Vizcaíno, Sebastián, Spanish navigator, xxxii, clxvi, 171, 377n

Vladivostok, clxxix, 271n, 374n

Voltaire, François-Marie Arouet dit, French writer, xviii, clxxix, ccii–cciii, 131n, 220n

Vries, Maarten Gerritszoon, Dutch navigator, clxxxiii, 273, 276n, 323n, 324–5, 337, 377, 514; — Strait, 325–6, 514

Wafer (Waffer), Lionel, buccaneer, and Davis Land, 54–5

Wales, Fort, Hudson's Bay, see Prince of Wales Fort

Wallis, Samuel, English navigator, cxxii, 6, 209, 213, 384n, 416, 418n, 427–8, 430, 440

Webber, John, artist on Cook's *Resolution*, befriends Monneron, xci, ci, ciii, cx, 413n

whales, hunting, cxiv, cxxxv, cl, 27, 37, 52, 168, 466; in Sakhalin, 320–1, 323, 335

White, John, surgeon of the First Fleet, account of the French in Botany Bay, cxciv–cxcv

Williams Sound, Alaska, 515–16

windmills on ships, xcviii

Woody Point (Cape Cook, British Columbia, 159

Yakutat Bay (Baie de Monti), Alaska, clxi, 98, 136n

Yesso, rumoured country north of Hokkaido, 273–4, 276, 284, 285–6, 314, 319, 323, 332–6, 337–8, 476, 510, 513

York Factory (Fort York), Hudson's Bay, lv–lvi, xci, 128

York Island (Moorea), Society Islands, 384

Zélée, Dumont d'Urville's ship, ccxxiv